The Climate Bonus

We urgently need to transform to a low carbon society, yet our progress is painfully slow, in part because there is widespread public concern that this will require sacrifice and high costs. But this need not be the case. Many carbon reduction policies provide a range of additional benefits, from reduced air pollution and increased energy security to financial savings and healthier lifestyles, that can offset the costs of climate action.

This book maps out the links between low carbon policies and their co-benefits, and shows how low carbon policies can lead to cleaner air and water, conservation of forests, more sustainable agriculture, less waste, safer and more secure energy, cost savings for households and businesses and a stronger and more stable economy. The book discusses the ways in which joined-up policies can help to maximise the synergies and minimise the conflicts between climate policy and other aspects of sustainability.

Through rigorous analysis of the facts, the author presents well-reasoned and evidenced recommendations for policy makers and all those with an interest in making a healthier and happier society. This book shows us how, instead of being paralysed by the threat of climate change, we can use it as a stimulus to escape from our dependence on polluting fossil fuels, and make the transition to a cleaner, safer and more sustainable future.

Alison Smith is an environmental policy consultant with 15 years' experience of advising the UK government and the European Commission, and has been a lead author for the Intergovernmental Panel on Climate Change.

'*The Climate Bonus* comes at a crucial time in national and international climate policy. It provides a compelling vision and strategy for a path forward to carbon reduction by linking climate change to the immediate concerns of policy makers: human health, land management, resource security and economic stability.'

Douglas Crawford-Brown, Director of the Cambridge Centre for Climate Change Mitigation Research, University of Cambridge, UK

'Policies to address climate change can yield many immediate benefits – cleaner air, improved health, a stronger economy, greater energy security, reduced waste and more robust ecosystems. *The Climate Bonus* documents these potential co-benefits of climate change policies. We need to act now to address climate change and to reap these numerous co-benefits.'

Erik Haites, President of Margaree Consultants Inc., Canada

'This book brings some fresh thinking to tackling climate change. It makes the key point that saving the planet is not rocket science and that there is much we can do now. It also brings out the often overlooked role of material resource efficiency, pointing out that it is just as important as energy efficiency for meeting climate targets.'

Marcus Gover, Director of the Waste and Resources Action Programme (WRAP), UK

'This book makes a very important contribution to the climate debate. It will enable both policy makers and the general public to understand that reducing our fossil fuel dependency won't just benefit the climate, but people's health across the board, as a result of cleaner air, healthier diets and more opportunities for walking and cycling.'

Génon Jensen, Executive Director of the Health and Environment Alliance and member of the World Health Organization's European Health and Environment Task Force, Belgium

The Climate Bonus

Co-benefits of climate policy

Alison Smith

Routledge
Taylor & Francis Group

LONDON AND NEW YORK

earthscan

from Routledge

First published 2013
by Routledge
2 Park Square, Milton Park, Abingdon, Oxon, OX14 4RN

Simultaneously published in the USA and Canada
by Routledge
711 Third Avenue, New York, NY 10017

Routledge is an imprint of the Taylor & Francis Group, an informa business

British Library Cataloguing in Publication Data
A catalogue record for this book is available from the British Library

Library of Congress Cataloging-in-Publication Data
Smith, Alison (Alison J.)
The climate bonus : co-benefits of climate policy / Alison Smith.
p. cm.
Includes bibliographical references and index.
1. Climate change mitigation. 2. Restoration ecology. 3. Restoration
ecology–Economic aspects. 4. Conservation of natural resources. I. Title.
QC903.S587 2013
363.738'74561–dc23
2012024557

ISBN: 978-1-84971-340-5 (hbk)
ISBN: 978-1-84971-341-2 (pbk)
ISBN: 978-0-203-10957-1 (ebk)

Typeset in Baskerville by
Keystroke, Station Road, Codsall, Wolverhampton

Printed and bound in Great Britain by
TJ International Ltd, Padstow, Cornwall

Contents

Figures

Colour figures in plate section

Figures

Tables

Boxes

Acknowledgements

I am extremely grateful to the reviewers who so kindly gave up their time to offer many helpful comments: Dr Doug Crawford-Brown of the Cambridge Centre for Climate Change Mitigation Research at the University of Cambridge; Marcus Gover, Director at the Waste and Resources Action Programme in the UK; Dr Philip Lawn, Senior Lecturer in Ecological Economics at Flinders University in Australia; Blake Alcott of the Sustainability Research Institute at the University of Leeds; and most of all, the indefatigable Debbie Ley of the Environmental Change Institute at the University of Oxford who reviewed every chapter. I am also very grateful to Dr Mike Holland of EMRC Associates, who was initially going to be my co-author until his consultancy work took over, for reviewing most of the chapters and making valuable contributions to Chapters 3, 5 and 9.

I regret that due to lack of space I was not able to incorporate all of the reviewers' suggestions; and any remaining errors are, of course, entirely my own responsibility. I would also like to thank my family and friends for their support and encouragement. Finally, I thank Helena Hurd, Louisa Earls and their colleagues at Routledge for their infinite patience in awaiting the delivery of the manuscript; my very thorough copy-editor Helen Lund; and proofreader Belinda Wakefield; and the production team at Keystroke.

A note on units

Dollars ($) are US dollars throughout unless stated otherwise. The year to which costs and prices refer varies depending on the source: it is stated in the endnotes where possible.

Billion and trillion are the English usage.

Billion	10^9; 1,000 million
Trillion	10^{12}; 1,000,000 million
ppb	parts per billion
ppm	parts per million
t	tonne (1,000kg)
kt	kilotonnes (1,000 tonnes)
Mt	megatonne (1,000,000 tonnes)
Gt	gigatonnes (10^9 tonnes); billion tonnes
µg	microgram (a millionth of a gram; 10^{-6}g)
micron	10^{-6} metres
ha	hectare (10,000m^2); 100ha = 1km^2
Mha	million hectares; 1 Mha = 10,000km^2
kWh	kilowatt hour (unit of energy equivalent to supplying one kilowatt of power for one hour)
MWh	megawatt hour (1,000 kWh)
GW	gigawatts; 1 GW = 1,000 MW (10^9 watts)
TW	terawatts; 1 TW = 1,000,000 MW (10^{12} watts)
mbd	million barrels per day. One barrel of oil is 42 US gallons, 159 litres. There are about 6.5 barrels of crude oil in a tonne, depending on density.
tcm	trillion cubic metres
Final energy	Energy received by the consumer, including refined fuels (gasoline, diesel, kerosene) and electricity after generation and transmission
Primary energy	Energy after extraction, but before conversion or transmission to the consumer. Includes coal, crude oil and gas, but excludes electricity and refined fuels

Greenhouse gas emission units

The three main man-made greenhouse gases are carbon dioxide (CO_2), methane (CH_4) and nitrous oxide (N_2O).

Emissions of carbon dioxide are expressed either as weight of carbon, or weight of carbon dioxide. Both are used in this book, depending on the reference cited. To convert from carbon to carbon dioxide, multiply by $44 \div 12$ (3.67).

Emissions of other greenhouse gases can be expressed in units of carbon dioxide equivalence (CO_2e), which is calculated by multiplying each greenhouse gas by its global warming potential (GWP) compared to carbon dioxide over a given period (because the lifetimes of the gases in the atmosphere vary). In calculations for this book, we have used 100-year GWPs of 25 for methane and 298 for nitrous oxide from the *Fourth Assessment Report* of the IPCC (IPCC, 2007).

Abbreviations

This is a selective list of abbreviations which appear frequently within the book. It does not aim to be comprehensive. Where other acronyms are used in the text, they are spelled out the first time they appear. See also 'A note on units' above.

CCS	Carbon capture and storage
CHP	Combined heat and power
EPA	Environmental Protection Agency (United States)
EROI	Energy return on investment
EU-27	The 27 countries of the European Union
FAO	Food and Agriculture Organization of the United Nations
FSC	Forest Stewardship Council
GDP	Gross Domestic Product
GHG	Greenhouse gas
HDI	Human Development Index
IEA	International Energy Agency
IIASA	International Institute for Applied Systems Analysis
IPCC	Intergovernmental Panel on Climate Change
IUCN	International Union for Conservation of Nature
LNG	Liquefied natural gas
NO_x	Nitrogen oxides
OECD	Organisation for Economic Co-operation and Development
OPEC	Organization of the Petroleum Exporting Countries
PCB	Polychlorinated biphenyl (toxic compound found in oil)
PINC	Pro-active investment in natural capital
PPP	Purchasing power parity (converting currency to reflect actual purchasing power in different countries)
PV	Photovoltaic
R&D	Research and development
REDD	Reduced emissions from deforestation and forest degradation
SCR	Selective catalytic reduction
UN	United Nations
UNEP	United Nations Environment Programme
UNFCCC	United Nations Framework Convention on Climate Change

US	United States of America
VOC	Volatile organic compound
WHO	World Health Organization
WWF	Worldwide Fund for Nature

1   Introduction

Climate change is widely held to be the most urgent issue facing humanity. After decades of study and debate, it is now generally accepted that we need to limit the global temperature rise to 2°C in order to avoid the prospect of catastrophic irreversible global warming, with many pushing for a lower target of 1.5°C.[1] According to the latest climate science, the two degree target means returning atmospheric greenhouse gas concentrations to 350 parts per million (ppm), though most climate policy is geared towards a target of 450ppm.[2] In 2009, concentrations were already at 399ppm.[3] To stabilise concentrations at 450ppm, global greenhouse gas emissions must be at least halved from 1990 levels by 2050, meaning that developed countries may need to cut emissions by 85 per cent.[4]

Saving the planet is not rocket science. We are not waiting for a technical breakthrough. Low-carbon energy sources and energy-efficient technologies have been around for decades, although performance is continually improving. So why has progress been so painfully slow? The reason is that there are political barriers to making the changes we need. Perhaps the most severe drawback is the lack of a universal mandate for change from an electorate confused by the science and told that the costs of achieving a stable climate are unaffordable.

Despite decades of research leading to a broad scientific consensus, many members of the public are still not convinced of the reality of climate change. A proliferation of websites, blogs and media articles is feeding a strong vein of denial that human activity is changing the climate. There is a fertile audience for these views – few people wish to believe that everyday actions such as driving, flying or powering their homes can influence the way that the planet works, or that individual action can have any impact.

It is all too easy to dismiss or ignore warnings of impending climate doom, because the impacts of greenhouse gases are invisible to most people. Unlike the lethal black smogs from coal burning that hung over many cities in the first half of the twentieth century, killing thousands, greenhouse gases are transparent and, at the concentrations found in the air, not directly harmful to human health. To assess the damage they cause, we have to rely on sophisticated computer models which attempt to replicate complex natural systems, giving scope for uncertainty. Also, the worst damages are projected to occur several decades into the future and in countries far away from those which are responsible for most of the emissions.

Indeed, for many in northern Europe, the main impacts of climate change are sometimes perceived to be a pleasant increase in balmy Mediterranean summers and a useful enhancement of crop yields.

For years, environmental campaigners have fought against the view that it is rash to spend huge sums fighting a threat that is not firmly proven to exist. We can't afford it, goes the refrain. It will damage the economy, reduce growth and threaten jobs. The money could be better spent tackling other global problems such as malaria or HIV infection. Underlying this scepticism is a deep-rooted conflict between the concepts of economy and environment. We have grown to regard the environment as a resource to be exploited in order to sustain and expand the economy. There is an apparent belief that human ingenuity will allow any depleted resource to be replaced with an alternative, provided the market is given free rein.

In reality, of course, the human economy is only a sub-system of the environment. Our true quality of life ultimately depends on maintaining healthy and vibrant ecosystems, clean water supplies and a stable climate. Yet the short-sighted culture of exploitation that has arisen makes it hard for democratic governments to act on climate change. Attempts to limit fuel use through taxation, for example, are met with determined resistance, and governments wishing to implement unpopular measures such as personal caps on carbon emissions might not survive the next election.

The climate debate has polarised into one of pressing immediate economic needs versus the long-term, uncertain and invisible threats of climate change. Yet this misses the bigger picture. Low-carbon policies often provide a whole range of additional environmental, social and economic benefits. These often overlooked co-benefits can help to offset the financial cost of the technology and boost its political acceptability. For example, switching from coal-fired electricity generation to renewable energy cuts air pollution from fine particles, acidic gases and ozone, significantly reducing death and chronic illness from respiratory diseases. Reducing the demand for imported oil and gas will improve national energy security and the balance of payments for most countries. Better insulation of houses can reduce winter deaths amongst the poor and elderly, and save money for householders. Energy-efficient technologies can boost the profitability and competitiveness of businesses. Many of these benefits are far more immediate and visible than the impacts of climate change, and can provide a much stronger motivation for supporting the move to a low-carbon society. For many low-carbon policies, we might argue that the co-benefits alone would justify their adoption even if climate change did not pose a threat.

This book evaluates the additional benefits that we can gain from strong action to tackle climate change: the Climate Bonus. Chapter 2 is a brief overview, showing how policies to reduce climate change are linked to a wide range of environmental, social and economic impacts. Chapters 3 to 8 then look in detail at how low-carbon policies can contribute to reduced air pollution, conservation of forests and other ecosystems, more sustainable agriculture, enhanced energy security, better use of finite resources, less waste, cost savings for households and businesses, employment benefits and healthier lifestyles, and Chapter 9 sums up.

Although low-carbon policies often provide sustainability benefits, there can sometimes be conflicts or trade-offs. In the popular media, two hotly debated examples are the visual impact of wind farms and the safety of nuclear power. Similarly, the current drive towards biofuels may lead perversely to the clearance of natural forests, which can have disastrous impacts both on biodiversity and on local communities. Often, there seems to be a lack of joined-up thinking by policy makers, exacerbated by the standard practice of treating issues such as climate change, air quality and energy security as separate policy issues handled by competing government departments. Consideration of the bigger picture can guide the development of an integrated strategy that can cut greenhouse gas emissions at the same time as delivering much wider benefits for society.

With sensible policy design, we can maximise the synergies and minimise the trade-offs between climate change and other aims, leading to a much more successful, popular and cost-effective programme of action. In this way, the threat of climate change can become a stimulus to escape our dependence on polluting and wasteful fossil fuel technologies and set a new course towards a cleaner, safer, healthier and more sustainable future.

2   The big picture

How climate policies are linked to co-benefits

This chapter gives an overview of the main sources of man-made greenhouse gas emissions, the policies used to reduce these emissions and how these policies are linked to co-benefits and conflicts.

Many of the co-benefits of climate policy are linked to a reduction in the use of fossil fuels, which produce 65 per cent of man-made greenhouse gases. The main source, of course, is carbon dioxide from burning fossil fuels for heat, power and transport, with smaller contributions from nitrous oxide produced during fuel combustion and methane emitted during coal mining and oil and gas production.

Important co-benefits are also linked to tackling emissions from land use: the carbon released when forests are cleared; and agricultural emissions of methane from livestock and nitrous oxide from fertilisers. Together these account for 24 per cent of man-made greenhouse gases. Finally, methane from decomposing organic waste and wastewater accounts for 3 per cent of emissions, and the remaining 8 per cent arises mainly from industrial processes such as cement making, chemical manufacture and the use of fluorinated gases[1] (Figure 2.1).

Fossil fuels and industrial processes

The main policies for cutting the climate impacts of fossil fuels are:

- cutting fossil fuel use through improving energy efficiency and promoting energy-saving behaviour – e.g. turning off unused lights and appliances, driving less and avoiding air travel;
- switching to low-carbon energy sources such as nuclear power and renewable energy, or switching from high-carbon coal to medium-carbon gas;
- carbon capture and storage (CCS) – separating CO_2 from flue gases at power stations, refineries or cement works, and storing it – e.g. in disused oil wells;
- reducing methane emissions from coal mines and oilfields by collecting the gas and using it for energy;
- reducing waste of materials – e.g. through reusing and recycling, to avoid the emissions associated with production of materials such as steel, chemicals, cement and plastic, and with manufacturing of finished goods;

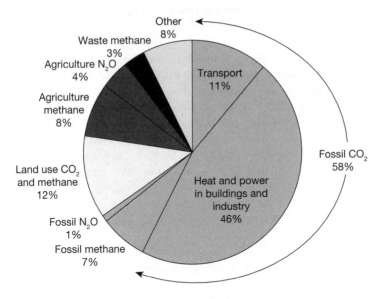

Figure 2.1 Global greenhouse gas emissions, 2008

Source: Data from EDGAR database (EC/JRC-PBL, 2011)

Notes: Other = CO_2 from cement (4.6 per cent), N_2O from chemicals (0.9 per cent) and F-gases from industrial processes (1.7 per cent).

Slight discrepancies between the percentages given here and those in the text are due to rounding up of figures.

- geo-engineering – e.g. through releasing sulphate aerosols into the stratosphere to reflect sunlight.

All of these options except CCS and geo-engineering will reduce the use of fossil fuels, giving rise to three major sets of co-benefits.

1. *Cleaner air*. Fossil fuel combustion is currently the main global source of air pollution. Cutting fossil fuel use will reduce emissions of fine particles, acidic sulphur dioxide and nitrogen oxides, and ozone-forming volatile organic compounds (VOCs), generating major co-benefits for human health and the environment.
2. *Cleaner, safer energy*. Cutting the use of fossil fuels will also reduce the impacts of extracting the fuels, such as water pollution and subsidence from coal mines; landscape impacts of open-cast coal mining; coal mine collapses; groundwater pollution from the extraction of shale gas; oil spills and blow-outs at oil and gas wells.
3. *Energy security*. Developing alternative energy sources can reduce dependence on increasingly scarce and expensive fossil fuels, which reduces the risks of price spikes, supply disruption and conflict. Certain renewable energy technologies can also improve access to energy for remote communities in developing countries.

In addition, cutting energy and material use by improving efficiency and changing behaviour can offer the following co-benefits.

- *Cost savings.* Cutting consumption of fuel, electricity or materials will generally save money, provided that the capital cost of the measure is not excessive.
- *Comfort.* Insulating and draught-proofing buildings increases comfort for the occupants, and efficient appliances and machinery can be quieter, lighter, cooler and generally more pleasant to use.
- *Conservation of scarce resources.* Material efficiency saves valuable resources such as metal ores, agricultural products, timber and water, and reduces the impacts of extracting the materials, such as air and water pollution and habitat loss.
- *Less waste.* Cutting the amount of material wasted reduces the costs and environmental impacts of waste disposal in landfills or incinerators.

Co-benefits are particularly high for low-carbon policies in the transport sector, which is the most damaging source of air pollution and is highly vulnerable to oil price shocks. Further co-benefits arise from measures that reduce the demand for motorised transport. These co-benefits include:

- *reduced congestion, noise and accidents*, as well as less vibration from heavy vehicles and lower costs for road maintenance;
- *time and cost savings* by videoconferencing or tele-working;
- *health and well-being benefits* from active transport (walking and cycling).

However, some low-carbon options are associated with potential conflicts:

- *Nuclear power* is associated with routine and accidental release of radionuclides, risks of nuclear proliferation and terrorism, the problem of disposal of radio-active waste and possible uranium supply issues.
- *Biofuels* can be associated with problems including biodiversity loss where biofuel crops replace natural ecosystems, displacement of local communities for biofuel plantations, the impact of biofuel crops on food prices and high use of energy, water, pesticides and fertilisers.
- *Renewable energy* faces issues over the intermittent supply of wind and solar power, high capital costs that can increase energy prices in the short term, landscape impacts of large wind or solar farms and associated power lines, loss of wildlife habitats as a result of tidal barriers or hydropower reservoirs and pollution and worker safety issues at some solar panel factories.
- *Reliance on rare metals* could be a problem for a number of low-carbon technologies such as wind turbines, solar panels and batteries for electric vehicles.
- *Geo-engineering* has potentially serious side effects, especially for ecosystems.
- *Energy and material efficiency* can stimulate greater economic growth and therefore lead to increased consumption and more emissions in the long term (the 'rebound effect').

- *Active travel* (public transport, walking and cycling) can be slower and less convenient, and discouraging air travel is likely to result in a loss of the enjoyment associated with foreign travel.

Land use and agriculture

A wide range of policies can be used to reduce climate impacts from agriculture and deforestation:

- forest protection and reforestation schemes, including payment to local communities to protect their forests;
- increasing carbon stored in soils – e.g. through reduced tilling (ploughing), adding manure or compost to the soil and building terraces to reduce soil erosion;
- agroforestry (growing trees on farmland);
- reducing over-application of fertilisers, to reduce nitrous oxide emissions;
- increased crop yields per hectare to reduce the pressure for deforestation;
- lifestyle choice by consumers to eat less meat;
- reducing waste of food, timber and paper.

The co-benefits are very significant, and include the protection of rapidly disappearing tropical forests as well as a major improvement in the sustainability of agriculture.

Forest protection

- protection of endangered species and increase in global biodiversity;
- reduction of floods, landslides and soil erosion, and protection of water supplies;
- social and economic benefits for local communities from forest resources, eco-tourism and carbon payments;
- aesthetic, cultural and recreational benefits from preservation of unique landscapes, habitats and species.

Agriculture

- Improved fertiliser management can cut air and water pollution such as nutrient overload that leads to algal blooms.
- Adding carbon-rich organic matter such as manure and compost to the soil can improve soil structure and water retention, boosting crop yields.
- Farmers benefit from reduced soil erosion, save money by using less synthetic fertiliser and diversify farm incomes through agroforestry.
- People enjoy health benefits from choosing to eat less meat and dairy produce, in cases where over-consumption was previously a problem.

However, there are some serious potential conflicts:

- Forest carbon payment schemes can lead to 'land grabs' in which local people lose access to forest land, or result in natural forests or grassland being replaced with monoculture commercial plantations, with adverse impacts on bio-diversity.
- Lower use of synthetic fertilisers could result in reduced crop yields per hectare, which could drive deforestation (to create more cropland) or lead to food security problems.

Waste management

Policies to tackle emissions from waste include:

- recovery of methane from landfill sites, preferably for use as a fuel;
- anaerobic digestion of organic waste to produce bio-gas for use as a fuel;
- composting of organic waste to avoid methane emissions if it was landfilled;
- energy recovery from waste incineration plants;
- recycling and waste prevention.

Co-benefits include:

- recovery of landfill gas reduces local air pollution and odour;
- composting and anaerobic digestion produce compost and digestate that can be used as soil improver and fertiliser;
- anaerobic digestion and energy recovery from waste and landfill gas displace fossil fuels and therefore reduce extraction impacts (oil spills, etc.) and increase energy security;
- recycling and waste prevention have major benefits attached to resource conservation, as described in the previous section.

Grouping the co-benefits

This wide range of climate policy co-benefits can be grouped under various headings, which will form the basis of Chapters 3 to 8:

- *cleaner air* (Chapter 3) – the health and ecosystem benefits of reduced air pollution resulting from a reduction in the use of fossil fuels;
- *greener land* (Chapter 4) – the benefits of climate policy in reducing deforestation and improving the sustainability of agriculture;
- *safe and secure energy* (Chapter 5) – how reducing our reliance on fossil fuels can improve energy security, provide safer and more affordable energy in the long term and reduce the impacts of fossil fuel extraction on health and the environment;
- *less waste* (Chapter 6) – the numerous benefits of moving to a more resource-

efficient economy, including conservation of scarce resources, financial savings
and reduction of the impacts of waste disposal;

- *a stronger economy* (Chapter 7) – the thorny question of whether we can afford
  to move to a low-carbon economy, and the all-important impacts on jobs,
  growth, productivity, fuel poverty and international trade and development;
- *health and well-being* (Chapter 8) – the surprisingly large benefits of low-carbon
  lifestyles for health and well-being, including the benefits of more walking and
  cycling, eating less meat and even, more controversially, a 'buy less, work less'
  lifestyle.

Within each chapter, there is also discussion of the key conflicts that can arise
between climate policy and other goals, and exploration of ways of avoiding or
minimising the conflicts while maximising the co-benefits. Finally, Chapter 9 pulls
together lessons learnt from the preceding chapters and suggests joined-up policies
to enable us to reap the full potential benefits of a low-carbon economy: the Climate
Bonus.

3 Cleaner air

Cutting fossil fuel pollution

Key messages

- *Air pollution causes serious health and environmental impacts.* Even developed countries with advanced pollution control technology are failing to meet air quality targets, and the problems are far worse in rapidly industrialising developing countries.
- *Climate change and air pollution are both mainly caused by our use of fossil fuels.* Climate policy based on cutting fossil fuel use can result in cleaner air, with benefits for health, ecosystems and the economy. In some cases, the air quality benefits can exceed the costs of climate action.
- *The most promising strategies* are based on cutting energy consumption and switching to cleaner energy sources. Cutting transport emissions is a priority, as traffic pollution causes severe health impacts.
- *Conflicts can arise* between climate change and air quality. Examples include switching to diesel for transport; air pollution from biofuel combustion; accidental release of radioactive pollution from nuclear power stations; and geo-engineering techniques that involve pumping sulphate into the atmosphere. These conflicts require careful management.
- *An integrated strategy* for addressing climate change and air quality can be cheaper and more cost-effective than separate strategies, because synergies can be exploited and conflicts avoided.
- *Reducing air pollution can provide a stronger motivation* for cutting fossil fuel use than climate concern, because the benefits are immediate, local and more easily quantified than climate benefits.

Acid, dust and ozone: the downside of fossil fuels

Exploiting fossil fuels has enabled dramatic improvements in the standard of living for many people. However, burning fossil fuels releases not only carbon dioxide but also pollutants including acidic gases, soot particles and toxic metals. Once in the air, these primary pollutants can react to form secondary pollutants such as ozone and fine particles. Both the primary and secondary pollutants are highly damaging to health and ecosystems.

Some fuels are more polluting than others. Natural gas is composed mainly of methane, with few impurities. It burns cleanly, with the only significant pollutants being the nitrogen oxides that are created during any combustion process. Coal and oil, however, contain complex mixtures of larger organic molecules together with impurities such as sulphur and toxic metals. They tend not to burn cleanly, resulting in the emission of unburnt hydrocarbons, soot, sulphur dioxide and toxic organic pollutants such as dioxins. Table 3.1 lists the main pollutants arising from fossil fuel combustion, with their sources and impacts.

Table 3.1 Air pollution from fossil fuels

Pollutant	*Source*	*Impacts*
Sulphur dioxide SO_2	From the sulphur contained in coal and oil, which is oxidised on burning.	• Causes lung irritation and coughing, tightening of the airways and damage to mucous membranes, especially in asthmatics. • Reacts with other pollutants such as ammonia to form fine sulphate particles, a key component of particle pollution (see below). • In foggy weather, dissolves readily in water droplets and can form acidic winter smogs. • Dissolves to form sulphuric acid which causes the acidification of soils and freshwater ecosystems.
Nitrogen oxides (NO_x) A mix of nitrogen oxide (NO) and nitrogen dioxide (NO_2), collectively known as NO_x.	Created during any high-temperature combustion process, from nitrogen in the air and in the fuel.	• NO_2 causes lung irritation and lower resistance to respiratory illness such as pneumonia and bronchitis, especially in young children and asthmatics. • NO_2 reacts with VOCs (see below) to form ground-level ozone. • Reacts with other pollutants to form fine nitrate particles. • Contributes to acidification and nutrient overload (see below).
Particulate matter (PM) Tiny solid or liquid particles suspended in the air. PM_{10} are particles less than 10 microns in diameter; $PM_{2.5}$ (fine particles) are less than	Natural sources include dust, volcanoes and forest fires. Man-made PM is mainly from fuel combustion. Coarse particles (over 2.5 microns) are mainly ash and mineral dust. Fine particles are generally secondary pollution formed by condensation or nucleation	• Damaging to health, especially fine particles as they can be inhaled deeper into the lungs. Particles tend to attract toxic and carcinogenic pollutants which stick to the particle surface. Ultrafine particles can penetrate into the bloodstream and be transported to other parts of the body, damaging blood vessels and the brain. $PM_{2.5}$ pollution is associated with increased illness and premature death from lung cancer,

Table 3.1 continued

Pollutant	Source	Impacts
2.5 microns; $PM_{0.1}$ (ultrafine particles) are less than 0.1 microns.	of the vapours and gases that result from combustion. They include sulphate and nitrate aerosols, organic carbon, black carbon and heavy metals. Around 85% of man-made particulate pollution is secondary.	irregular heartbeat, heart failure, bronchitis, asthma attacks, low birth weight, premature birth and reduced lung development in children. • PM influences climate in different ways. Sulphate aerosols reflect sunlight, causing cooling, but black carbon particles absorb sunlight, contributing to warming.
Volatile organic compounds (VOCs)	A wide range of organic compounds typically emitted in vehicle exhausts, from oil refining and from industrial processes that use organic solvents.	• React with NO_x to form ground-level ozone. • Some VOCs such as benzene and 1,3-butadiene are health hazards. Benzene is carcinogenic and is known to cause leukaemia.
Methane (CH_4)	Agriculture, waste disposal, coal mining and oil and gas production.	• Reacts with NO_x to form ground-level ozone. • A powerful greenhouse gas with a 12-year lifetime: 25 times stronger than CO_2 over 100 years; 72 times stronger over 20 years.
Ozone (O_3)	A secondary pollutant, caused by complex reactions involving nitrogen oxides, volatile organic compounds, methane and carbon monoxide, especially in the presence of sunlight.	• Stratospheric ozone (above 10km) is the naturally occurring ozone layer that protects us from ultraviolet radiation. • Tropospheric ozone (i.e. below 10km) is a greenhouse gas. • Ground-level ozone, caused by pollution, is a major health hazard, causing lung irritation and breathing difficulties including asthma attacks. It restricts plant growth, reducing crop yields and damaging ecosystems. Ground-level ozone does not contribute to stratospheric ozone.
Heavy metals including arsenic, cadmium, chromium, mercury, copper, selenium, tin, lead, nickel and vanadium.	Coal and oil burning.	• Can cause deterioration of the immune system, nervous system and metabolic system, and some are suspected to be carcinogenic. • Accumulation in the environment can disturb nutrient uptake and be toxic to plants, wildlife and livestock.
Dioxins	Combustion of organic substances containing	Persistent toxins which accumulate in fat deposits in humans and animals, with

Pollutant	Source	Impacts
	chlorine, including coal, oil, waste incineration, forest fires and industrial processes.	adverse effects on the immune system, nervous system, reproductive system and sexual development. Thought to be carcinogenic.
Polycyclic aromatic hydrocarbons (PAHs)	Combustion of carbon-based fuels.	Carcinogenic pollutants.

The following sections briefly describe the main impacts of fossil fuel pollution on health and ecosystems.

Health effects

Air pollution, especially from ozone and fine particles, is linked to a range of health problems including heart, lung and respiratory diseases and lung cancer. Concern over air pollution dates from the time when the use of coal first became widespread. In thirteenth-century London, for example, people turned from wood to coal as the surrounding forests shrank. The smoke from thousands of coal fires would hang over the city in a dense sulphurous fog, especially in cold, still conditions, causing lung irritation. Edward I banned coal burning in 1306, but even though the offence carried the death penalty at first, people had little choice but to continue using coal. The problem worsened during the Industrial Revolution of the mid-1700s to mid-1800s, leading to the infamous 'pea soup' fogs of Victorian London, and culminated in the premature deaths of over 4,000 people in a week during the Great Smog of 1952.[1] Similar problems with smog occurred in industrialised cities around the world.

Governments responded by introducing laws to control pollution, such as the US Air Pollution Control Act of 1955 and the UK Clean Air Act of 1956, which established smokeless zones in many cities. Further legislation was introduced to address the problem of acid rain which became apparent during the 1970s and 1980s, leading to the installation of pollution control systems at power stations. This had a significant but limited effect – few power stations are currently (2012) fitted with the best available technologies. In Europe and North America, however, a significant reduction in emissions was achieved almost by accident, as market liberalisation led to the 'dash for gas' in which coal and oil-fired power stations were replaced by cleaner natural gas turbines. At the same time, the spread of gas and electricity grids led to the replacement of domestic coal fires with more convenient gas or electric heating. The end result is that coal smogs in developed countries are now largely a thing of the past. In rapidly industrialising countries such as China, however, where large numbers of coal power stations are being built with low levels of pollution control, they are a serious and growing problem.

Meanwhile, the rise of the car has resulted in transport pollution overtaking coal pollution as the major threat to health in developed countries. Winter smogs have been replaced by summer smogs, which are caused by the action of sunlight on a mix of transport fumes and industrial pollution (Box 3.1).

Box 3.1 Smog

Coal smogs (winter smog)

Coal is mainly carbon, but it also contains small amounts of sulphur and variable amounts of rock-forming minerals such as silicates and carbonates. When coal is burnt, most of the carbon reacts with oxygen to form carbon dioxide, and the sulphur forms sulphur dioxide. The minerals become ash, much of which remains in the grate as 'bottom ash', but many fine ash particles become entrained in the smoke as 'fly ash', together with particles of unburnt carbon (soot).

The term 'smog' was coined in 1905, meaning a cross between smoke and fog. Winter smog forms when coal smoke meets damp, cold air, especially under the cold, still conditions that often develop during winter. Moisture in the air condenses around the ash particles in the smoke to form a fog, and the sulphur oxides then dissolve in these droplets of water to form a fine mist of sulphuric acid. When inhaled, this acidic mist causes irritation and damage to the mucous membranes lining the lungs, and can cause constriction of the bronchial airways. Children, the elderly and the sick are most at risk – children because their respiratory systems are not fully developed, and the sick and elderly because theirs are already damaged through age or illness.

Photochemical smogs (summer smog)

Photochemical smogs tend to form during periods of hot, sunny, still weather. Sunlight causes nitrogen oxides from fossil fuel combustion to react with volatile organic compounds from industrial and vehicle pollution, forming a mixture of ground-level ozone and fine particles that irritate the lungs and can cause asthma, bronchitis, heart disease and early death.

The process of ozone formation can take several days, so high ozone levels can often be found many miles downwind of the source of pollution. Paradoxically, ozone levels can be higher in rural areas than in the cities that generate the pollution. This is partly because ozone (O_3) reacts readily with the nitric oxide (NO) in vehicle exhaust to form nitrogen dioxide (NO_2). This 'scavenging' process reduces ozone levels close to roads. Certain volatile organic compounds, such as isoprene, can be emitted by vegetation in hot weather, and this can also enhance ozone formation in the countryside. Finally, the haze of pollution that hangs over many cities cuts down the amount of sunlight reaching the ground, which can also slow down ozone production until the pollution diffuses out into surrounding areas.

Ecosystem effects

Although human health is the leading cause for concern, air pollution also damages ecosystems through acidification, nutrient overload (eutrophication), the accumulation of toxic substances in the food chain, and ozone damage to plants. These effects are briefly described in the following sections.

Acidification

Acid rain forms when sulphur dioxide and nitrogen oxides dissolve into rainwater. Fine dry particles of sulphates and nitrates can also be deposited directly onto soil or plants. Both of these processes contribute to the acidification of soils and freshwater ecosystems. As acid pollutants can travel hundreds of miles, large regions can be affected, often far from the original sources of pollution. Apart from causing direct damage to plant foliage and aquatic organisms, acid conditions can make it harder for plants to absorb soil nutrients such as calcium, making plants weaker and more vulnerable to other environmental stresses. Acid rain can also cause extensive damage to crops and buildings.

During the 1970s and 1980s, many streams and lakes in Europe and North America became acidic and unable to support life. This led to a major effort to reduce emissions from power stations through legislation such as the Large Combustion Plant Directive of 2001 in Europe and the 1990 amendment to the Clean Air Act in the United States. Acid emissions were greatly reduced, but many ecosystems have been slow to recover. In the Adirondack Mountains of New York State, for example, between a third and two-thirds of the lakes have acidified due to air pollution, and between a tenth and a quarter are highly acidic, with pH levels below 5.0. This has led to a reduction in fish diversity, with the extinction of some species of fish and many completely fishless lakes.[2] Recovery is also slow in Europe, with soil acidity remaining high in many areas and a fifth of trees still showing signs of damage.[3] Acid rain is an increasing problem in the developing world, with a third of China's land area being affected.

Eutrophication

Eutrophication occurs when excess nutrients (mainly nitrogen, phosphorous and potassium) are added to soil or water, stimulating plant growth that can damage the balance of natural ecosystems. The main cause of eutrophication is run-off of fertiliser from farmland (see Chapter 4) and discharge of liquid effluents such as sewage, but the deposition of nitrogen oxides and ammonia from the air also plays a part.

Plant species are very sensitive to the availability of nitrogen. Some require nitrogen-rich soils, while others are adapted to live in places where there is very little nitrogen (an extreme example being the carnivorous plants that need to supplement their nitrogen uptake with insects). Extra nitrogen means that nitrogen-loving plants can displace the existing flora, leading to significant loss of species. In forests, for example, the typical woodland flora may be displaced by brambles and nettles. In aquatic ecosystems, too much nitrogen can lead to excessive growth of algae, leading to the algal blooms often seen on lakes and in coastal waters. These can use up all the oxygen in the water, leading to the death of fish and other organisms.

Toxicity

Toxic pollutants such as dioxins, mercury and other heavy metals tend to accumulate in the bodies of animals at the top of the food chain, with adverse effects on behaviour, reproductive capacity and health. In the United States, for example, where 50 per cent of mercury and 62 per cent of arsenic emissions come from coal- and oil-fired power plants, high levels of mercury have been found in many species of birds, fish and mammals, with 75 per cent of trout sampled in the state of New York having mercury concentrations above the level of concern.[4] There is also a health risk to humans from eating fish contaminated with mercury. Around 6 per cent of American women of child-bearing age have unsafe levels of mercury in their blood.[5] A study in 2010 estimated that the reduction in children's IQ due to ingestion of mercury in the food chain resulted in an economic cost to society of almost $3 billion due to reduced productivity.[6]

Ozone damage

As well as causing health problems in humans, high ozone levels lead to restricted plant growth which damages ecosystems and reduces crop yields. Ozone enters plants through the pores (stomata) in their leaves. It is highly reactive, and can cause direct damage to plant tissue as well as reducing the rate of photosynthesis and accelerating leaf ageing. A high ozone episode in Greece caused 100 per cent loss of lettuce and chicory crops, for example.[7] Ozone also damages polymers such as the rubbers used for many car parts.

Ozone levels have doubled since pre-industrial times, due mainly to fossil fuel burning. One study estimated that ozone damage could reduce global vegetation growth by between 14 and 23 per cent between pre-industrial times and 2100, and by up to 30 per cent in the worst-affected regions, which include North America, Europe, India and China. For Asia, some sources suggest crop yield losses of 10–30 per cent under current levels of ozone pollution, with a particularly high risk for the Indo-Gangetic plain, one of the most important agricultural regions in the world. Global crop losses for wheat, rice, soybean and maize could be worth $16–30 billion per year, with 40 per cent of this loss in China and India.[8]

Ozone damage to plants also has the effect of reducing the amount of carbon absorbed from the atmosphere as plants grow, which leads to an increase in carbon dioxide in the atmosphere.[9] In fact, it has been suggested that this damage to the carbon sink could have a greater warming effect than the effect of ozone as a greenhouse gas in the troposphere.[10]

The big clean-up: is it working?

Great progress has been made in reducing air pollution over recent decades, especially through regulations introduced in developed countries. Many power stations are now fitted with electrostatic precipitators to trap ash particles, low-NO_x burners and catalytic reduction units to reduce nitrogen oxide emissions, and flue gas desulphurisation equipment to remove sulphur dioxide. Vehicles are becoming cleaner, with catalytic converters and filters to reduce emissions of nitrogen oxides, carbon monoxide and soot particles, while new regulations restrict the amount of sulphur in transport fuels.

Unfortunately, these technical advances have been partly offset by the rapid growth in vehicle use and electricity demand. In Europe, freight transport has increased by 30 per cent and passenger travel by 20 per cent since the early 2000s. Increased shipping activity is offsetting the emission reductions arising from new legislation on marine pollution.[11] The increasing use of household electrical appliances ensures that electricity demand continues to grow by around 2 per cent a year, despite efficiency improvements. Nevertheless, sulphur dioxide emissions in Europe were halved between 1999 and 2009, while nitrogen oxide emissions fell by 27 per cent, PM_{10} by 16 per cent and volatile organic compounds (which lead to ozone formation) by 34 per cent.[12]

But despite this dramatic fall in emissions, air pollution limits are still being exceeded. Twelve of the EU-27 countries are expected to have missed their 2010 emissions targets for nitrogen oxides, volatile organic compounds and ammonia.[13] In addition, most countries of the EU failed to meet the 2005 and 2010 air quality targets for particle pollution and nitrogen oxides (Figure 3.1, see plate section).[14] Many cities are highly polluted – the proportion of the urban population of the EU exposed to pollution levels above the World Health Organization guidelines is 80–90 per cent for particle pollution and more than 95 per cent for ozone.[15]

In 2011, air pollution was estimated to cause at least 570,000 premature deaths across Europe, corresponding to over 6 million years of life lost, and total health impacts were valued at €678 billion.[16] An earlier study estimated that Europe loses 200 million working days a year to pollution-related illness and that over 6 per cent of deaths and illnesses in young European children are caused by air pollution.[17] There is a similar picture in other developed nations. In the United States, more than half of all citizens live in places that fail to meet the air quality standards set by the Environmental Protection Agency,[18] and health damages are estimated at $120 billion.[19]

Crop and ecosystem damage is also significant. The fall in sulphur emissions in Europe has reduced the area affected by acidification, but nitrogen emissions continue to cause problems with acidification, eutrophication and ozone damage. Between 1996 and 2008, 30–69 per cent of crops in Europe were exposed to ozone concentrations above the target level. In 2010, 69 per cent of sensitive European ecosystems were at risk of eutrophication and 54 per cent were thought to exceed critical levels of mercury pollution.[20] More than half of all natural habitats in the UK have harmful levels of acidity or eutrophication.[21]

The situation in developing countries is even worse. While developed countries generally have strict emission standards but high energy consumption per person, developing countries tend to have the opposite problem: energy use *per capita* is only a fifth of that in developed countries,[22] but there are very low standards of pollution control. Many rapidly industrialising countries such as China face huge challenges from pollution, both from the growing use of coal for power generation and from increasing vehicle use. It is estimated that more than 70 per cent of the urban population of Asia live in cities where levels of particulate pollution exceed 60 micrograms per cubic metre, which is the most lenient standard used in Western Europe or North America. More than half of China's large cities fail to achieve national standards for air quality, and the health impacts were estimated to cost $23 billion in 2003.[23] The rate of premature death from outdoor air pollution is now higher in developing than in developed countries, and the problem is expected to worsen as urbanisation increases and the population ages.

Globally, this all adds up to a trend of increasing levels of air pollution around the world. The World Health Organization estimates that outdoor air pollution from particles kills 1.3 million people each year, and indoor air pollution (largely smoke from open fires and stoves in developing countries) kills a further 2 million.[24] Other estimates are far higher – the European Environment Agency estimates seven times more premature deaths from outdoor particle pollution in Europe (500,000 compared to 70,000 in the WHO study), and a study for the OECD estimates global deaths from air pollution as 4.5 million per year in 2000, increasing to 13 million per year by 2050 under a business-as-usual scenario.[25] Another study estimates that ozone pollution alone will cause 2 million premature deaths world-wide by 2050, with associated health costs of $580 billion.[26] Air pollution has been estimated to cost around 2–5 per cent of GDP in China, Europe, Russia and the United States, and costs in China could increase to 13 per cent of GDP by 2020.[27]

Air pollution has been traditionally considered as a local or regional problem, because many of the common air pollutants tend to be removed from the atmosphere within a few days or weeks by rain or by chemical reactions, meaning that pollution would not travel far from its source. However, recent studies have shown that air pollution is fast becoming a global issue due to the increasing use of fossil fuels worldwide, and the use of taller chimney stacks for factories and power stations which means that pollution travels further. Ozone, for example, can travel up to 500 kilometres in a day, and lasts for several weeks before being broken down to oxygen. Plumes of pollution from highly polluted megacities[28] have been shown to travel large distances around the world, with plumes from Asia reaching the

United States, and plumes from the United States extending as far as Europe.[29] Background ozone levels in the northern hemisphere have been rising by up to 10 micrograms per cubic metre ($\mu g/m^3$) per decade over the last 20 to 30 years, reaching 80 $\mu g/m^3$ in Western Ireland, where there is relatively little local pollution. This high background level makes it very difficult to achieve the UK air quality target of 100 $\mu g/m^3$.[30]

Will further technical advances succeed in cleaning up the air? End-of-pipe measures such as filters and catalysts can be very effective at cutting emissions from individual sources such as vehicles and power stations, and there is certainly considerable scope for extending the use of pollution control in developing countries. However, there are limits to how much pollution can be abated in this way. Nitrogen oxide pollution, for example, can be reduced by up to 95 per cent using catalytic reduction technology, but the last 5 per cent remains an intractable problem. Other problems with end-of-pipe measures are discussed in the section on 'Conflicts' below.

Even as we try to reduce air pollution, climate change is expected to exacerbate the problem. As summers in many places become hotter, drier and sunnier, photochemical smog will worsen. The high temperatures will increase emissions of methane from soils, wetlands and melting permafrost, and emissions of volatile organic compounds from vegetation, leading to more ozone formation in areas where nitrogen oxide emissions are high. Although changes in rainfall patterns are hard to predict, the frequency of droughts is expected to increase, so that less pollution will be removed by rain in the affected areas (though other areas may have extra rainfall). Droughts also mean that less pollution will be removed from the air by plants, as their leaf pores (stomata) close to preserve water. A glimpse of what this could mean was offered during the exceptionally hot summer of 2003 (Box 3.2 and Figure 3.2, see plate section).[31] By 2040, summers like this could be the norm in Europe.[32]

Box 3.2 The 2003 ozone bubble

In the first 2 weeks of August 2003, Europe was hit by an extreme heat wave and drought. Record high temperatures up to 40°C were reached, and slow air movement from the east allowed extra pollution to drift into Western Europe and the UK. In the UK, ozone levels were more than twice the air quality target. The poor air quality was estimated to account for around 700 of the 2,000 heat-wave-related premature deaths over a 7-day period. The UK deaths were concentrated in the London area and South East England, where ozone levels were highest.

The high ozone levels were partly related to the effect of the high temperatures on vegetation. Plants respond to heat stress by producing large amounts of isoprene, which is a volatile organic compound and ozone pre-cursor. High isoprene levels were thought to be responsible for around a third of the UK ozone production during the heat wave.

Across Europe, temperatures and ozone levels were even higher, resulting in an estimated 45,000 extra deaths, of which around a third were probably due to air pollution.[33] In the Paris area, the death rate doubled; and on the worst day, mortality was six times higher than normal, with half of these deaths attributed to high ozone levels.[34] Increases in death rates were highest in elderly people and in lower socio-economic groups.

Relevant climate policies

There are strong synergies between reducing air pollution and tackling climate change. Both result largely from our reliance on fossil fuels, which supply 81 per cent of world primary energy (Figure 3.3) but account for two-thirds of greenhouse gas emissions (see Chapter 2) and are the main source of outdoor air pollution worldwide. In the UK, for example, fossil fuels produce 91 per cent of sulphur dioxide emissions, 88 per cent of carbon monoxide, 97 per cent of nitrogen oxide and 58 per cent of particle emissions, as well as being a major source of carcinogens such as benzene and heavy metals such as mercury and nickel (Figure 3.4).

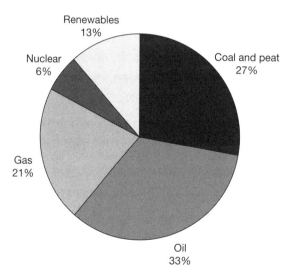

Figure 3.3 Fossil fuels supplied 81 per cent of world primary energy in 2009
Source: Data from IEA, 2011b

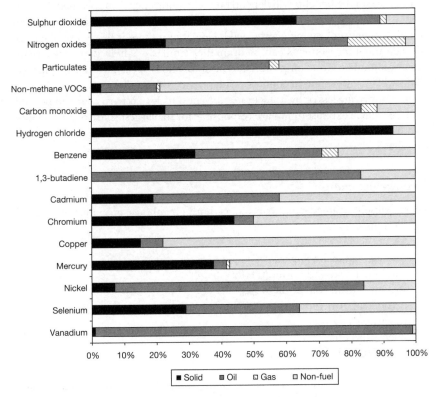

Figure 3.4 Proportion of air pollutants arising from fossil fuel use in the UK, 2006
Source: Data from NAEI, 2008

A number of policies can tackle both climate change and air pollution at once, the first three of which exploit the fact that both are linked to our use of fossil fuels:

1   cutting total energy consumption;
2   switching to cleaner and lower-carbon energy sources;
3   reducing emissions of air pollutants that are also greenhouse gases (black carbon, methane and ozone precursors);
4   increasing forest cover (forests can filter out pollutants from the air as well as storing carbon).

Cutting energy consumption

Cutting overall energy use is one of the quickest and most cost-effective ways of cutting both greenhouse gas emissions and air pollution. Developing clean energy infrastructure will take several years, but switching off an unused appliance or installing an energy-efficient light bulb achieves instant savings – and usually also saves money for businesses and consumers.

The scale of the benefits will depend on the existing energy supply mix: if a large share is from renewable technologies or nuclear power, then the carbon and air pollution benefits of saving energy will be low; but if there is heavy use of coal or oil, then the improvements can be dramatic. Similarly, where there are already strict pollution controls on power stations and vehicles, the air quality improvements will be lower than in countries with poor control. Cutting transport fuel use – either by driving less or by using more efficient vehicles – is particularly important, because vehicle exhaust fumes are responsible for the largest share of air pollution health impacts. Methods of cutting energy use are discussed in more detail in Chapter 5.

Low-carbon energy

Energy supply strategies that can reduce both greenhouse gas emissions and air pollution include:

- cleaner coal technologies;
- combined heat and power (CHP);
- switching from coal or oil to gas;
- carbon capture and storage (CCS);
- renewable energy;
- nuclear power;
- fuel cells, hydrogen and electric cars.

These technologies have very different implications for air pollution: some offer a large benefit compared to conventional fossil fuels, and some offer only a small improvement, or even worsen some emissions.

Cleaner coal technologies

Most coal-fired power plants in operation worldwide are conventional steam cycle (CSC) plants in which pulverised coal is burnt to raise steam, which powers turbines to generate electricity. However, considerable effort has been directed at developing alternative 'clean coal' technologies with higher efficiency and lower emissions. The main options are:

- *Supercritical (SC) and ultra-supercritical (USC) pulverised coal.* These plants operate at high pressures and temperatures, thus raising the efficiency from 39 per cent for the best CSC plants to 42–47 per cent. If advances in materials technology permit even higher operating temperatures, efficiency could reach 50 per cent in future.[35]
- *Integrated gasification combined-cycle (IGCC).* Coal is gasified, by partially burning it at high pressure, to produce a fuel gas that is a mixture of carbon monoxide and hydrogen. After removal of particles, sulphur and nitrogen oxides, this gas then fuels a 'combined-cycle' gas turbine generator, using surplus heat to raise

steam for a secondary steam turbine. Emissions of sulphur and nitrogen oxides are very low, and although the efficiency of the four existing plants is 40–43 per cent, there is the potential for future designs to reach efficiencies of 50 per cent. There is also the potential to integrate the system with carbon capture and storage (CCS) technology, by separating out the carbon monoxide from the fuel gas and converting it to carbon dioxide for storage. The remaining hydrogen can be burnt to generate heat or power, or used in fuel cells or for transport fuel. Alternatively, the fuel gas can be used to synthesise liquid transport fuel. Although this possibility has led to renewed interest in IGCC, in view of the increasing scarcity of conventional oil (see Chapter 5), significant technical obstacles remain to be overcome.

- *Fluidised bed combustion (FBC).* High-pressure jets of air are used to suspend the fuel in a 'fluidised bed', thus allowing more efficient mixing of air and fuel. The higher combustion efficiency means that operating temperatures can be lower, thus substantially reducing nitrogen oxide emissions. Adding limestone to the fuel allows 95 per cent of sulphur emissions to be removed in the fuel bed; this is more efficient than removing sulphur by 'scrubbing' the flue gases. Fluidised bed combustion is suitable for a wide variety of solid fuels including waste and biomass.

To date (2012), there has been little uptake of these technologies. Under a business-as-usual scenario, the International Energy Agency (IEA) expects pulverised coal technology to continue to dominate through to 2050, though supercritical and ultra-supercritical would gradually replace subcritical generation. Fluidised bed combustion is expected to remain a niche technology, mainly used for low-quality solid fuel, and there will be a small uptake of IGCC. Under the IEA's climate-friendly BLUE Map Scenario, however, most coal plants would be fitted with CCS technology. This would lead to a greater uptake of IGCC because this is slightly cheaper than pulverised coal when CCS is fitted.[36]

Despite the potential for cleaner coal technologies to reduce emissions, the synergy between climate and air quality objectives is limited. Supercritical and USC plants can offer climate benefits, but little improvement in air quality; FBC and IGCC plants, on the other hand, offer significant air quality benefits, but little improvement in carbon emissions. The main area of interest is the potential combination of IGCC plants with CCS, which could offer both air quality and climate benefits in the future, if the considerable technical obstacles to both technologies can be overcome.

Combined heat and power (CHP)

In conventional coal or gas power plants, 45–65 per cent of the energy contained in the fuel is wasted as heat that is lost in the cooling towers. The idea behind combined heat and power (CHP) is that the waste heat from the power generation process is recovered and used to provide space heating or cooling for nearby buildings.[37] This reduces the demand for fossil fuels or grid electricity for space heating and cooling, and hence reduces emissions of greenhouse gases and air pollutants. The overall efficiency of the process is typically 75–80 per cent, and can even reach 90 per cent in the best modern plants.[38]

Any kind of fuel can be used in a CHP system: coal, gas or biofuels. The system is usually set up to meet the required demand for heat; any surplus electricity is sold back to the grid. The main limiting factor is the availability of nearby buildings where the heat can be used: it has been mainly applied in new developments where a CHP plant is integral to the design. Micro-CHP systems that heat and power a single building are now being developed, and can overcome this limitation, although they are less cost-effective. Combined heat and power is widespread in Denmark, Finland and the Netherlands, as well as in Russia and China, and currently provides 10 per cent of world electricity generation. There is potential for a much wider uptake in other countries, especially if regulations are reformed to make it easier for CHP operators to sell electricity back to the grid. However, concern has been raised that replacing large rural power plants with local CHP could increase air pollution in urban areas.[39]

Fuel switching: from coal or oil to gas

Natural gas burns more efficiently and is less carbon-intensive than coal and oil. When burnt in an efficient combined-cycle gas turbine, it produces around half of the total greenhouse gas emissions for each unit of electricity produced and very little sulphur dioxide, particulates or heavy metals. However, it does produce significant amounts of nitrogen oxide, typically at around 500g/MWh compared to 700g/MWh for coal.[40]

In the UK, the 'dash to gas' for power generation reduced carbon dioxide emissions from 187 million tonnes in 1970 to 150 million tonnes in 2000, and caused sulphur dioxide emissions to plummet from over 6 million tonnes to 1.2 million tonnes. Similar trends were seen across Europe, where switching from coal or oil to gas accounted for about a third of the 70 per cent fall in emissions of sulphur dioxide from European power stations between 1990 and 2008.[41]

Although natural gas is cleaner than coal and oil, the potential for fuel switching is constrained by its availability, with conventional reserves becoming depleted in many countries, although shale gas production is increasing. Also, the increased use of gas (without carbon capture) is not sufficient to limit the global temperature rise to 2°C, especially since recent work suggests that methane leaks during production and distribution may offset the carbon benefits more than was previously thought, especially for shale gas (see Chapter 5). Natural gas is therefore generally viewed as a transition fuel: useful in the short term, but ultimately needing to be replaced by more sustainable energy sources.

Carbon capture and storage (CCS)

Carbon capture and storage (CCS) is widely viewed as an essential strategy to combat climate change. It involves capturing the carbon dioxide produced during fossil fuel combustion, and storing it indefinitely, typically by injecting it into porous rock formations such as exhausted oil and gas reservoirs or saline aquifers. Trials involving the storage of gas in depleted oil reservoirs are under way in several locations, but technology for efficiently capturing carbon dioxide has only been demonstrated in small-scale test plants. There are three main types of carbon capture.

1   *Post-combustion.* Carbon dioxide is separated out from the exhaust gases of a power station or industrial process, usually by passing the gases through a liquid solvent such as monoethanolamine (MEA), where the carbon dioxide (which typically forms 3–15 per cent of the gases) dissolves and is retained. The carbon dioxide is then extracted by heating the solvent: 87–90 per cent can be captured. This technology can be retrofitted to existing power stations. It has been demonstrated, but not at a commercial scale.

2   *Pre-combustion.* Carbon dioxide is removed from the fuel before combustion. The fuel is heated in a reactor with steam and air or oxygen to produce a synthetic gas ('syngas') consisting mainly of carbon monoxide and hydrogen. The carbon monoxide then reacts with steam in a second 'shift' reactor to produce carbon dioxide and more hydrogen. The CO_2 can then be separated out, leaving the hydrogen to be used as a carbon-free fuel. This technique can be applied at IGCC plants (see above) or at gas-fired plants, but it is not possible to retrofit it at existing conventional coal power stations. The technology is not yet fully developed, but it could eventually (in 10–20 years) be cheaper and more efficient than post-combustion capture, extracting 89–95 per cent of the CO_2 and with a lower energy penalty.

3   *Oxyfuel combustion.* Fuel is burnt in oxygen instead of air, to produce a flue gas that is 80 per cent CO_2, with the remainder being mainly water vapour. It is then relatively easy to remove the water vapour by cooling and compressing the flue gas, so that the remaining CO_2 can be sent for disposal (after removal of pollutants such as sulphur dioxide). The CO_2 removal rate could be as high as 95–98 per cent. The main disadvantage is the cost and energy penalty involved in separating oxygen out of the air. This technology is not yet operational.

The process of separating, compressing, transporting (usually by pipeline) and storing the carbon dioxide increases fuel consumption by 25–40 per cent for conventional coal power stations, 15–25 per cent for IGCC and 10–25 per cent for gas combined-cycle plants.[42] This 'energy penalty' leads to extra environmental impacts: direct emissions of air pollutants from fuel combustion; and indirect impacts from extracting, processing and transporting the additional fuel to the power plant. The energy penalty offsets the carbon savings, so that although around 90 per cent of the carbon dioxide can be captured, the overall greenhouse gas reductions per unit of electricity produced are around 72 per cent for conventional coal plants, 59 per cent for gas-fired power stations and 81 per cent for IGCC.[43]

The impact on air pollution is not yet clear, but for post-combustion separation, it seems that emissions of NO_x will increase in line with the amount of extra fuel consumed. Emissions of sulphur dioxide will fall, because the sulphur content of the flue gases must be reduced to very low levels to avoid degradation of the amine solvent and corrosion of equipment. Particle emissions must also be reduced to avoid contaminating the solvent, although the reductions will be offset by increased emissions from coal mining and transport so that the net effect is uncertain. Emissions of ammonia could increase by a factor of 10 to 25, due to degradation of the amine solvent, although total emissions would still be relatively low compared to existing emissions from the agriculture sector. For pre-combustion removal and the oxyfuel process, emissions of all these pollutants will be lower than for conventional combustion. Emissions of methane from extraction of coal and gas will increase for all CCS options. Further environmental impacts will arise from the manufacture, use and disposal of solvents, the landscape damage from additional mining of coal and limestone, and the additional disposal of ash and flue gas treatment residue.[44]

Although storage of carbon dioxide in well-characterised oil and gas reservoirs is not expected to result in significant routine leakage, it is not possible to rule out leaks of carbon dioxide from pipelines and from unexpected events such as earthquakes, groundwater movements, leaks from unmapped faults and fractures (especially in large saline aquifers) or well blow-outs. As well as reducing the effectiveness of CCS as a carbon removal option, leaks could have safety implications in inhabited areas as atmospheric concentrations of over 10 per cent of carbon dioxide can cause suffocation. Another disposal option is injection of carbon dioxide into the deep sea, but this will alter ocean chemistry and could have significant consequences for marine organisms and ecosystems, especially as deep-water fish are more sensitive to environmental changes than shallow-water fish.[45]

Although many development projects are planned, it is not yet clear whether CCS technology can be deployed quickly enough to play a major role in reducing carbon emissions within the timescale needed to limit warming to 2°C. It also may not be cost-effective to retrofit expensive CCS equipment to older coal-fired plants that are nearing the end of their life. There are also questions over the size of the available underground storage capacity, with estimates varying wildly.[46] One study estimated that the storage available in the EU is only enough for 20 years of carbon emissions at current rates.[47] Despite these uncertainties, one promising application for CCS could be in combination with biomass combustion, where storage of the CO_2 would effectively act as a carbon sink, locking away CO_2 that has been removed from the atmosphere during plant growth.[48]

Renewable energy sources

Most renewable energy sources – wind, wave, tidal, solar, hydro and geothermal energy – produce no emissions during operation, though emissions do arise when the equipment is manufactured and installed, especially during production of energy-intensive raw materials such as steel, concrete and silicon. Biofuels, on the other hand, produce emissions when they are burnt, as well as during the production of the fuel and feedstock.

Biofuel emissions are highly variable, depending on the feedstock, the fuel production method, the combustion method and the degree of pollution control. Feedstocks can include energy crops such as sugar cane or corn; wood and tall grasses; waste material such as forestry waste, food waste, dung and crop residues; and algae. These can be converted into solid, liquid and gaseous fuels, which can be used for household heating and cooking, industrial heat, electricity generation or transport. In almost all of these applications, the biofuel is ultimately burnt, generating direct emissions to air. The one exception would be if a gaseous biofuel were used to power a fuel cell, which generates no direct emissions.

Most biofuels are low in sulphur, except for bio-gas derived from certain sources such as protein-rich waste, where high sulphur levels can be reduced by wet or dry scrubbing before the gas is used. Direct emissions of particles, carbon monoxide and trace elements also tend to be low compared to fossil fuels using the same combustion technology. Emissions of nitrogen dioxide, however, can be as high as or even slightly higher than those from fossil fuels. For biofuels from crops, emissions arise from fertiliser manufacture and use, and from the energy used by agricultural machinery for irrigating and harvesting the crop. For all biofuels, further emissions will be associated with the energy used to process and transport the fuels.

The greenhouse gas impact of biofuels is complex. Burning fossil fuels releases carbon that was removed from the atmosphere millions of years ago and has been stored underground ever since, thus adding carbon to the atmosphere. Biofuels, however, absorb carbon from the atmosphere as they grow, so that when the carbon is released on combustion there is no net change. Biofuel combustion emissions can therefore be considered to be carbon-neutral, although the upstream emissions from fuel production must still be taken into account. In Chapter 4 we will also discuss emissions arising from land-use change, such as when natural forests are cleared to grow biofuels.

A further complication arises for biofuels derived from waste. In landfill sites, biodegradable waste such as food and paper will decay to form methane – a greenhouse gas that is 25 times more powerful than carbon dioxide. Even though modern landfill sites generally collect methane and flare it or use it for fuel, thus converting it to carbon dioxide, collection rates can be as low as 20 per cent.[49] If waste is diverted from landfill and burnt as a biofuel, however, or used to generate bio-gas, all the carbon will eventually be converted to carbon dioxide instead of methane, with a much lower climate impact. Biofuels derived from waste that would otherwise have been landfilled can therefore be credited with avoiding methane emissions.

Figure 3.5 (see plate section) shows the full life-cycle emissions of greenhouse gases, sulphur dioxide, nitrogen oxides and particle pollution for various electricity generation technologies, including emissions from fuel production, plant construction, operation and decommissioning. Renewable energy sources generally produce far lower emissions of both air pollutants and greenhouse gases per unit of energy generated than fossil fuels. For example, wind power produces around 10 grams of carbon dioxide equivalent per kilowatt hour of electricity generated (gCO_2/kWh), compared to around 500 gCO_2/kWh for natural gas and 1,000 gCO_2/kWh for coal and oil power.

For photovoltaic solar power, the life-cycle emissions are higher, because the manufacture of silicon solar cells is energy-intensive. The emissions are highly dependent on the type of solar cell used. Polycrystalline silicon cells produce 50–100 gCO_2/kWh, amorphous silicon cells around 40 gCO_2/kWh and thin film solar cells just 20 gCO_2/kWh.[50] Solar thermal technology has lower emissions than solar photovoltaic power. For example, parabolic trough systems, in which the sun's rays are concentrated by a curved mirror and used to raise steam to drive a generator, produce only 14 gCO_2/kWh.[51]

Figure 3.5 shows two biofuel technologies: burning forest wood in a steam turbine to generate electricity, and burning bio-gas from anaerobic digestion of waste to generate electricity. The wood combustion produces very low emissions except for nitrogen oxides, where the emission level is extremely high. However, the coal power plant is assumed to use selective catalytic reduction, which cuts nitrogen oxide emissions by 80 to 90 per cent, whereas no pollution control is assumed for the wood plant. Emissions of nitrogen oxides from wood combustion could be reduced by using more advanced technology such as gasification. The bio-gas combustion example has comparatively high levels of nitrogen oxides and sulphur dioxide, but a large negative contribution to greenhouse gas emissions, arising from the avoidance of methane emissions had the waste been disposed of to landfill.

Life-cycle studies have also been used to estimate the 'external costs' of different power generation technologies – in other words, the costs that fall on society (such as health or ecosystem damage), but which are not included in energy prices. Figure 3.6 shows the estimated external costs of air pollution and climate change for different energy sources, from the European Commission's ExternE project (ExternE, 2005). Lignite and oil-fired plant were estimated to have external costs of 5.8 c€/kWh and 4.8 c€/kWh respectively, followed by hard coal at 4.1 c€/kWh and natural gas with 1.5–2.2 c€/kWh. These external costs are very significant compared to typical EU electricity prices of around 10–20 c€/kWh. Renewable and nuclear technologies have far lower external costs, below 0.4 c€/kWh. It should be noted that the results are largely based on power plants with advanced pollution control systems. In countries with lower pollution control, external costs could be larger.

As energy supply systems become cleaner and more efficient, the impact of renewable energy systems could be reduced even further in a 'virtuous circle'.[52] For example, if more electricity is generated from renewable energy, the emissions from production of metals in electric-arc furnaces would decrease, reducing the impact of wind turbine manufacture. Fossil fuel systems, on the other hand, owe most of their impact to pollutants produced during fuel combustion, and opportunities for further abatement are limited unless CCS becomes viable.

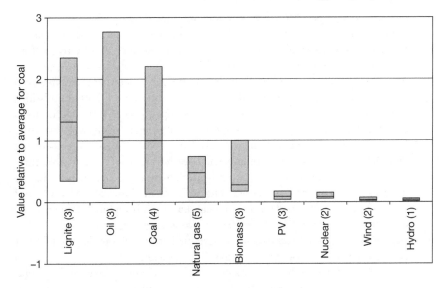

Figure 3.6 External costs of different fuel cycles relative to coal

Source: ExternE, 2005

Note: The number of case studies for each fuel chain is indicated in the x-axis labels. The horizontal bar within the ranges indicates the average of all studies considered.

Nuclear power

Nuclear power is increasingly viewed as an essential part of a low-carbon energy future. The life-cycle energy impact arises from building, operating and decommissioning the plant; treating and disposing of nuclear waste; and mining, processing and enriching the uranium fuel. Most estimates show total emissions of greenhouse gases and major air pollutants to be very low compared to fossil fuels – comparable with renewable energy at around 10 gCO_2/kWh (Figure 3.5). However, another study claimed higher impacts of 100 to 200 gCO_2/kWh, of which almost half was due to the energy used during uranium fuel production.[53]

Although emissions of conventional pollutants are low compared to fossil fuels, nuclear power plants do release radioactive substances into the air and water, both through routine discharges and through occasional accidental releases. This is discussed in the section on 'Conflicts' below.

Cleaner vehicles

Vehicles are a major source of both greenhouse gases and local air pollution. Exhaust fumes are emitted at breathing level in crowded streets, giving rise to severe health effects. Cleaner and more efficient vehicles can help to reduce both air pollution and carbon emissions, although reducing the use of motor vehicles would have a greater effect. Various options exist, with differing climate change and air pollution impacts.

- *Increased fuel efficiency.* More efficient vehicles, including hybrids, consume less fuel per kilometre driven, giving benefits for both climate and air quality. Efficiency is increasing in most world regions, driven largely by regulations and standards, although this is partly offset by an increase in vehicle size.
- *Electric or plug-in hybrid vehicles.* Vehicles operating in electric mode have zero emissions at the point of use, although there may be additional emissions from power stations. The emission benefits depend on how the electricity is generated, and are greatest where there is a large share of clean, low-carbon electricity. Even if generation is purely from coal, however, there are net greenhouse gas savings.[54] Air quality will improve because it is easier to control emissions from power stations than from vehicles, and health benefits are magnified because power stations tend to be located away from towns and cities. However, uptake of electric vehicles has so far been limited, mainly due to their restricted range and the need to provide recharging points.
- *Gas- or hydrogen-fuelled vehicles.* Natural-gas-fuelled vehicles emit about 25 per cent less CO_2 per kilometre than gasoline vehicles. Bio-gas generated from anaerobic digestion of organic waste can even have negative greenhouse gas emissions, if the waste would otherwise have emitted methane in a landfill site. Bio-gas can also be generated from gasification of solid biomass, in which case there may be emissions related to the production of the biomass feedstock. All these fuels produce less air pollution than gasoline and diesel (provided that bio-gas is pre-treated, if necessary, to remove sulphur). Hydrogen vehicles, using either fuel cells or internal combustion engines, also have very low exhaust emissions, but additional emissions can arise during the production of the hydrogen fuel. Producing hydrogen by electrolysis of water can actually increase emissions if the typical US generating mix is used, but emissions will substantially decrease if renewable electricity is used.[55] Uptake of gas-fuelled vehicles is limited by the need to provide a network of refuelling infrastructure – at present (2012), deployment is mainly by fleets, such as buses, that return to a central depot for refuelling.

Figure 3.7 (see plate section) shows lifetime emissions for various light-duty vehicles. Hybrid and electric vehicles offer lower emissions for most pollutants even compared to the most advanced gasoline and diesel vehicles. Emissions for hydrogen fuel cells depend heavily on the hydrogen generation method. Renewably generated electricity or hydrogen-fuelled vehicles would offer near-zero emissions.

However, some vehicle technologies can give rise to conflicts between air quality and climate, as we will see later, including the impact of particle filters on fuel efficiency, and the impacts of liquid biofuels and diesel vehicles on air quality.

Emission controls

Some air pollutants are also greenhouse gases: these include black carbon (see Box 3.3) and ozone. Curbing emissions of black carbon and the ozone precursors

– nitrogen oxides, carbon monoxide, methane and other volatile organic compounds – can therefore tackle air pollution and climate change simultaneously. Methane is particularly significant as it is both a greenhouse gas and an ozone precursor. The concentration of methane in the air has increased from 700ppm in pre-industrial times to 1,800ppm today (2012), due largely to emissions from agriculture, oil and gas production and disposal of waste in landfill sites. Because it is long-lived, with a lifetime of around 12 years, it has become a global pollutant, although concentrations are slightly higher in the northern hemisphere.[56]

Strategies for limiting these emissions include:

- recovering methane from landfill sites;
- reducing venting of methane during oil production, and cutting leaks from gas distribution;
- controls on emissions of black carbon and ozone precursors from vehicles, industry, and power stations;
- reduced burning of traditional biofuels on inefficient stoves or fires, to cut black carbon;
- reducing methane emissions from agriculture (see Chapter 4).

Box 3.3 Black carbon

Black carbon is the main component of soot. It consists of short chains of pure carbon released during combustion. Around 40 per cent of global black carbon pollution comes from burning fossil fuels (coal and diesel), 40 per cent from burning of forests or grasslands and 20 per cent from traditional biofuels, mainly wood and animal dung burnt in domestic stoves.[57] Black carbon typically forms around 5–10 per cent of fine particle pollution in Europe, although it can be up to 20 per cent in some roadside samples due to the high level of black carbon in diesel exhaust.[58] Black carbon emissions have increased dramatically in Asia since the mid-2000s due to the growth in vehicle use and coal-fired power generation.

As well as being a health hazard, black carbon has recently emerged as a strong contributor to global warming. When airborne, it absorbs the sun's radiation and warms the atmosphere, with a warming effect over 1,000 times stronger than the same amount of carbon dioxide. However, this effect is short-lived as black carbon usually only remains airborne for around 2 weeks before settling or being washed out by rain. For this reason, black carbon was not considered to be a significant greenhouse gas, and was not included in the 1997 Kyoto Protocol.

More recently, a UNEP study (2011) has found that considerable quantities of black carbon are being deposited on snow and ice in the polar regions and on glaciers in Asia and elsewhere. This decreases the albedo (reflectivity) of the glaciers, so that less of the sun's radiation is reflected back into space, thus contributing to global warming. As the black carbon absorbs heat from the sun, it accelerates the melting of the glaciers and ice caps. This is becoming an increasing problem in Asia, where millions of people depend on rivers fed by glacial meltwater for their fresh water supplies. Excessive glacial melting is also contributing to the formation of large, unstable glacial lakes, which pose a flood threat to communities downstream.[59] In addition, black carbon is thought to play a role in disrupting regional weather patterns such as monsoon rains in Asia and rainfall in central Africa.

Black carbon is now acknowledged as a major short-term contributor to global warming, with estimates showing a warming effect equivalent to 55 per cent of the impact of current levels of carbon dioxide.[60] Reducing black carbon could have major benefits both for climate change and health.[61]

Forests and other land-use changes

It is well known that trees store carbon, and increasing forest area is therefore a key climate policy. This has major benefits for biodiversity, as we will see in Chapter 4, but trees can also help to improve air quality. They can absorb gaseous pollutants through the pores in their leaves, and they can also trap particles of pollution. Forests can also increase local rainfall, which helps to wash pollution out of the air. One study estimated that trees in the United States remove over 700,000 tonnes of pollution from the air each year, avoiding over $3 billion worth of damage.[62] Urban trees are particularly beneficial, as they can remove traffic pollution from the areas where it does the most damage, as well as providing shade and aesthetic pleasure. Trees in London are estimated to remove between 800 and 2,000 tonnes of particle pollution each year.[63]

Set against this benefit, it is true that vegetation emits volatile organic compounds (VOCs) in hot weather, which can play a part in increasing ozone production in areas where nitrogen oxide pollution is high. However, this is a complex relationship: in areas where nitrogen oxide pollution is low, VOCs can actually remove ozone from the air. Also, trees exert a cooling effect, which can help to reduce ozone production. The overall impact depends on local weather conditions and on the species of tree – for low VOC-emitters, there is usually a net ozone reduction.[64]

Burning forests, grasslands and crop residues has negative effects on both climate change and air quality. Reducing these practices can reduce emissions of black carbon and result in more stored carbon in the soil and vegetation (see Chapter 4).

Co-benefits

We have seen that many climate policies, especially those that reduce our use of fossil fuels, can also reduce air pollution. Benefits for health, ecosystems and the economy include:

* lower incidence of premature death and illness from heart and lung diseases and cancer;
* lower health costs and less working time lost due to pollution-related illness;
* healthier forests, streams, lakes and other ecosystems, and savings in the cost of liming to reduce acid damage;
* reduced damage to buildings from acid rain and soot;
* increased crop yields due to reduced ozone concentrations;
* savings in the economic and environmental costs of installing and operating pollution control equipment – e.g. less quarrying of limestone for de-sulphurisation.

The scale of the benefits will depend on a number of factors, including the size of the cut in carbon emissions, the choice of policies and technologies, the existing energy supply mix and the degree of pollution control already in place. To take account of all these factors, estimates of air quality co-benefits generally model different scenarios using an approach similar to that outlined in Box 3.4.

Box 3.4 Air quality scenario modelling

1 *Develop scenarios* including a baseline 'business-as-usual' scenario based on continuation of current trends in energy use and fuel mix; and alternative scenarios that might, for example, involve cutting carbon emissions by a given amount.

2 *Calculate the emissions* from transport, power stations, buildings and industry for each scenario.

3 *Use an 'atmospheric dispersion model'* to generate maps of air pollution concentrations, using predicted weather patterns to model the way pollutants move through the air, are washed out by rain, or react to form other compounds.

4 *Estimate the damage* caused by the pollution, such as increases in premature death or decreases in crop yield. This is done using 'dose-response functions', which are derived from observed relationships such as the link between hospital admissions and pollution levels.

5 *Estimate the cost of the damage.* This makes it easier to compare the costs and benefits of different policies, but estimates should be treated with

> caution because they are usually based on limited data and also depend on complex and controversial assumptions about, for example, the value of a life.

Although few studies have addressed ecosystem impacts such as acid rain, eutrophication and crop damage, there is a large body of evidence concerning the health impacts of air pollution. The clearest evidence comes from cases where public health is monitored after action to reduce pollution. For example, when the burning of bituminous coal was banned in Dublin, Ireland in 1990, there was an immediate and permanent improvement in public health. Deaths from respiratory disease fell by 15 per cent and deaths from cardiovascular disease fell by 10 per cent.[65] Studies in the United States have shown an increase in life expectancy across the population of 0.6 years for every 10 $\mu g/m^3$ reduction in exposure to fine particles.[66] A major study in Canada has just demonstrated (2012) that long-term exposure to fine particles increases mortality even at very low concentrations, down to around 1 $\mu g/m^3$.[67] This new finding has not been incorporated into the assessment models discussed below, but it would tend to increase the benefits of pollution reduction still further.

The next section summarises some of the key studies to illustrate the scale of the air quality co-benefits that could result from climate action.

Health benefits in the UK

Although emissions in the UK have decreased dramatically over the last few decades, air quality limits are still exceeded, with hot-spots of nitrogen oxide and particle pollution in large cities such as London. The health costs of particle pollution alone are estimated to be £15 billion per year.[68] However, a 2007 study estimated that if optimal policies were chosen to achieve the target of reducing UK carbon emissions by 60 per cent in 2050, based on strong uptake of zero-carbon energy sources and zero-carbon vehicles, levels of fine particle pollution and nitrogen oxide in London could be cut by around 55 per cent.[69]

The carbon target has now been tightened to a cut of 80 per cent by 2050, which could yield air quality benefits worth £775 million annually by 2022.[70] A 2010 study emphasises the importance of taking an integrated approach to climate and air quality, which could increase the cumulative air quality benefits by 2050 from £15 billion to £40 billion.[71] The extra benefits come from using hydrogen or gasoline-hybrid vehicles rather than diesel; gas or nuclear power rather than coal with CCS; and restricting the use of wood for home heating to rural areas.

Health and ecosystem benefits in Europe

A number of studies have looked at the air quality co-benefits of climate policy in Europe. The European Commission estimated the benefits of meeting the EU climate target to cut greenhouse gas emissions by either 20 per cent or 30 per cent from 1990 to 2020, and to increase the share of renewable energy from 8.5 per cent to 20 per cent. The cuts would be achieved by providing incentives for renewable energy, and by vigorously promoting energy efficiency. Importantly, the study assumed that the cuts are to be achieved entirely within the EU and not by buying carbon credits from non-EU countries.

This EC study and a follow-up report by Holland revealed considerable health and ecosystem co-benefits. For a 30 per cent cut in greenhouse gas emissions, EU countries would save over 300,000 life-years every year and gain over 6 million working days, giving total health benefits valued at between €20 billion and €77 billion (Table 3.2). Ecosystem benefits were quantified only for a 20 per cent cut, which would reduce the area of forest affected by critical levels of acidification by 22 per cent, and bring 22,000 square kilometres of forest out of the critical zone for eutrophication.[72]

More recently, the ClimateCost study (2011) extended the analysis to 2050, and looked at a more ambitious target to restrict climate change to 2°C. Around half of this target would be achieved through cutting energy use, with the rest from switching to nuclear and renewable energy and installing CCS. This could cut EU-27 sulphur dioxide emissions by 60 per cent, nitrogen oxides by 46 per cent and fine particles by 19 per cent. Across the whole of Europe (including Russia and Turkey), this would save over a million life-years, cut deaths from particle pollution

Table 3.2 Co-benefits of a 20 per cent or 30 per cent cut in greenhouse gases for the EU by 2020

	Baseline	20% GHG cut	30% GHG cut
Share of renewable final energy in 2020	12.5%	20%	
Energy intensity improvement per year	1.8%	3%	
Primary energy demand change 2005–2020	+9%	−10%	
Share of EU energy imported by 2030	67%	58%	

	Baseline	Change compared to baseline	
Air pollution compared to baseline (sum of SO_2, NO_x and $PM_{2.5}$)		−14%	−21%
Life-years lost amongst people over 30	2,800,000	−218,000	−323,000
Loss of work days	56,531,000	−4,405,000	−6,528,000
Cost of lost work days, million Euros	4,975	−388	−574
Health benefit, million Euros, low	172,000	−13,000	−20,000
Health benefit, million Euros, high	666,000	−52,000	−77,000
Forest with critical nitrogen levels (km^2)	893,000	871,000	Not available
Forest with critical acidification levels (km^2)	130,000	102,000	Not available

Source: Holland, 2009; European Commission, 2008a

by almost 40 per cent and also prevent 6,000 deaths from ozone pollution, with total health benefits valued at €93 billion (Table 3.3 and Figure 3.8, see plate section). These savings are equivalent to around €24 per tonne of carbon dioxide mitigated.[73]

The study also demonstrated benefits for ecosystems, with the area of forest suffering from critical levels of acidification and nitrogen deposition falling significantly (Figures 3.9 and 3.10, see plate section). Nitrogen deposition affects a larger area than acidification, but the impact of cutting nitrogen oxide emissions is limited because a large part of the nitrogen comes from the agriculture sector in the form of emissions of ammonia and nitrous oxide from fertilisers and manure

Table 3.3 Health benefits for Europe in 2050 of meeting a 2°C target for climate change

	Baseline	Mitigation	Co-benefit	Value (M€)
Ozone effects				
Acute mortality (all ages), deaths	35,800	29,800	6,000	360
Respiratory hospital admissions (age 65+)	37,900	31,700	6,170	14
Minor restricted activity (age 15–64), days	57,000,000	46,900,000	10,200,000	451
Respiratory medication use (age 20+)	25,000,000	20,500,000	4,280,000	4
Fine particle effects				
Chronic mortality, life-years lost	2,760,000	1,680,000	1,080,000	64,700
Infant mortality (1 mo.–1yr), deaths	636	363	273	1,560
Chronic bronchitis (age 27+), new incidence	133,000	81,600	51,200	11,100
Hospital admissions	81,900	50,300	31,600	72
Restricted activity days (age 15–64)	230,000,000	140,000,000	90,100,000	8,650
Respiratory medication use, days	26,200,000	16,100,000	10,100,000	10
Lower respiratory symptom, days	371,000,000	228,000,000	143,000,000	6,290
Total value of health co-benefits				**93,100**
Ecosystem benefits				
% of forest suffering critical acidification	3%	1.1%	1.9%	
% of forest suffering critical nitrogen levels	34%	28%	6%	
EU-27 only				
Acid damage to buildings				264
Ozone damage to crops				638
Forest with critical acid levels (km²)	78,000	36,000	42,000	
Forest with critical nitrogen levels (km²)	926,000	783,000	145,000	

Source: ClimateCost study (Holland *et al.*, 2011a, 2011b)

Note: For clarity, values have been rounded to three significant figures. Based on the value of a life-year in the EU being €60,000, and applying this value to all countries in Europe regardless of differences in GDP. Values not discounted.

(see Chapter 4 for a discussion of how climate policy can help to reduce these emissions). The economic value of the improvement in eutrophication was not quantified, but could be considerable.

A similar study showed that the EU target to extend life expectancy by 3.8 months by cutting air pollution could be largely achieved through cutting greenhouse gas emissions by 20 per cent by 2020, which would improve life expectancy by 3.5 months. In contrast, the business-as-usual scenario of just implementing existing air quality legislation would only improve life expectancy by 2.6 months, and would increase greenhouse gas emissions by 3 per cent.[74]

A study for the European Environment Agency looked at a climate action scenario under which total greenhouse gas emissions would be cut by 40 per cent between 1990 and 2030; of this, around half could be achieved by trading carbon with other countries.[75] The study predicted that nitrogen oxide emissions in 2030 would be 10 per cent lower, sulphur dioxide 17 per cent lower and fine particles 8–10 per cent lower than in the baseline, and breaches of air quality limits in urban hot-spots such as street canyons would fall considerably. In 2030, there would be 23,000 fewer premature deaths from ozone and fine particle pollution (288,000 compared to 311,000) and 200,000 extra life-years, resulting in health benefits worth €14–46 billion per year. Ecosystems would also benefit: the area of forest affected by acidification would fall by 20,000 square kilometres and the area affected by eutrophication would fall by 30,000 square kilometres.

Health and ecosystem benefits worldwide

For developed countries worldwide, benefits are estimated to be similar to those predicted by the European studies. For example, a study of the industrialised countries which signed Annex 1 of the UN Framework Convention on Climate Change found that cutting greenhouse gas emissions to 20 per cent below the baseline level by 2020 would cut sulphur dioxide emissions by 30 per cent and nitrogen oxide and particle emissions by 10–15 per cent, mainly as a result of reduced coal use.[76]

For developing countries, the issues are slightly different. The severe pollution problems in rapidly industrialising cities mean that cutting fossil fuel use can give large public health benefits. However, the social and economic costs can also be high because energy consumption per person is already very low, giving rise to fears that curbing fossil fuel use could stifle economic development and poverty reduction.

Modelling for the IEA confirms that developing countries stand to gain the most from the air quality co-benefits of climate action. If carbon emissions were reduced in line with achieving a target of 450ppm greenhouse gas concentrations, global sulphur dioxide emissions would fall by a third by 2035, nitrogen oxide emissions by 27 per cent and fine particle emissions by 8 per cent compared to the business-as-usual scenario.[77] Years of life lost through exposure to fine particles in China, India and Europe in 2035 would fall by 23 per cent, from 3.2 billion to 2.5 billion, saving 740 million life years. The health benefits are greatest in China and India, due to their low levels of pollution control, high coal use and rapidly growing transport emissions. The model suggests that particle emissions in OECD countries would actually rise by 19 per cent due to the greater use of solid biomass (see the section on 'Conflicts' below).

A study for the OECD looked at how air quality co-benefits could provide an incentive for climate action. If global greenhouse gas emissions were cut to 50 per cent of 2005 levels by 2050, the study predicted that premature deaths from chronic exposure to outdoor fine particle pollution would be reduced by 42 per cent compared to the business-as-usual case, avoiding more than 5 million early deaths per year by 2050.[78] An integrated strategy tackling climate change and air pollution together would be even more effective, achieving a 67 per cent reduction in premature deaths (Figure 3.11).

The study predicted higher health benefits in developing countries than in developed ones, but also higher costs. This was because it was assumed that the greenhouse gas cuts were achieved by imposing a global carbon price, which would hit developing countries hardest as they are starting from very low energy consumption, making further carbon reduction difficult and expensive. For all countries, the health benefits would compensate for a large part of the costs of climate action in the short term: and for OECD countries, the benefits would exceed the costs from 2030 onwards; whereas for non-OECD countries, the benefits would exceed the costs only after 2050.[79] This delayed effect for the non-OECD countries is because the study predicted that the initial greenhouse gas reductions in these countries would come from improving the efficiency of power stations, which has

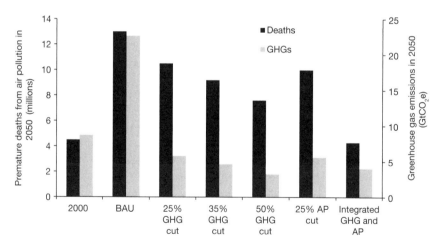

Figure 3.11   Premature deaths from air pollution in 2050 are reduced by 42 per cent with a 50 per cent cut in global greenhouse gas emissions, and by 67 per cent with an integrated climate and air pollution strategy

Source: Based on data from Bollen *et al.*, 2009a

Note: BAU = business-as-usual; AP = air pollution.

a comparatively small effect on public health. Later, deeper cuts would require abatement in the transport sector, where the public health impacts are larger. The study considered only the impact of premature deaths from particulate pollution – if the costs of illness, time off work, ozone pollution and crop damage were also included, then the co-benefits would be even higher.

It is also worth noting that the value of a life was assumed to be proportional to the average GDP *per capita* of the country – a somewhat controversial practice. If the value of a life was assumed to be the same in all countries, the health co-benefits in developing countries would be assigned a higher value. To illustrate the scale of the differences, we can return to the ClimateCost study. Table 3.4 shows that the study estimated very high health co-benefits in China and India in 2050, reflecting predicted fine particle pollution levels that are ten times higher in China and up to 15 times higher in India than the WHO guideline level of 10 μg/m³ achieved in Europe. Climate mitigation cuts fine particle concentrations by 47 per cent in China and 63 per cent in India compared to the baseline scenario, increasing life expectancy by 20 months and 30 months respectively (Figure 3.12, see plate section). However, Table 3.5 shows that when EU values are used for the value of life-years, the co-benefits are around four times higher in China and almost nine times higher in India than when local values scaled by the difference in GDP *per capita* are used.

Other studies from around the world reinforce these results, consistently estimating substantial health co-benefits from climate policy.[80] For Asia and Latin

Table 3.4 Health co-benefits for China and India in 2050 of meeting a 2°C target for climate change

	Baseline	*Mitigation*	*Co-benefit*
China			
Ozone: deaths	55,000	34,000	22,000
Fine particles: life-years lost	63,000,000	34,000,000	29,000,000
India			
Ozone: deaths	90,000	35,000	55,000
Fine particles: life-years lost	71,000,000	27,000,000	44,000,000

Source: ClimateCost study (Holland *et al.*, 2011a, 2011b) (figures rounded to nearest 1,000)

Table 3.5 Value of health co-benefits in 2050 for meeting a 2°C climate target (€ million)

	EU valuations	*Local valuations*
EU-27	43,730	43,730
Europe	94,037	77,273
China	2,488,882	616,520
India	3,787,796	423,865

Source: ClimateCost study (Holland *et al.*, 2011a, 2011b)

America, a moderate reduction of greenhouse gases by 10–20 per cent over the next 10 to 20 years is estimated to reduce sulphur dioxide emissions by 10–20 per cent and nitrogen oxide emissions by 5–10 per cent, with health benefits of the order of tens of thousands of lives saved. For North America and Korea, the benefits are lower, but still substantial – of the order of thousands of premature deaths avoided. In India, greenhouse gas emissions are expected to triple between 2005 and 2030, and the average loss of life expectancy from particle pollution could grow from 14 months to 59 months, but climate action could reduce the growth in greenhouse gases by a third and cut the increase in health impacts by a quarter.[81]

Estimates of the value of co-benefits vary widely between studies. A 2010 survey found estimates varying from a low of $2 per tonne of carbon dioxide abated to a high of $196, with a mean value of $49.[82] The lower estimates generally do not consider the full range of pollutants or impacts, whereas higher benefits are predicted in developing countries with little existing pollution control, and for high-impact measures such as reducing transport emissions in cities and improving domestic stoves in developing countries.

Despite these differences, co-benefits are generally estimated to be of the same order of magnitude as mitigation costs, with some studies indicating that co-benefits could exceed costs in certain cases. For example, various studies estimate that greenhouse gas emissions could be reduced by 13–23 per cent in India, 15–20 per cent in China and 20 per cent in Chile (compared to the business-as-usual case for 2010) at no net cost, because of the value of the air quality co-benefits.

The studies tend to confirm the finding of the European and OECD reports that integrated policies to address climate change and air quality can be more cost-effective than separate policies. For example, one study showed that a 50 per cent reduction in health impacts from fine particle pollution in China could be achieved at a 40 per cent lower cost and with a 9 per cent reduction in carbon dioxide emissions if climate change measures such as combined heat and power (CHP) and energy efficiency were included, compared with a package based just on end-of-pipe pollution control.[83]

The OECD study highlights an important issue: co-benefits can often be realised on a much shorter timescale than climate benefits. The worst impacts of climate change are projected to occur many decades into the future, and even if emissions today are reduced, the long lifetime of carbon dioxide in the atmosphere means that temperatures will continue to rise for some time. Health benefits, on the other hand, can occur instantly after cutting local air pollution, and this can provide a much stronger motivation for action.

Similarly, although developed countries emit the most greenhouse gases per person, the worst effects of climate change (in the short term at least) will occur in developing countries. This can reduce the motivation to cut emissions, at least among the less altruistic members of society. Health co-benefits, on the other hand, are local. Cutting transport pollution will save lives in the town where you live, which can provide a much stronger incentive for action.

Local benefits, however, are undermined by the use of 'flexibility mechanisms', which are policies that allow countries to pay for emission cuts in other countries instead of reducing their own emissions. They include the Clean Development Mechanism (CDM), carbon trading and offsetting, and the Reduced Emissions from Deforestation and Forest Degradation (REDD) initiative (see Chapter 4) where countries are paid to preserve forests. The study of co-benefits in UN-FCCC Annex 1 countries, mentioned above, estimated that using the CDM would slash the sulphur dioxide reduction by 75 per cent and halve the nitrogen oxide and particle reduction.[84]

Finally, a 2010 study found that combining climate and air quality policy with energy security policy could have even greater benefits. The study found that the added incentives of improved health (with 3 million lives per year saved globally) and reduced oil dependence could drive a switch away from fossil fuels that achieves almost zero carbon emissions by 2100, as well as almost eliminating air pollution by 2050 and cutting oil consumption in Europe to a tenth of its current level by 2100.[85]

Methane, ozone and black carbon

Although the studies described above tend to focus on cutting carbon dioxide emissions from fossil fuels, there can also be significant air quality benefits from reducing the greenhouse gases that are also air pollutants: black carbon and ozone precursors including methane.

One study estimated that cutting global methane emissions by 20 per cent would prevent around 30,000 premature deaths from ozone pollution annually by 2030, and around 370,000 between 2010 and 2030. If avoided deaths are valued at $1 million each, the benefit would be equivalent to $240 for every tonne of methane removed, far greater than the estimated cost of $100 per tonne for reducing methane emissions.[86] In other words, the health benefits alone make methane abatement cost-effective, even before considering the climate change benefits.

Cutting ozone levels can also give considerable economic benefits from improved crop yield. One study showed benefits to crop yields for China to be comparable to health benefits in 2010, with the combined health and crop yield benefits allowing a 15–20 per cent reduction in CO_2 at no net cost.[87] Another study showed that crop losses from ozone damage in China could increase by 10–20 per cent by 2030 under business-as-usual, but by only 2–4 per cent under a climate change scenario.[88] Increased crop yields provide benefits to rural populations, complementing the health co-benefits that mainly accrue to urban areas.

The United Nations Environment Programme (UNEP) estimated in 2011 that fast action to control all three of these pollutants – black carbon, methane and ozone – could slow near-term warming by 0.4–0.5°C by 2050, as well as avoiding 2.4 million premature deaths from outdoor air pollution and saving between 32 and 52 million tonnes in crop yields per year.[89] Key actions included reducing soot emissions from cook-stoves and diesel vehicles, and stopping methane leaks from landfill sites, coal mines and oil and gas infrastructure. Half of the measures aimed at reducing black carbon and methane emissions also saved money; for example, through improved fuel efficiency of cook-stoves or reduced leakage of gas from pipelines, and a further third of the measures could be achieved at moderate cost. The health and crop yield benefits of measures to reduce methane emissions were valued at $700–$5,000 per tonne, far greater than the typical cost of the measures which was estimated as under $250 per tonne.[90]

Indoor air pollution

In developing countries, over 3 billion people use coal, kerosene or traditional biofuels such as wood, dung, charcoal and crop residues for cooking and heating.[91] Many of these households use open fires or inefficient and smoky stoves, often in poorly ventilated rooms, leading to severe indoor air pollution that causes lung problems and eye irritation, as well as black carbon emissions that contribute to global warming (Box 3.3 above). The World Health Organization has estimated that 2 million premature deaths a year can be attributed to indoor air pollution.[92] Poor families often spend as much as a quarter of their household income on fuel wood, or alternatively women and children spend many hours a day gathering wood, which can be a dangerous occupation and also cuts the time available for education or income generation. Firewood collection can also lead to deforestation, biodiversity loss and soil erosion. Where animal dung or crop residues are used as fuel, they are unavailable for improving soil structure and fertility (see Chapter 4).

These problems can be tackled by providing efficient, well-designed stoves that use less fuel and also emit less pollution. This is harder than it sounds – some early designs managed to save fuel wood but actually increased air pollution, because the enclosed stove reduced air flow to the flame. Enclosed stoves with chimneys can reduce indoor pollution to almost zero, but some designs can increase fuel use (due to heat being absorbed by the body of the stove) and therefore increase outdoor pollution. Other initiatives have failed for cultural reasons, such as failure to adapt stoves to local cooking methods. Nevertheless, a number of stoves have now been developed that can halve fuel use compared to the best-managed three-stone open fire, as well as cutting particle emissions by 75 per cent, or even by 99 per cent for new designs of fan-assisted stove (though at $20–$50 each, these are relatively expensive).[93]

A modelling study (2009) estimated that a 10-year programme to install 150 million efficient biomass cook-stoves in India could save 1.8 million premature deaths in adults and 240,000 deaths in children under five, as well as cutting emissions of greenhouse gases (including black carbon) by 0.5–1 billion tonnes of carbon dioxide equivalent.[94]

Despite these benefits, climate policy must be carefully designed to avoid conflicts with the goal of reducing indoor air pollution. Gas or electric stoves could improve indoor air quality, but increase greenhouse gas emissions, for example (see next section on 'Conflicts'). There is even a potential issue in developed countries, where complete sealing of buildings to eliminate heat loss through draughts can increase indoor air pollution unless adequate ventilation is installed.

Conflicts

Although there is generally a strong synergy between cutting climate emissions and improving air quality, there are some exceptions: certain climate policies are bad for air quality, and some pollution-control methods increase climate emissions. In this section we will look at some of the areas in which conflicts can arise:

- end-of-pipe solutions for cutting air pollution or capturing carbon;
- gasoline versus diesel for transport;
- the air pollution impacts of biofuels;
- radioactive pollution from nuclear power;
- geo-engineering.

End-of-pipe solutions

We have seen that efforts to address climate change, which usually centre on cutting the use of fossil fuels, tend to give big benefits for air quality. However, policies to reduce air pollution have traditionally focused on end-of-pipe control technologies which often tend to increase greenhouse gas emissions, because extra energy is needed to operate the equipment. For example, operating flue gas desulphurisation equipment uses around 2 per cent of the power output of a coal-fired power station, resulting in more fuel being burnt and more greenhouse gas emissions for each unit of power produced. Carbon capture and storage increases energy use by 10–40 per cent, as we saw earlier, which can increase emissions of some pollutants.

Many end-of-pipe technologies require the use of chemicals which react with the pollutant in order to remove it from the waste gas stream. For example, flue gas desulphurisation involves injecting powdered lime or calcium carbonate into the power station flue, where the calcium reacts with sulphur dioxide to form gypsum (calcium sulphate). There will be visual impacts and habitat loss resulting from the quarrying of large quantities of limestone, which is often in sensitive and ecologically important areas such as the Peak District in the UK and the Karst region of China. In addition, extra fuel will be needed to quarry and transport limestone to the power station. The absorption of sulphur dioxide also generates additional carbon dioxide which is released from the limestone reagent.

Similarly, removing nitrogen oxides from power station flue gases through selective catalytic or non-catalytic reduction requires the use of an ammonia-based reagent such as urea. Manufacturing these reagents is energy-intensive and can generate emissions of nitrous oxide which is a greenhouse gas. Carbon capture and storage requires the use of amine-based solvents, resulting in emissions of various pollutants during manufacture.[95] End-of-pipe solutions also generate solid or liquid waste. For example, coal-fired power stations with desulphurisation units generate large quantities of waste gypsum sludge, and ash removal systems produce ash that is contaminated with highly toxic substances including mercury and dioxins. Disposal of these waste streams can pose problems with groundwater pollution, and habitat and amenity loss. For carbon capture, the solvent sludge must be incinerated as hazardous waste.

Similar conflicts occur in the transport sector. Catalytic converters and particulate filters on vehicles decrease fuel efficiency by a few per cent, increasing carbon dioxide emissions. Selective catalytic reduction to control nitrogen oxide emissions can increase emissions of the greenhouse gas nitrous oxide. Cleaning up transport fuel also has an energy penalty: reducing the sulphur content of diesel increases refinery emissions of CO_2 by 5–10 per cent.[96]

Cutting air pollution by fitting end-of-pipe controls to power stations can lead to another unintended consequence – the removal of the aerosol cooling effect. The aerosols of fine sulphate and nitrate particles which form as secondary pollutants when fuel is burnt are light-coloured and reflect solar radiation back into space. The cooling effect is estimated to be around 30 per cent of the warming effect of carbon dioxide in the atmosphere, and could be reducing global average temperatures by about 0.7°C.[97] This means that a drastic reduction of sulphur emissions without simultaneously cutting greenhouse gas emissions could enhance global warming. For example, one study modelled the impact of a policy to reduce deaths from air pollution globally by 25 per cent by 2050, which meant cutting sulphur dioxide emissions by over 80 per cent. Although the policy also cut greenhouse gases by 40 per cent, global temperatures did not fall in the short term, due to the reduction of the sulphate cooling effect.[98]

However, end-of-pipe controls will also cut emissions of black carbon and ozone precursors, which will partly compensate for the loss of the sulphate cooling effect. The IPCC estimated that soot and ozone together produced a warming effect that was around half of the sulphate cooling effect.[99] A study in 2011 estimated that cutting black carbon emissions by bringing vehicle emission standards in developing countries up to European levels could not only save between 120,000 and 280,000 premature deaths annually by 2030, but would also reduce warming in the northern hemisphere by 0.2°C.[100]

Another conflict can arise from reductions in emissions of nitrogen compounds. If nitrogen is deposited to soil or water in areas where lack of nitrogen is limiting plant growth, it can stimulate growth and thus lead to uptake of carbon from the atmosphere. So reducing nitrogen pollution could, in this case, lead to increased concentrations of carbon dioxide in the atmosphere. This effect is partly offset by the fact that areas where there is excess nitrogen in the soil tend to emit nitrous oxide, a powerful greenhouse gas. The interacting carbon and nitrogen cycles are very complex and poorly understood.

With all these complex interactions and competing effects, it should be no surprise that estimates of the impact of air pollution on climate change are highly uncertain and hard to model. However, it is clear that the relative balance of different pollutants such as soot and sulphates will be crucial in determining the net effect. Further research is needed to understand and quantify these issues.

The potential removal of the aerosol cooling effect should not be seen as an argument against fitting end-of-pipe controls to power stations. Air pollution is highly detrimental to health and the environment and therefore needs to be tackled. But it does show that air pollution and climate change should be tackled as part of an integrated strategy. For example, black carbon and ozone are not included in existing climate change targets, despite the vital role their reduction could play in offsetting the removal of the sulphate cooling effect. A policy that tackles multiple pollutants – sulphur dioxide as well as black carbon, methane and ozone precursors – can deliver air quality benefits while reducing adverse effects on the climate. Better still, by cutting energy demand and moving to clean, low-carbon energy sources, we can cut carbon and air pollution at the same time.

Diesel for transport

Some countries have adopted policies to reduce greenhouse gas emissions by encouraging drivers to switch from gasoline to diesel vehicles, typically through applying lower taxes to diesel fuel. Diesel engines produce 10–25 per cent less CO_2 per kilometre driven than gasoline engines, and no nitrous oxide; but they produce more fine particles, nitrogen oxides, black carbon and other local air pollutants.[101] Diesel particles are particularly damaging to health because they are very small, and so can be inhaled deep into the lungs, and they tend to be coated with other toxic pollutants. They have been shown to be linked to a higher cancer risk.[102]

The UK formerly had a policy of setting lower taxes for diesel than for gasoline, in order to reduce greenhouse gas emissions, but this was suspended following consideration of health impacts. One study showed that between 2001 and 2020, there would be 1,850 extra deaths due to increased diesel use in the UK.[103] Modern diesel vehicles equipped with particle filters and selective catalytic reduction units have much lower emissions,[104] but their introduction has been limited because filters can only be used with ultra-low-sulphur diesel fuel. The planned introduction of tighter vehicle emission standards in China was recently postponed because ultra-low-sulphur diesel is still not widely available.[105] Options such as gasoline hybrids or electric vehicles could offer greater benefits for both climate and air quality.

Air pollution from biofuels

The move towards biofuels has generated widespread concern over the impacts of energy crops on food security and ecosystems (see Chapter 4), but air pollution is also an issue. Data are poor and often contradictory, but it seems that although biofuels generally offer benefits over fossil fuels, there are some exceptions. In addition, extra emissions can arise during fuel production, such as black carbon from forest clearance or sugar cane burning (a common practice to make harvesting easier), or emissions of nitrogen compounds from fertiliser application.[106]

A 2011 review found that transport biofuels tend to produce lower exhaust emissions than gasoline or diesel for most pollutants (probably because the higher oxygen content of the fuels leads to more complete combustion), but similar or slightly higher NO_x emissions, probably due to higher combustion temperatures (Table 3.6). Nitrogen oxides (NO_x) emissions could be reduced by fine-tuning the engine or by use of catalytic pollution control units. Virgin plant oil, however, appeared to have higher emissions for some pollutants, possibly because the engine must be tuned differently. For bio-ethanol, there is an increase in acetaldehyde (a

Table 3.6 Ratio of emissions from road transport biofuels to emissions from gasoline or diesel

Biofuel	Compared to	NO_x	PM	CO	HC	Notes
Bio-ethanol E15	Gasoline	1.0	0.4	0.7	0.9	Low benzene, 1,3-butadiene and other toxics; high acetaldehyde, formaldehyde and evaporative ethanol
Bio-ethanol E85	Gasoline	1.0	0.8	1.0	1.0	
Bio-diesel (esterified) B100	Diesel	1.08	0.62	0.66	0.31	Low polyaromatic hydrocarbons (PAHs) and other toxics
Virgin plant oil	Diesel	1.0	1.5	1.5	1.5	
Bio-gas	Diesel	0.5	0.3	0.83	0.65	
Synthetic diesel from waste gasification	Diesel	lower	lower	lower	lower	

Source: AQEG, 2011

Note: E15, E85 = blends of 15 per cent or 85 per cent bio-ethanol in gasoline.

VOC) and evaporative emissions. This was not expected to lead to a significant increase in ozone levels for low blends (up to 15 per cent bio-ethanol), but could become a problem if there was significant uptake of high blends.[107] For the UK, the study concluded that a realistic scenario of biofuel uptake could lead to a 0.5 per cent increase in NO_x emissions from the transport sector by 2020, but decreases of 17 per cent in particle emissions, 8 per cent in VOCs and 23 per cent in carbon monoxide (CO).

For wood burnt in high-quality household boilers, emissions are lower than those from coal but higher than those from natural gas. In the UK, government policy therefore encourages biomass boilers only in rural areas, where pollution impacts are lower, and only if coal or oil (not gas) is being displaced.[108] Similarly, for biomass burnt in power stations, emissions would generally be lower than those from coal but greater than those from gas, if the same level of pollution control was applied.[109] However, small biomass power stations may not apply the same level of pollution control as larger coal power stations – partly because smaller plants may be exempt from regulations.

'Energy-from-waste' plants that burn municipal waste also attract concern, especially over emissions of dioxins from the combustion of plastic and organic waste. Although dioxin emissions have been reduced significantly by strict regulations for waste incinerators in developed countries, public concern remains high and there are frequent objections to proposals for new incineration plants.

In developing countries, 'blunt instrument' climate policies such as a global carbon price or tax could result in higher gas or electricity prices, which could deter households from switching away from smoky traditional biomass stoves or fires to cleaner cooking methods. Options for dealing with this conflict include the use of cleaner and more efficient biomass stoves, solar cookers or bio-gas stoves – which can provide considerable benefits both for climate and health – as well as a shift towards cleaner and lower-carbon sources for electricity generation, which would make electric cookers a less damaging option.

These examples emphasise that air quality must be protected if climate policy encourages greater use of biofuels. Solutions include more efficient household stoves, pollution control technology for biofuel combustion plants and vehicles, and production of 'second-generation' biofuels and bio-gas from waste materials to minimise emissions from energy crop cultivation (see Chapter 4).

Nuclear pollution

Although nuclear power is a low-carbon energy source with low emissions of conventional air pollution, it carries the potential for the release of small amounts of radioactive material which can have severe health consequences. This section focuses on air and water pollution arising from routine discharges, accidents and waste disposal. Impacts related to safety and energy security, including the availability of uranium ore and the threat of proliferation or terrorist activity, are discussed in Chapter 5.

Pollution during routine operation

Nuclear power involves the routine discharge of radioactive material into nearby air, land and water. Radioactive carbon and hydrogen are released to the atmosphere when pressure vessels are opened to replace nuclear fuel, approximately once per year, and radioactive water from the cooling system can be discharged into lakes, rivers or the sea. Fuel reprocessing is responsible for a large proportion of routine discharges, accounting for 83 per cent of the dose to the public from nuclear activities in the EU.[110] Groundwater pollution also occurs during uranium mining (see Chapter 5). The amounts involved are small, because national safety legislation generally aims to ensure that the average dose to members of the public is well below background radiation levels, but there has been concern that over time they could accumulate and become concentrated in the food chain or in areas where human exposure could occur. Discharges from the Sellafield reprocessing plant in the UK (2003) led to high levels of radioactive technetium in lobsters, for example.[111] Studies have found a doubling of childhood cancer rates near nuclear reactors in France and Germany, and some studies found leukaemia clusters near nuclear sites in the UK; but other studies found that clusters in the UK, France, Finland and Spain were no more likely near nuclear sites than elsewhere.[112] There is a theory that leukaemia clusters near nuclear sites could be caused by infections introduced by an influx of workers from outside the area, rather than by radioactive releases, but this is unproven.

It is important to note that burning coal and oil also releases small amounts of radioactive material, as do industrial activities such as phosphate manufacture.

Accidents

Perhaps the most emotive issue is the potential for catastrophic accidents (see Box 3.5 for some notable examples). Although rare, the consequences of nuclear accidents are severe and long lasting because they can result in the contamination of large areas of land. Two of the main problems come from release of iodine-131, which can become concentrated in the thyroid gland, and caesium-137, which is readily absorbed into the food chain and has a half-life of 30 years.[113] Indeed, upland sheep farms in the UK are still subject to restrictions on sale of produce as a result of caesium pollution from Chernobyl in Ukraine, more than 25 years after the event.

The accidents at Three Mile Island in the United States (1979) and Chernobyl (1986) contributed to a period of stagnation for nuclear power during the 1980s and 1990s, and although a 'nuclear renaissance' was under way, boosted by nuclear's low-carbon credentials, the Fukushima disaster in Japan (2011) has led to a number of countries reconsidering their options. Germany, Italy and Switzerland have since rejected nuclear energy, and Japan faces public opposition to reopening any of its 54 reactors. Nevertheless, plans to expand nuclear power remain strong in most countries (see Chapter 5).[114]

Box 3.5   Major nuclear accidents

Nuclear accidents and incidents are rated 0–7 on the International Nuclear Event Scale (INES).

Windscale 1957 (INES level 5)

Overheating of a plutonium pile at a military reactor led to the dispersal of radioactive material into the surrounding area. The pile was written off, and modelling studies (Wakeford [2007], citing Clarke [1990]) suggest that the accident may have caused 100 deaths from cancer.

Kyshtym 1957 (INES level 6)

A nuclear waste storage tank at the Mayak reprocessing plant in Russia exploded, killing at least 200 people and heavily contaminating a wide area, leading to the evacuation of 272,000 people.

Three Mile Island 1979 (INES level 5)

A faulty valve combined with operator errors caused loss of cooling water leading to core meltdown. Operators eventually regained control, with some venting of radioactive gas to the atmosphere. There were no casualties, but the brand-new reactor was written off.

Chernobyl, 1986 (INES level 7)

A malfunction during a safety test led to an explosion and fire, with a major release of radioactive material over large areas of Belarus, Ukraine, Russia and Europe. There was a permanent evacuation of 350,000 people, a 30-kilometre exclusion zone remains in place, and 7,000 square kilometres were severely contaminated with caesium-137 at levels over 555 kilobecquerels per square metre, meaning that there are strict restrictions on agriculture and other activities. Forty-seven workers were killed during the explosion or died from radiation poisoning while they tried to contain the damage. As of 2005, there have been an estimated 7,000 cases of thyroid cancer and nine children have died. Estimates of the eventual death toll from cancer are highly controversial, ranging from 4,000 among the most highly exposed population (The Chernobyl Forum, 2005) to 16,000 across Europe (Cardis *et al.*, 2006) to 985,000 globally (Yablokov *et al.*, 2009).

Fukushima, 2011 (INES level 7)

Following a devastating tsunami, cooling systems at the Daiichi nuclear plant failed, causing overheating at four reactors and partial meltdown at three of these. Hydrogen explosions damaged the containment systems, releasing a plume of radioactive material extending north-west over the surrounding area, and large amounts of radioactive water from emergency cooling operations were discharged into the sea. There was an evacuation of 130,000 people living within 20 kilometres of the plant. Over 30,000 square kilometres have been contaminated with caesium-137; of this, 600 square kilometres are severely contaminated with over 555 kilobecquerels per square metre, the level that triggered compulsory evacuation during the Chernobyl incident. It has been estimated that 1,000 deaths from cancer might result from the accident.[115]

Since 1952, there have been around 100 accidents at nuclear power plants that have led either to death or to property damage exceeding $50,000 (though some of these did not involve radioactive releases).[116] Modern reactors do offer improved levels of safety, and some new third- and fourth-generation reactor designs incorporate 'passive safety' features which aim to shut down the reactor safely without human intervention. But construction of these new designs has run into problems, with third-generation reactors in Europe being well behind schedule and over budget (see Chapter 5).

The ExternE project of the EU estimated that the health impacts of accidents add just €0.0001/kWh to the cost of nuclear power, but the Japanese Atomic Energy Commission made a post-Fukushima estimate of 1.1 yen/kWh (€0.01/kWh).[117] This could be an underestimate, as later work suggests that the release of caesium-137 from Fukushima was twice the original estimate.[118] One problem with these cost estimates is that they use a discount rate, which makes future costs smaller than present costs. After about 100 years, any discounted cost becomes practically zero, making long-term impacts such as nuclear pollution essentially disappear. It has also been argued that nuclear power receives hidden subsidies through governments effectively underwriting the costs of insuring against accidents. One source argues that these costs would add at least €0.14/kWh and perhaps as much as €2.36/kWh to the cost of nuclear power, making it uncompetitive.[119]

Waste

Nuclear power produces very long-lived radioactive waste. Spent fuel contains various fission products together with uranium and plutonium isotopes with long half-lives, such as plutonium-239 with a half-life of 24,000 years. Further waste is generated from fuel reprocessing, uranium mining, plant decommissioning and fuel fabrication.

Disposing of high-level waste is problematic. Underground waste repositories are the main option, but these are unpopular with local communities. Despite decades of research, no deep repository for high-level waste is yet operating anywhere in the world, though plans are well advanced in Sweden and Finland. All the high-level nuclear waste produced to date (roughly 270,000 tonnes as of 2007) is currently in temporary storage, 90 per cent of it in cooling ponds at power stations.[120]

The challenge with underground disposal lies in ensuring that radionuclides do not contaminate water supplies or enter the food chain. This is difficult because it is impossible to construct a repository that will keep out groundwater permanently. Typical designs involve encasing the waste in concrete-filled metal drums or vitrifying it into glass canisters, then placing it in a deep repository which is back-filled with clay or concrete. Over a period of hundreds of years, groundwater will eventually penetrate and flow through this repository, and safety studies tend to focus on the speed at which radionuclides will be transported away from the repository and the likelihood of them ending up in human water supplies. The risk depends on factors such as future rainfall levels, groundwater fluctuations, sea-level changes and geological movements.

Fast breeder reactors produce less waste and they can also burn reprocessed fuel, which can reduce existing spent fuel stockpiles. However, their development has been hindered by technical challenges (see Chapter 5).

Geo-engineering

Various geo-engineering schemes have been proposed to tackle climate change. The most widely discussed is the concept of spraying sulphuric acid out of aeroplanes to form an aerosol of sulphate particles in the stratosphere, thus reflecting the sun's radiation and cooling the earth. The sulphate particles gradually return to the earth's surface, and must be constantly replenished by more spraying.

Although there are clear implications for air quality, preliminary modelling has suggested that the amount of sulphate used would only add a few per cent to current sulphate emissions, and that any effects on human health and ecosystems would be small.[121] However, others contend that effects could be more severe, and that sulphate injection would involve at least doubling the current atmospheric concentration of particulate matter, and perhaps increasing it by a factor of between 5 and 10 in future years.[122] There are also concerns that there could be damage to the ozone layer and that local and regional weather patterns such as monsoon rains could be disrupted, leading to drought in Asia and Africa.[123]

Other geo-engineering schemes such as iron fertilisation of the oceans (to encourage the growth of algae that would take up carbon dioxide) also have potentially serious and poorly understood impacts for ecosystems. Iron fertilisation can result in ocean acidification, a shift in the balance of species, and oxygen deprivation. Its effectiveness is unclear, as in order permanently to sequester carbon dioxide, the algae would have to sink to the deep ocean and remain there without decomposing. It could take an area the size of the Southern Ocean to absorb just 3 per cent of man-made carbon dioxide emissions, and this could be offset by an increase in nitrous oxide emissions.[124]

The way forward

A strong climate policy, based on cutting our use of fossil fuels, offers a tremendous opportunity to reduce air pollution. Cleaner air will improve health (especially for children and the elderly), save lives, enhance crop yields and reduce damage to forests, buildings, rivers and lakes. Studies show that these air quality benefits offset much of the climate policy cost, and can even exceed it. Air quality benefits can also provide a stronger incentive for action than climate benefits, because the benefits are immediate, local and obvious.

To date, however, policies to address climate change and air quality have been separate. Air quality has typically been seen as the responsibility of local or regional governments, whereas climate change is a global issue. As our scientific understanding improves, it is becoming increasingly obvious that this is a false division. Pollution is no longer a local issue: long-range pollution increases the background concentration of ozone across the globe, making it harder to achieve air quality targets through national measures alone. Also, many local pollutants are now known to be climate-altering gases, including ozone, sulphate particles and black carbon.

There is an urgent need for a more integrated strategy to maximise the synergies and minimise the conflicts between the two goals, with the potential to achieve cleaner air and lower greenhouse gas emissions at a lower cost than with separate policies. For example, there is little point in building a new power station with expensive pollution control equipment if it would be more cost-effective to invest in clean renewable energy technologies, thus avoiding the pollution altogether. Inappropriate investments carry the risk of locking us into unsustainable technology. There are also technical limits to the pollution cuts that can be achieved through end-of-pipe methods. Ultimately, we cannot meet air quality targets without cutting our use of fossil fuels.

Despite the compelling reasons for developing an integrated strategy, governments around the world continue to exclude air quality co-benefits from their decision-making process. A survey in 2010 showed that out of 13 major climate policy assessments carried out by the IPCC and the UK and US governments, only two assessed air quality co-benefits and only one of these included these benefits in the final cost assessment. Most of the models focused narrowly on achieving a specific carbon reduction goal whilst minimising the cost per tonne of carbon abated. The study concluded that if air quality benefits were included in the assessments, stronger and earlier climate action would be justified.[125]

A good starting point for an integrated strategy is to prioritise win-win options that are good for both climate and air quality. Some examples are listed below.

Win-win (good for both climate and air quality)

- energy-efficient technology;
- energy-saving behaviour change;
- reducing transport activity – e.g. through better public transport;
- renewable energy;
- switch from coal or oil to natural gas;
- switch from coal or oil to biofuels;
- nuclear power (except for radionuclide pollution);
- hybrid, electric and hydrogen vehicles – provided that hydrogen and electricity are produced with clean, low-carbon technologies;
- emission controls to cut black carbon and NO_x emissions from vehicles;
- cutting methane emissions from agriculture, waste and fossil fuel production;
- more efficient biofuel stoves.

Win-lose (good for climate, bad for air quality)

- switch from gasoline to diesel vehicles (end-of-pipe particle filters and NO_x controls can reduce this conflict);
- switch from gas to biofuels;
- traditional biofuel stoves and fires in developing countries;
- carbon capture and storage (borderline: good for some pollutants, but worse for NO_x)
- geo-engineering (sulphate injection to the stratosphere).

Lose-win (good for air quality, bad for climate)

- end-of-pipe abatement technologies such as flue gas desulphurisation on power stations;
- reducing the sulphur content of fossil fuels;
- switch from traditional biofuels to gas or electric stoves.

Lose-lose (bad for both air quality and climate)

- use of fossil fuels without emission control and CCS.

Although we can maximise benefits by focusing on the win-win options, the other options may still have a role to play, especially with careful policy design to minimise the conflicts. For example, end-of-pipe solutions and the use of low sulphur fuel will still be essential to minimise emissions from vehicles and power stations – and the energy penalty could be offset or even exceeded by the resulting reductions in black carbon or ozone precursors.

Using an integrated approach, several important issues emerge.

- The transport sector is a priority, as vehicles are the fastest growing source of greenhouse gases and produce the most severe health impacts from air pollution. Key solutions include support for public transport, cycling, walking and cleaner vehicles.
- Biofuels have a role to play provided that they are produced from sustainable sources and that adequate pollution control is applied. Second-generation bio-fuels from waste are particularly promising, especially bio-gas from anaerobic digestion.
- Climate change policies must address the indoor air pollution problems from traditional biofuel use in developing countries; for example, by promoting cleaner and more efficient biofuel stoves, bio-gas or renewable electricity.
- Cutting air pollution can reduce the cooling effect of sulphate and other inorganic aerosols. This can be mitigated with integrated strategies that also reduce black carbon, methane and ozone precursors – thus also improving health and crop yields, especially in Asia.
- Carbon capture and storage (CCS) could achieve both air quality and climate benefits in theory, but the large energy penalty would cause some impacts to increase, including NO_x emissions and impacts related to coal mining and waste disposal.
- Geo-engineering based on injecting sulphate aerosols to the stratosphere fails to achieve air quality benefits, fails to address ocean acidification and has potentially serious consequences for the ozone layer and local rainfall patterns.
- Nuclear power is a win-win solution for climate and air quality (for conventional air pollution), but faces problems in other areas (e.g. waste disposal and safety) which must be solved to the satisfaction of the general public if nuclear power is to play a key role in climate policy.

4 Greener land

Forests, food and farming

Key messages

- *Protecting forests* saves carbon emissions and has many co-benefits: safeguarding biodiversity; preventing floods and soil erosion; increasing local rainfall and providing livelihoods for local people. However, carbon payment schemes must be designed very carefully to protect the rights of local people and avoid replacement of natural forests with plantations.
- *Agriculture* is currently a major source of greenhouse emissions, but the sector could become carbon-neutral by increasing the carbon stored in soils and reducing emissions from fertilisers and livestock, using methods such as conservation tillage, adding organic matter to the soil, avoiding over-application of synthetic fertilisers, planting trees, and eating less animal produce. Many of these methods have important co-benefits, such as improving soil structure and fertility, reducing air and water pollution, protecting biodiversity and enhancing resilience to climate change. However, there is an ongoing debate over whether the best way forward lies with more intensive agriculture based on chemical inputs, or agro-ecology based on organic principles. There are important implications for food security and the demand for farm land.
- *Biofuels* are widely viewed as a key climate strategy, but some biofuels have higher life-cycle greenhouse gas emissions than fossil fuels. Biofuels can also result in deforestation, loss of land for food production (and thus increased food prices), displacement of local people and high use of water, fossil fuels, fertilisers and other agrochemicals. Strict guidelines are needed to avoid these problems and ensure that biofuels are produced sustainably.

Population growth and soaring consumption are placing increasing burdens on the land. Natural forests and other important wildlife habitats are being converted to farmland and plantations to satisfy our growing demand for food, fibre, timber and biofuels. We have lost 40 per cent of the world's original forests, 50 per cent of wet-lands since 1900 and 20 per cent of mangroves since 1980.[1] A combination of habitat loss, pollution, over-fishing, over-grazing and climate change is now leading to what has been referred to as the 'sixth great mass extinction', with rates of species loss that are 100 to 1,000 times the natural background levels.[2] At the same time, the ability of our agricultural system to feed the world sustainably is in doubt. Intensive farming has increased crop yields, but also caused water pollution, soil erosion and groundwater depletion. Shortages of water and fertile land, together with the increasing cost of fertilisers and other agrochemicals, are now contributing to rising food prices and fears about food security.

Climate policy can play a major role in addressing these issues. Forests, soils and other ecosystems store vast amounts of carbon which are emitted to the air when vegetation is cleared or soils disturbed. This accounts for an estimated 12 per cent of man-made greenhouse gas emissions, mainly from deforestation. Agriculture produces a further 12 per cent of global greenhouse gas emissions in the form of nitrous oxide from synthetic nitrogen fertilisers and manure, and methane from livestock farming and rice paddy fields (Figure 4.1).[3] Climate policy is therefore increasingly looking for ways of protecting the 'green carbon' stored in soils and ecosystems, and reducing emissions from agriculture. This is particularly important because climate change itself is likely to accelerate losses of green carbon through a series of feedback loops: drought could lead to the loss of much of the Amazon rainforest, for example; and higher temperatures can speed up the breakdown of soil carbon into carbon dioxide. This type of feedback loop could eventually transform land-based ecosystems from a carbon sink, absorbing an estimated 30 per cent of man-made greenhouse gas emissions at present (2012), to a carbon source.[4]

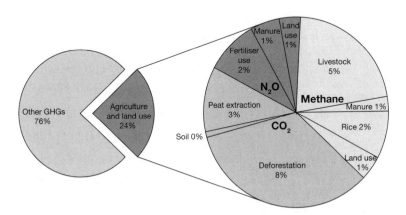

Figure 4.1 Global greenhouse gas emissions from agriculture and land use, 2008

Source: Based on data from EDGAR database (EC/JRC-PBL, 2011) (figures rounded)

Note: Excludes carbon dioxide from fuel used in agricultural machinery and for fertiliser manufacture. N_2O = nitrous oxide.

Preventing deforestation and moving to more sustainable agriculture can reduce greenhouse gas emissions at the same time as providing multiple co-benefits, such as preserving valuable ecosystems, improving soil fertility and cutting water pollution. In developing countries, forests can provide food and firewood for local people, and trees on steep slopes can greatly reduce the risk of floods and landslides. These benefits can generate strong public support, particularly in view of the widespread concern over the fate of rainforests and endangered species.

However, there are also some severe potential conflicts. Poorly designed climate policy can lead to the felling of rainforest for biofuel cultivation, or to 'land grabs' as powerful elites seize control of carbon-rich forests in anticipation of future payments for preserving them. There is also a heated debate over the future of agriculture: should we move to more intensive techniques (including controversial options such as genetic modification) in order to reduce the expansion of agriculture into wilderness areas; or to less intensive techniques in order to minimise other impacts such as pollution from agrochemicals? This chapter will try to identify a way through this complex ethical and environmental maze, in order to exploit the strong potential co-benefits and minimise the conflicts.

Relevant climate policies

A wide range of climate policies can provide co-benefits (or conflicts) for ecosystems and agriculture: protecting forests; increasing carbon stored in soils; agroforestry; reducing emissions of methane and nitrous oxide from farming; using bioenergy; improving crop yields; and eating less meat.

Protecting and increasing forest carbon

It is difficult to measure the amount of carbon stored in soils and vegetation, but it is believed to be well over 2,000 billion tonnes – three times the amount in the atmosphere, which is 750 billion tonnes. Of this, about 600 billion tonnes is in vegetation and 1,600 billion tonnes is in the top metre of soil.[5] Further large stores of carbon are held in deeper soils, including peat and permafrost where decomposition of organic matter is very slow.[6] All these carbon stores are dwarfed by the 38,000 billion tonnes stored in the ocean.

Forests cover about 30 per cent of the land surface, yet they store about 60 per cent of all the carbon held in ecosystems (down to the first metre of soil depth). The most carbon-rich ecosystems are tropical forests, where most of the carbon is held above ground in the dense vegetation, followed by boreal forests where more of the carbon is stored in the soil as accumulated organic matter (Table 4.1). However, significant amounts are also held in grasslands and wetlands.

Oceans and land-based ecosystems are thought each to have absorbed about 30 per cent of the carbon dioxide emissions from fossil fuels since the Industrial Revolution. Land-based ecosystems are thought to absorb 2–3 billion tonnes of carbon each year, much of which is taken up by forests, while oceans absorb another 2 billion tonnes, set against current emissions of around 8 billion tonnes from fossil fuels.[7] Yet this vital carbon sink is dwindling, as forests are felled and peatlands drained to clear land for agriculture and plantations.

It was once thought that only newly planted or regenerating forests could remove carbon from the atmosphere, and that mature 'old-growth' forests were at

Table 4.1 Estimates of carbon stored in ecosystems and oceans

Ecosystem	Tonnes per hectare in vegetation	Tonnes per hectare in soil	Total tonnes per hectare	Total stored carbon worldwide, Gt carbon	Risks
Estimates to 1m depth					
Tundra	40	220	260	155	Warming
Boreal forests	60–90	120–340	180–440	385	Logging, mining, draining of peat
Temperate forests	90–190	60–130	150–320	315	Historic losses, currently expanding
Temperate grassland and shrubland	8	130	140	185	Agriculture (mainly historic)
Tropical and sub-tropical forests	170–250	90–200	260–450	550	Agriculture, biofuel production, logging
Tropical and sub-tropical savanna	9–80	170	180–260	285	Fire, agriculture
Deserts and dry shrubland	2–30	14–100 (desert); 270 (dry shrubland)	16–300	180	Soil erosion, over-grazing
Total terrestrial to 1m depth				**2,055**	

Estimates to greater depths (note that these cannot be added to the figures above as the permafrost and peatland estimates partially overlap with estimates above and with each other)

Northern permafrost region				1,670	Warming
Peatland			1,450	550	Drainage for biofuels and pulpwood plantations

Table 4.1 continued

Ecosystem	Tonnes per hectare in vegetation	Tonnes per hectare in soil	Total tonnes per hectare	Total stored carbon worldwide, Gt carbon	Risks
Oceans					
Oceans – surface				1,020	
Oceans – dissolved organic carbon				700	
Oceans – deep				38,100	

Source: Based on data in Trumper *et al.*, 2009 (figures rounded)

equilibrium, with the carbon absorbed by photosynthesis being offset by the carbon released as dead matter decayed. However, research has shown that even old-growth forests continue to absorb carbon dioxide, partly in response to increasing levels of carbon dioxide in the atmosphere, which stimulates extra plant growth.[8] A 2011 study estimates that existing forests absorb 2.4 billion tonnes of carbon each year, but that 2.9 billion tonnes are released through deforestation. This is partly offset by the absorption of 1.6 billion tonnes as newly planted forests (mainly commercial plantations) grow, so that forests still act as a net carbon sink for 1.1 billion tonnes of carbon each year.[9] Draining of peatlands, especially through clearing of tropical swamp forest for pulpwood and biofuel plantations in Malaysia and Indonesia, is estimated to release a further 0.5–0.8 billion tonnes of carbon each year.[10]

The carbon stored in forests and other ecosystems can be protected by:

- Avoiding deforestation.
- Planting new forests (afforestation) and restoring felled areas (reforestation). It is estimated that up to 2 billion hectares of degraded former forest land is suitable for reforestation, of which 0.5 billion hectares is sparsely populated and could be allowed to regenerate to continuous natural forest, and 1.5 billion hectares is moderately populated land and could be suitable for 'mosaic' restoration, which is where patches of natural forest are interspersed with farming and agroforestry.[11]
- Better forest management, such as 'reduced-impact logging' (selective removal of individual trees, leaving the rest of the forest undisturbed), replanting after felling, and avoiding fires. Reduced-impact logging can reduce carbon losses by around 30 per cent compared with conventional techniques.[12]

- Avoiding the draining of peat soils for forestry, agriculture, biofuel plantations or peat harvesting, and restoring drained peat lands through re-wetting.

These objectives can be supported through a variety of policies including:

- setting up protected areas to preserve existing forests;
- certification to help promote the use of sustainably produced timber and alternatives to peat;
- paying landowners to preserve their forests or to plant new trees via 'payment for ecosystem services' (PES) schemes – these can include schemes funded by national or regional governments, voluntary carbon offsets purchased by individuals or companies, and international schemes such as the proposed REDD mechanism (see Box 4.1).

Box 4.1 REDD, PINC and green carbon

As awareness of the amount of 'green carbon' stored in ecosystems grew, a group of developing countries led by Costa Rica and Papua New Guinea proposed that there should be a system for paying developing countries not to cut down their forests. This concept has gradually evolved through the following stages:

CDM	Limited carbon credits for afforestation and reforestation are allowed under the Clean Development Mechanism of the Kyoto Protocol
RED	Reduced emissions from deforestation
REDD	Reduced emissions from deforestation and forest degradation
REDD+	REDD plus conservation, sustainable management of forests and enhancement of forest carbon stocks (this allows funding for new planting and sustainable felling as well as protection of existing forests)
PINC	Pro-active investment in natural capital. As REDD would pay more to countries with high historic deforestation rates, PINC was therefore proposed as a complementary system to allow payment to countries with large forests but low deforestation rates, which would not benefit from REDD. The approach is based around the recognition that forests provide a range of valuable ecosystem services, not just carbon storage.

Reduced emissions from deforestation and forest degradation (REDD) is thought to be a very cost-effective way of reducing carbon emissions. The

Eliasch Review (2008) estimated that annual payments of $17–33 billion under a carbon trading scheme could halve deforestation rates by 2020 and (with afforestation) make the forest sector carbon-neutral by 2030, with long-term net benefits of $3.7 trillion in avoided climate damage. A more ambitious target of reducing deforestation by 90 per cent would have even higher net benefits of over $6 trillion.[13]

At present, proposals for a REDD+ mechanism are well advanced, and many pilot projects are under way, mainly funded under the UN-REDD Programme and the World Bank Forest Carbon Partnership Facility (FCPF). However, the prospects for large-scale funding as part of an international climate agreement are currently (2012) in doubt (see 'The way forward' below).

Increasing carbon stored in agricultural soils

Large amounts of carbon are lost from agricultural soils during tilling, which is when the soil is broken up by ploughing or harrowing. Tilling is used to kill weeds, mix in fertiliser and crop residues (the remaining stems and roots of the previous crop), level the soil surface and create a fine soil texture ready for planting, but it also increases soil erosion and speeds up the oxidation of carbon by soil microbes. As a result, cultivated soils typically contain only 50–75 per cent of the carbon content of undisturbed natural soils.[14] It has been estimated that between 42 and 78 Gt of carbon could have been lost from agricultural soils to date, leaving many soils depleted in organic matter and nutrients, especially in hot, humid regions where organic matter is broken down rapidly. However, it could be possible to restore 50–66 per cent of this carbon using techniques such as conservation tillage, cover crops and addition of organic matter to the soil.[15]

Low-till or no-till agriculture, also collectively called 'conservation tillage', avoids disturbing the soil as far as possible. The field is not tilled, or only tilled occasionally (rotational tilling), or only tilled in thin strips for planting, leaving undisturbed soil in between (strip tilling). The crop residue from the previous crop is left in place (rather than being burnt, removed or ploughed in) in order to protect the soil from erosion and add nutrients, and seeds are planted using special equipment to cut a narrow slot or drill single holes through the crop residue and soil crust.[16] Low-till methods conserve soil carbon at the same time as reducing the use of fossil fuels in farm machinery – the number of tractor passes needed to cultivate a crop can be reduced from around ten to just one or two, saving time and effort for the farmer. Conservation tillage is practised on around 5 per cent of the world's cultivated area,[17] especially in the United States – where it was introduced to address severe soil erosion that formed the Great Plains 'dustbowl' in the 1930s – and Brazil, where it is used on half of the farmed area. It is estimated that greater worldwide uptake of no-till agriculture could sequester 1.5–4.5 $GtCO_2$ per year, which represents 3–9 per cent of total greenhouse gas emissions.[18] It is common for conservation tillage to be used in conjunction with the use of cover crops and crop rotation: this grouping of practices is termed 'conservation agriculture'.

Cover crops are grown in the periods between commercial crops (such as during the winter, or as part of a crop rotation cycle), in order to suppress weeds, prevent soil erosion and add nutrients. Nitrogen-fixing cover crops such as vetch, clover, peas, beans and other legumes are often used as a 'green manure', as they absorb nitrogen from the air and thus reduce the need for fertilisers. Prior to planting, cover crops and weeds are ploughed into the soil or, for no-till systems, killed with herbicides. Organic farmers wishing to use no-till techniques have experimented with alternatives to herbicides including mowing, rolling with a 'crimper' to crush the plants, or (for smaller-scale agriculture and horticulture) covering with cardboard to suppress light. The regular use of cover crops can increase soil carbon by 15–28 per cent in the long term.[19]

Soil carbon can also be increased by adding organic matter such as compost and manure. Long-term use of organic fertilisers can increase soil carbon levels by 10–100 per cent compared to synthetic fertilisers.[20] All these methods of increasing soil carbon may eventually reach a saturation point where further additions of carbon are balanced by losses, but studies have shown that carbon levels can continue to increase for several decades provided that more organic matter is added regularly.[21]

Soil carbon can also be increased by a variety of other methods:

- use of terracing, 'bunds' (low stone walls or earth banks built along slope contours) and trees to reduce soil erosion on steep slopes, especially 'contour planting' of trees and shrubs along terrace edges;
- rotational grazing (moving livestock from place to place, leaving the pasture elsewhere to recover) or reducing grazing density;
- allowing set-aside (fallow) periods to allow degraded soils to recover;
- increasing crop diversity and using integrated pest management (e.g. use of natural predators) instead of pesticides;[22]
- choosing species with higher carbon storage, such as deep-rooted grasses;
- burying biochar in soil (see Box 4.2).

There are complex interactions between some of these options, which can lead to conflicts and trade-offs. For example, the herbicides commonly used as part of no-till agriculture can kill beneficial soil organisms, and thus reduce soil carbon, as well as having other undesirable environmental effects. Similarly, fertilisers can increase carbon uptake in vegetation by promoting growth, but can also speed up the decomposition of soil carbon by micro-organisms, as well as leading to nitrous oxide emissions and increased use of fossil fuels for fertiliser manufacture. These issues will be explored further in the section on 'Conflicts'.

Box 4.2 Biochar

Biochar is simply charcoal, which is created by burning biomass in the absence of oxygen – this is called 'pyrolysis'. Water and organic impurities are driven off, leaving a fuel consisting mainly of pure carbon, which burns very cleanly and with a high temperature. Charcoal has been produced for centuries as a fuel for metal smelting, until it was largely replaced by coke made from pyrolysis of coal. Traditional charcoal production in earth-covered wood piles could take many days, but modern pyrolysis in furnaces can take just a few minutes or seconds, and allows the recovery of useful heat or electricity as well as by-products of oil and gas which can be used as fuel.

The difference between biochar and charcoal is largely in the application – biochar is charcoal that is deliberately buried in soil in order to sequester the carbon and improve the structure and fertility of the soil. Modern interest in biochar was stimulated by the observation of the rich, fertile 'terra-preta' soils in Latin America, left by past civilisations who buried charcoal as a soil improver. Biochar is rich in nitrogen, potassium, phosphorous and trace elements, and can slowly release these nutrients over time. The porous structure of biochar also aids water and nutrient retention, especially in poor-quality sandy soils. Research to date shows that biochar can dramatically increase yields in tropical soils, especially where water is scarce, but achieves more modest improvements in the fertile soils of temperate regions such as Europe.[23] However, biochar can contain toxic heavy metals, dioxins and polycyclic aromatic hydrocarbons (PAHs) which could pose a problem for food crops, grazing livestock, nearby water supplies and farm workers (especially if dust from biochar is dispersed by wind).

Attention is increasingly turning to the possibility of using biochar as a method of carbon sequestration. The idea is that biomass plantations such as coppice woodland would absorb carbon dioxide from the air as they grow, and the harvested biomass would then be used to create charcoal which would be buried in agricultural soils. In theory, the relatively inert carbon should stay in the soil for hundreds or even thousands of years, thus locking it out of the atmosphere. Further climate benefits could be gained if the biochar improves soil structure and fertility sufficiently to reduce the need for synthetic fertilisers, thus avoiding nitrous oxide emissions and carbon emissions from fossil fuels used in fertiliser manufacture. Heat and electricity generated during the pyrolysis stage of biochar production could also displace fossil fuel emissions.

Research into biochar is at a relatively early stage, and questions exist over how long the carbon would remain in the soil, to what degree biochar can improve soil fertility and productivity, whether it speeds up or slows down the breakdown of soil organic carbon, how much land is suitable for biochar burial and whether biomass can be produced sustainably. At present, it seems that biochar may be best suited to poor or degraded tropical soils, and only where there is a sustainable source of biomass such as waste material that is not needed for other purposes such as composting or animal fodder.

Two studies demonstrate that biochar may not always offer major benefits compared to using biomass directly in a boiler, or using it to produce biogas in an anaerobic digester. A UK study (2011) estimated that a pyrolysis–biochar system could save 7–30 tCO_2e per hectare of biomass grown per year, compared to savings of 1–7 tCO_2e per hectare if the biomass was burnt in a boiler to displace fossil fuels. About half of the carbon removal was due to carbon locked up in the biochar, with the rest being indirect effects from avoided fertiliser use and avoided use of fossil fuels (from the heat and electricity produced during pyrolysis). However, with less optimistic assumptions about fertiliser displacement and biochar carbon storage times, the biochar system was less beneficial than the conventional biomass systems.[24]

The second study (2010) estimated that large-scale use of biochar could offset 12 per cent of annual global carbon emissions, whereas conventional combustion of biomass (displacing fossil fuels for heat and power generation) could reduce emissions by up to 10 per cent. However, once again the advantage of biochar over conventional biomass combustion depends on unproven benefits such as an increase in soil productivity, and may not apply in soils that are already fertile.[25]

Finally, it is unclear how the use of biochar would be funded, as the cost of production could well exceed the amount that farmers would be likely to pay on the basis of its use for soil improvement. Wide-scale use could therefore depend on including the use of biochar in a carbon trading mechanism.

Agroforestry and perennial crops

Agroforestry involves planting trees and shrubs on farmland. This covers a wide range of options, including scattered trees amongst crops or on pasture; hedgerows along field or terrace edges; strips of woodland on steep slopes or stream banks; and orchards or plantations. The trees and shrubs can provide fruit, timber or fuel for the farmer. Perennial crops are those which can be harvested and will then regrow from the roots the following year, including certain fodder crops such as alfalfa and hay.

Agroforestry and perennial crops have a number of climate benefits compared to annual crops:

- more carbon is stored and absorbed in roots and vegetation;
- soil carbon accumulates as leaf litter falls to the ground;
- soil erosion is decreased because the ground is stabilised;
- less soil carbon is lost because the soil is not disturbed through tillage;
- less nitrogen fertiliser is generally required than for annual crops.

Depending on the exact conditions, agroforestry can increase annual carbon uptake to 1.5–3.5 tonnes per hectare, compared to a typical value of 0.2 tonnes per hectare for annual crops,[26] and further climate benefits arise from reduced fertiliser use. However, at present there are few options for perennials to replace the staple grain crops that form the bulk of the world's food supply.[27] Perennials also make it difficult to practise crop rotation, so that there can be a build-up of pests and diseases over time.

Reduced emissions from nitrogen fertilisers

The use of nitrogen fertilisers – both organic and inorganic – increases yields, but also leads to emissions of nitrous oxide as the fertilisers degrade in soils, or after the nitrates are washed out of soils by rain. For synthetic fertilisers, there are also carbon dioxide emissions from manufacture, which is very energy-intensive. Emissions can be reduced by various methods:

- avoiding over-application of fertilisers;
- using slow- or controlled-release fertilisers to minimise periods when there is surplus nitrogen in the soil;
- adjusting the timing of fertiliser application to coincide with plant needs;
- applying fertiliser in the best location for plant roots to take it up more efficiently (e.g. adding to the planting hole rather than spraying the whole field);
- rotational grazing or set-aside periods to reduce the need for fertiliser application by allowing the land to recover naturally;
- using winter cover crops to soak up excess nitrogen from the soil;
- using organic fertilisers, such as manure, compost, or nitrogen-fixing cover crops, instead of inorganic fertilisers.

Using organic fertilisers avoids carbon dioxide emissions from fertiliser manufacture, but the effect on nitrous oxide emissions is less clear. Manure may emit more nitrous oxide than synthetic fertilisers do, but it can also be argued that the manure already exists as a consequence of livestock production and therefore the emissions would have occurred anyway.[28]

Reducing methane emissions

Various methods exist for controlling methane emissions from agriculture:

- better management of rice paddy fields – e.g. by draining them during the off-rice season;

- improved feed or pasture quality for livestock, to reduce the methane produced in their digestive systems;
- better manure management, such as covering manure and animal slurry when stored in tanks and lagoons, composting manure in well-aerated small piles instead of large heaps, or collecting methane emissions in an anaerobic digester for use as bio-gas.[29] However, some techniques to minimise methane emissions can lead to higher emissions of nitrous oxide, and vice versa.

Bioenergy

Bioenergy already provides 13 per cent of the world's final energy. Most of this is traditional biomass – wood and dung burnt in cooking stoves and on open fires in developing countries – but the use of 'modern' bioenergy is growing. This includes solid biomass such as wood chips and pellets burnt in power stations, factories, domestic boilers or CHP plants; anaerobic digestion of food or farm waste to produce bio-gas for heat, power or transport; and liquid biofuels produced from crops, woody material or algae (Box 4.3).[30]

Sixty governments now have targets and incentives to support bioenergy, mainly in the transport sector but also for heat and power generation.[31] For transport, targets generally specify that a certain proportion of biofuel must be blended into all fuel sold. In Brazil, for example, gasoline must contain 20–25 per cent bio-ethanol and diesel must contain 5 per cent bio-diesel. The US government sets mandatory targets for the amount of biofuel that must be blended into gasoline and diesel each year: 48 billion litres in 2010, rising to 136 billion litres by 2022.[32] In the European Union, there was a target for 5.75 per cent of all road transport fuel to be derived from biofuels by 2010, now replaced by a target of 10 per cent of transport fuel to be renewable (including renewable electricity or hydrogen) by 2020.[33]

Globally, biofuels provide almost 3 per cent of transport fuel, and production is expected to double over the next decade (2010–2020) under current trends.[34] With strong climate policy, future production could be even greater: the IEA envisages that biofuel production could increase from 1.1 million barrels per day in 2009 to 8.1 million barrels per day in 2035, when it would supply 14 per cent of global transport fuel demand, and this could grow to 27 per cent of transport fuel demand by 2050.[35] There is particular interest in the aviation sector, where other options for reducing emissions are limited. However, the way in which the biomass feed-stock is produced is all-important: some biofuels can have worse greenhouse impacts than conventional fuels, as well as having adverse impacts on land use, biodiversity and greenhouse gas emissions. This will be discussed in the section on 'Conflicts' below.

Box 4.3 Algal biofuels: panacea or mirage?

Algae are fast-growing, efficient at converting sunlight to biomass, and can be grown in wastewater or seawater, avoiding the need for productive land and freshwater. However, algal biofuels have not yet been produced on a commercial scale or at competitive prices. Yields are currently low, and there are limited sites with the right combination of a wastewater source, a carbon dioxide source to stimulate growth (such as power station flue gas), adequate sunlight (which cuts out much of northern Europe, Russia and Canada) and a warm and stable climate (above 15°C average monthly temperatures in winter) or a suitable waste heat source.

Although yields per hectare could potentially be considerably higher than conventional biofuels (see Table 4.2), most production methods still need large areas of flat land on which to site tanks or ponds in which to grow the algae. Temperature, carbon dioxide, pH, light exposure and nutrient concentrations must be carefully controlled to provide optimum conditions for growth, and open ponds (the cheapest production method) need to be stirred so that all algae receive enough light and nutrients, while guarding against invasion by foreign strains of algae or bacteria. Finally, converting the algae into usable biofuels can be complex, costly and energy-intensive, depending on the method chosen.[36]

Despite these challenges, over 200 companies are involved in research and development of algal biofuels. Innovative solutions include the possibility of growing algae at sewage treatment plants, using the nutrients and water already present in the sewage.[37] The residue left after oil has been extracted from the algae can be fed into an anaerobic digestion plant to produce biogas, which can then be burnt to provide heat and electricity to operate the plant, with the carbon dioxide from combustion and the nitrogen-rich digestate being used to fertilise the algae.[38] Algae can also be grown in closed bio-reactors rather than open ponds, which reduces the land and water requirement but increases the capital cost considerably. There is also an

Table 4.2 Oil yields per hectare from algae can be far higher than from crops

Crop	Potential oil yield (litres/hectare/year)
Soybean	450
Sunflower	955
Jatropha	1,890
Oil palm	5,940
Algae	3,800–50,800

option of growing seaweed (macroalgae) in coastal areas, such as on the foundations of offshore wind turbines. Production costs for various demonstration plants currently range from $200 to $1,200 per barrel of crude oil equivalent, compared to crude oil prices of just over $100 per barrel, but it is estimated that costs could fall to as little as $60 per barrel in future.[39] The IEA anticipates that algal biofuels will become commercially viable between 2020 and 2030.[40]

Improving yields

Improving crop and livestock yields per hectare means that less land is needed for agriculture, which can reduce deforestation. However, some methods of improving yields involve trade-offs with other climate impacts: irrigation increases the use of fossil fuels to power pumps; synthetic fertilisers emit nitrous oxide; and biocides can kill soil organisms and thus reduce soil carbon levels.

Eating less meat and dairy produce

Meat and dairy production are thought to be responsible for 80 per cent of agricultural greenhouse gas emissions and 18 per cent of total man-made greenhouse gas emissions, including 35–40 per cent of methane emissions, 65 per cent of nitrous oxide emissions and about 9 per cent of carbon dioxide emissions.[41] About a third of this is carbon dioxide from deforestation to clear land for pasture and feed crops; a quarter is methane from enteric fermentation in the digestive systems of cows, sheep and goats; and the rest is mainly nitrous oxide emissions from manure and fertiliser used on pastures and feed crops (Figure 4.2). Cattle and dairy farming tend to have the highest impacts because of the high methane emissions from the digestive systems of ruminants. Poultry and pigs produce less methane and are more efficient at converting plant energy into animal energy.

Globally, demand for meat is predicted to grow from an average 37 kilograms (kg) per person per year in 2000 to over 52kg by 2050.[42] The technical options for reducing emissions from livestock, such as improving feed quality or manure storage, are limited, but there is considerable scope for demand to be reduced. Although animal produce is a valuable source of nutrition, especially for the undernourished, many affluent consumers could cut down on their consumption with no ill effects – in fact, there could even be health benefits (see Chapter 8). Recent modelling suggests that in order to meet a 2°C climate target, nitrous oxide concentrations must be stabilised by 2050, which could be achieved if consumers in developed countries cut their annual meat consumption from 78kg (the 2002 level) to 37kg, while those in developing countries increased their consumption from 28kg to 37kg. This would cut overall global meat consumption by 21 per cent, and provide major environmental co-benefits, including reduced deforestation and pollution.[43]

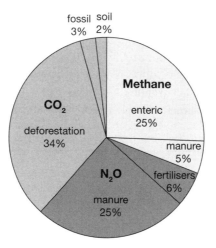

Figure 4.2 Breakdown of greenhouse gas emissions from meat and dairy production
Source: Data from FAO, 2006

Co-benefits

It is clear that better land management is essential if we are to meet climate targets. Deforestation and agriculture currently contribute around 24 per cent of greenhouse gas emissions, yet with best practice this could be turned around so that both agriculture and forestry are carbon-neutral by 2030.[44] At the same time, there could be a wide range of very significant co-benefits from protecting forests and shifting to more sustainable agriculture.

Forests

Forests are more than just a carbon store: they are rich in biodiversity, help to prevent floods and landslides, maintain clean water supplies, reduce air pollution (see Chapter 3) and provide livelihoods, recreational opportunities and aesthetic benefits for billions of people. Yet over 15 million hectares (150,000 km²) of forest are being cleared each year – up from 14 million during the 1990s – and roughly the same area is being degraded through logging. Although 9 million hectares of new forest are created each year, mainly in boreal and temperate regions of Europe, Russia and North America, these are mainly low-biodiversity monoculture plantations.[45]

Old-growth forests, which now form only a third of the remaining forest area of 3.7 billion hectares,[46] are a unique and irreplaceable resource. Tropical forests are the most at risk: 10 million hectares are cleared each year (Figure 4.3). Climate policy can play a vital role in helping to safeguard these forests and the numerous services they provide, which are described below.

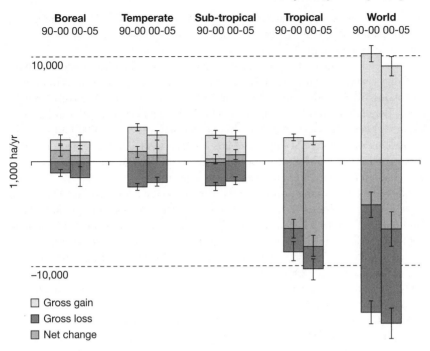

Boreal	Temperate	Sub-tropical	Tropical	World
90-00 00-05	90-00 00-05	90-00 00-05	90-00 00-05	90-00 00-05

Figure 4.3 Average annual forest area cleared, planted and net change for 1990–2000 and 2000–2005, showing that tropical forests are most at risk

Source: FAO, 2011b

Biodiversity

Biodiversity is simply the variability of living organisms, including the number of different species, the genetic variation within a single species, and the variety of different ecosystems. There is an ethical argument that biodiversity is valuable in its own right, as all species have an equal right to exist, but there are also economic benefits. Most obviously, biodiverse ecosystems provide a wide range of species that can be exploited for food, medicine and materials. But biodiversity also enables ecosystems and agriculture to adapt to change. If rising temperatures or a new pest or disease threaten to wipe out wheat crops, for example, we can turn to wild varieties to help breed new strains that are resistant.

The Economics of Ecosystems and Biodiversity (TEEB) study produced some useful indicative figures for the economic value of biodiversity. For example, 25–50 per cent of the $640 billion pharmaceutical industry, much of the $70 billion biotechnology industry, and all of the $30 billion agricultural seed industry are based on natural genetic resources such as medicinal plants, wild seeds and naturally occurring enzymes.[47] In addition, the natural world is invaluable as a source of knowledge – through 'bio-mimicry' we are increasingly copying the way plants and animals work to derive useful products including medicines and advanced materials.[48] But ultimately, biodiversity is worth even more than this: we depend on healthy ecosystems for our very existence. With a changing climate, increasing pollution and the spread of pests and diseases, we will need the resilience of bio-diverse ecosystems more than ever.

Biodiversity is currently declining at an alarming rate. The International Union for Conservation of Nature (IUCN) estimates that around 25 per cent of mammals and plants, 40 per cent of amphibians and 13 per cent of birds are at risk of extinction.[49] Populations of wild vertebrates fell by 31 per cent between 1970 and 2006.[50] Figure 4.4 (see plate section) shows that global biodiversity in 2000 had fallen to 73 per cent of its potential value, and it is predicted to fall to 62 per cent by 2050 under current trends. Most of the decline is in species-rich ecosystems such as forests and grasslands, where human activities are concentrated.

The main driver of this decline in biodiversity is habitat loss. Around 37 per cent of the earth's land area has already been converted to farmland, plantations or infrastructure. Under a business-as-usual scenario, a further 11 per cent (7.5 million km^2) of natural land will be lost by 2050, and 40 per cent of the area currently used for low-impact agriculture, such as semi-natural pasture, could be converted to intensive agriculture.[51] Other pressures include over-fishing, over-hunting (especially for traditional Asian medicines based on animal parts), pollution, climate change and the spread of invasive species.[52]

Against this grim picture, climate policy can play a major role in protecting threatened habitats. The most carbon-rich ecosystems – tropical forests and wetlands – are also those with the highest biodiversity, and are most at risk from habitat loss. The Amazon rainforest alone is thought to contain a quarter of all terrestrial species. Tropical forests are home to a large number of endemic species (those not found in any other country), many of which are at high risk of extinction. The populations of wild vertebrate species in the tropics fell by 59 per cent between 1970 and 2006,[53] and over 1,000 rainforest animal species are endangered or critically endangered, including gorillas, orangutans, chimpanzees, Javan rhinoceroses, Asian elephants, giant pandas, tigers and many species of monkeys, lemurs and gibbons.[54] Even within particular regions, there is generally a very strong synergy between areas of forest with high carbon storage and those with high biodiversity, as can be seen in the UNEP World Conservation Monitoring Centre's *Carbon and Biodiversity: A Demonstration Atlas.*[55] These areas are also most at risk – 46 per cent of deforestation in Indonesia is in peat forest.[56]

Under forest carbon payment schemes such as REDD (see Box 4.1 above), forest owners and users are compensated for giving up the short-term profit they could obtain by logging or clearing forest for agriculture or plantations. In Kalimantan in Indonesia, for example, it has been estimated that if carbon credits could be sold for $10–$33 per tonne of carbon dioxide (current prices are around $10 per tonne), it would be more profitable to conserve the forest than to clear it for oil palm plantations.[57] This could safeguard the habitat of 40 threatened mammals, including the orangutan and Borneo pygmy elephant. If only the most carbon-rich areas were targeted – peat forest with twice the diversity of mammal species of other areas – the price would drop to as little as $2 per tonne CO_2.

Although the highest biodiversity benefits come from conserving natural forest, there can also be benefits from reforestation, especially where strips of new forest can act as wildlife corridors linking existing areas of natural forest. The benefits will depend strongly on the tree species chosen, and on the previous use of the land. Planting of mixed native species on converted farmland, felled areas or degraded land can improve both carbon storage and biodiversity. However, commercial monoculture plantations on natural grassland or wetland tend to reduce biodiversity, providing green cover but not the range of food and prey species needed to sustain indigenous forest wildlife.

For reforestation schemes, the best option for biodiversity is to allow natural regeneration, by simply protecting the area from threats such as fires, grazing or invasive species. However, this is a slow process and also depends on there being enough natural forest nearby to allow dispersal of seeds. Regeneration can be speeded up by planting a few 'perch trees' for birds so that they can drop seeds, 'nurse trees' to provide shade for seedlings and help to stabilise soil, and green manure crops such as legumes to provide nutrients and replenish the soil. Direct replanting of trees is even faster, but ultimately leads to a forest with lower biodiversity, even if native species are chosen.[58]

Water catchment, flood prevention and soil protection

Trees perform a vital function in stabilising soil, providing clean water supplies and preventing flooding. Tree roots prevent soil from being washed away during heavy rain, especially on steep slopes. A good root network enables the soil to soak up water like a sponge, and this water can then be released more slowly over the following days and weeks, helping to maintain river levels. In addition, the soil and tree roots help to filter out pollution and suspended sediment from water flowing downhill, allowing clean water to enter the river system.

The devastating consequences of losing these ecosystem services are clearly illustrated when forests on steep slopes are clear-felled. Subsequent heavy rainfall frequently leads to landslides and flooding with widespread loss of life, and leaves the slopes denuded of soil and vegetation. One 10-year study of 56 developing countries found a clear correlation between deforestation and flood risk.[59] The problem is especially acute in tropical regions where rains are often heavy and soils tend to be thin and low in nutrients, with most of the nutrients stored in above-ground vegetation. Once the trees are gone, soils are quickly washed away and the land is of little use for farming.

Widespread deforestation in China's mountainous western provinces, to clear land for farming, eventually culminated in the deaths of thousands of people when record monsoon rains triggered severe floods and landslides in 1998. The government set up the Sloping Land Conversion Program (SLCP) in response, paying farmers to convert cropland on steeply sloping land back into forest, and also to plant trees on 'waste' land in order to combat desertification. Farmers were supplied with saplings and paid both in grain and cash, with payments continuing for up to 8 years provided that the survival rate of the saplings was acceptable. The programme had resulted in the conversion of 9 million hectares of cropland to forest by the end of 2006, and it has reduced soil erosion by up to 68 per cent in some areas, as well as helping to improve the habitat of the giant panda in the Wolong Nature Reserve. However, it is not yet clear whether farmers will continue to maintain the forests when payments cease.[60]

In Hiware Bazar in India, deforestation caused severe soil erosion and flooding, with only 12 per cent of farmland remaining usable. Reforestation of the surrounding area, together with the building of banks to prevent soil erosion, doubled the number of working groundwater wells and the area irrigated. Grass production went up from 100 tonnes in 2000 to 6,000 tonnes in 2004, the number of livestock increased, and milk production soared from 150 litres to 4,000 litres per day. Poverty was cut by 73 per cent in less than 10 years.[61]

Flooding and soil erosion also lead to high sediment levels in streams and rivers, so that water treatment systems have to be installed to make the water drinkable. Recognising the role that forests play in providing clean, high-quality drinking water, around a third of the world's 100 largest cities have set up protected areas to safeguard the forests in their water catchment areas. In New York, for example, the city spent $2 billion buying and restoring land in the forested Catskill water catchment area, rather than spending $7 billion on a water treatment plant. In Ecuador, local communities in the Pisque watershed area receive $11–16 per hectare per year for maintaining forests that ensure clean water supplies.[62] In Quindío, Colombia, reforestation of river banks and exclusion of livestock led to a rapid improvement in the cleanliness of the water supply and a decrease in water-borne diseases.[63]

Soil erosion also leads to silting up of water courses, irrigation channels and hydroelectric reservoirs. In Madagascar, for example, deforestation in upland areas led to silting up of streams and rivers that reduced water flows to thousands of rice farmers further downstream.[64]

Cloud forests, which are often enveloped in mist, can help to increase the water supply by scavenging droplets of moisture out of the humid air. In Mexico, where water scarcity is a major problem, the Payment for Hydrological Environmental Services Program pays communities to preserve their forests, with a premium for cloud forests in recognition of their role in enhancing the water supply during the dry season.[65] And in Ecuador, where removal of the cloud forests in the Lorna Alta watershed was estimated to cost local households over $600 per year in lost water supply, the community decided to establish a reserve of 30 square kilometres to protect the remaining forest.[66]

Forests also help to recycle rainfall, as trees take up water through their roots and return surplus water vapour to the atmosphere through leaf pores – a process called transpiration. Around 25–50 per cent of rainfall is recycled in this way. In the Amazon, for example, water vapour recycled by the forest travels in 'flying rivers' down to the major agricultural regions of Southern Brazil and Paraguay, and also feeds the glaciers and snowfields in the Andes that provide melt-water for upland communities.[67] Large-scale clearing of forests can change regional rainfall patterns, reducing overall rainfall by 10–25 per cent, although some areas may receive more rain and others less.[68]

There may be a tipping point when a certain proportion of forest has been removed, when reduced rainfall causes forests to die back and be replaced with savannah or grassland. This can be accelerated by climate change and by logging (which increases the risk of fires because the removal of trees allows the lower undergrowth to dry out), potentially leading to a vicious circle of forest degradation.[69] It is thought that the Amazon rainforest could be tipped into this state if 30–40 per cent of the forest is removed. This could release over 100 Gt of carbon, equivalent to around 12 years of global fossil fuel emissions, and significantly reduce rainfall in La Plata River Basin, Brazil's main farming region, as well as in important agricultural regions of Paraguay, Argentina and Bolivia. Reduced rainfall, together with increased sedimentation in reservoirs, would also reduce the generation of hydroelectric power, which provides much of the electricity supply in Brazil and Paraguay.

In coastal areas, forests and natural wetlands help to protect the land from coastal erosion and also from the risk of storm surges and tsunamis – often providing a far cheaper and more effective alternative to man-made infrastructure such as sea walls. Mangrove forests are particularly important, as the trees grow semi-submerged in shallow coastal waters, providing a strong physical barrier to waves as well as stabilising the sea floor. Not only are mangroves important stores of carbon, but they are also high in biodiversity, sheltering a large number of unique species, and are important as breeding grounds for fish. In Northern Vietnam, an investment of $1.1 million in restoring mangroves along the coast saved over $7 million a year in building and maintaining sea walls. The replanted area suffered far less damage during Typhoon Wukong in 2000 than neighbouring areas with no mangroves.[70] A 2012 study found that mangroves in Thailand are worth $11,000 per hectare if replaced by shrimp farms, but $18,000 per hectare if retained for flood protection.[71]

Planting trees can also help to combat desertification. In Mauritania, trees are being planted to halt the advance of shifting sand dunes that now cover two-thirds of the country. Fences woven from branches and twigs harvested sustainably from local forests are used to deflect the wind and create artificial dunes surrounding the capital city of Nouakchott. The dunes are then stabilised by planting them with drought-tolerant indigenous species of trees, shrubs and grass. Constant watering and maintenance are required to get plants established in these conditions, but the local community will eventually benefit from a new source of firewood and animal fodder as well as protection from sand encroachment.[72] Another ambitious experimental project aims to combat the southwards advance of the Sahara Desert by establishing a 'Great Green Wall' – a strip of forest 30 kilometres deep that runs across the entire width of Africa, and which could provide fruit and timber for local communities as well as enhancing biodiversity.[73]

Despite the wide range of potential benefits from reforestation, local impacts must be considered carefully. Planting new forests on grassland, for example, can reduce local water flow into streams and rivers as the trees suck water out of the ground, as well as leading to an initial loss of soil carbon when the ground is disturbed for planting. However, there will generally be an increase in rainfall on a regional scale as the trees return moisture to the air through transpiration.[74]

Livelihoods

Forests provide livelihoods not just for the 60 million indigenous people who live entirely on forest resources, and the further 60 million who work in the forestry sector, but for many households who gather fuel wood, fruit, nuts, timber, roofing materials, traditional medicines, animal fodder and other products such as rubber, coffee or rattan from forests. Around 20,000 wild tropical plants are used as food, and forest animals are also hunted for meat and hides.[75] The World Bank estimates that more than 1.6 billion people depend on forests to some extent for their livelihood, and that forests help to support around 90 per cent of the 1.2 billion people living in extreme poverty.[76]

International trade in 'non-wood forest products' is estimated as worth $11 billion per year,[77] plus over $100 billion for medicinal products,[78] but far more forest resources are informally harvested and do not appear in national statistics, even though they can form between a fifth and a half of household income.[79] Forests are thought to be important as back-up resources during hard times, such as when harvests fail, when people can turn to wild food. In fact, it has been estimated that although the official forestry, fishing and agriculture sectors provide only 6–17 per cent of the formal GDP of Indonesia, India and Brazil, forests and other ecosystems provide 47–89 per cent of the income of poor households, termed the 'GDP of the poor' (Figure 4.5).

Forests can also generate revenue from eco-tourism, which is a rapidly growing sector. The number of international tourists grew from 500 million to 900 million between 1995 and 2008, and 40 per cent of all trips were to developing countries. Tourism is the main source of foreign exchange earnings for many developing

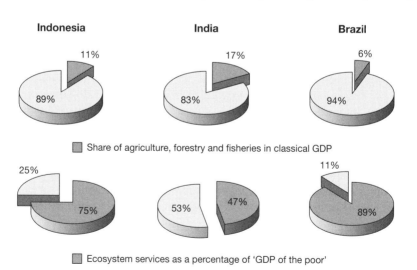

Figure 4.5   Forests and other ecosystems form a large part of the livelihoods of low-income households, termed the 'GDP of the poor'

Source: Adapted from TEEB, 2011b

countries, providing revenue worth three times the amount of international gov-ernment aid. For countries such as Kenya, Tanzania, Madagascar and Costa Rica, eco-tourism is the main tourist attraction.[80] However, there is a conflict between local economic benefits and the global impacts of additional air travel.

Forests also provide essential habitat for insects, birds and bats, which are responsible for the pollination of 35 per cent of all food crops.[81] Bees are especially important for pollination; yet wild bees are in decline in many countries due to a combination of factors including habitat loss, poisoning by pesticides and infection by diseases transmitted by the Varroa mite.[82] In the United States, for example, the wild population of the native honey bee has plummeted, and pollination services are now mainly carried out by mobile hives transported from field to field by commercial beekeepers. This dependence on a small number of commercially bred bee species means that essential pollination services are vulnerable to loss of the bee colonies through disease or pests. In developing countries without a com-mercial bee-keeping sector, wild bees and other insects are essential for pollination. In Sulawesi in Indonesia, for example, deforestation is expected to reduce yields on local coffee plantations by 18 per cent between 2001 and 2020. In Costa Rica, pollinating insects from forests increased coffee yields by 20 per cent and also improved the quality of the coffee beans, providing a service estimated to be worth almost $400 per hectare, or 7 per cent of farm income – similar to the revenue that can be obtained from cattle or sugar cane production, which are leading causes of forest clearance.[83] As well as pollinators, forests also shelter a range of natural predators that can help to control pests and diseases.

On top of these livelihood benefits, local communities can receive income from carbon payment schemes such as REDD. Some schemes are specifically designed to target those on low incomes, including Mexico's Payment for Hydrological Environmental Services Program and Ecuador's Socio Bosque programme.

A study published in 2012 concludes that ecosystems provide over $1 trillion per year in essential local services to poor communities, including food, fuel, water, pollination, soil protection and medicine, but excluding climate protection, bio-diversity and recreation (where the benefits are distributed more globally). Revenue from Payment for Ecosystem Services (PES) schemes could provide them with another $500 billion per year, assuming that revenues were distributed evenly from the people receiving the service, such as fresh water, to those living in areas providing the service. Total benefits could amount to over $1 per day for 331 million people living in poverty; for comparison, the world's poorest billion people live on $1 a day or less, so this could double their income. The study demonstrated a strong overlap between areas of high and threatened biodiversity and those providing the most valuable ecosystem services: the top 25 per cent of areas in terms of conservation importance delivered about half of the ecosystem services (taking into account the number of people benefiting from the services). In these areas, the benefits from preserving the forest are three times the income that would be received from clearing the forest for agriculture.[84]

Of course, there can be conflicts between local use of forests and environ-mental goals. Extraction of forest resources will often reduce carbon storage and biodiversity, even if done sustainably. Local communities can also be responsible for unsustainable forest use, such as slash-and-burn agriculture; over-hunting of bush meat (forest animals and birds) for food, hides or medicine; or over-harvesting of wood for fuel. There can also be conflicts where local people are denied access to reserves set aside for conservation or carbon storage, or where forest animals such as elephants cause damage to crops. Later in this chapter we will explore ways of reconciling the needs of local communities with sustainable use of forests.

Well-being

This chapter has already mentioned a number of ways in which forests contribute to human health and well-being, by providing food, clean water and medicine. Well-known examples of natural medicines include aspirin, from the salicylic acid in willow bark; quinine, the anti-malarial drug from the cinchona tree; and Taxol, the anti-cancer drug from the bark of the Pacific yew tree. Although there are now sophisticated methods for making artificial drugs by randomly combining different molecules, it is much quicker to let nature do it for us: millions of years of evolution have produced a vast array of complex molecules that perform useful biological functions, and these naturally generated compounds are more readily applicable to treatment of human diseases. In fact, around 70 per cent of new medicines introduced in the United States in the 25 years between 1981 and 2006 originated from natural substances.[85] Rainforest plants, in particular, have provided two-thirds of all anti-cancer drugs. The rosy periwinkle from Madagascar – now extinct in the wild due to deforestation – was used to develop two of the most important anti-cancer drugs in use today, vincristine and vinblastine, which have increased the survival rate for childhood leukaemia from 20 per cent to 80 per cent. Yet only a few per cent of all known plant species have been tested for medicinal use, and hundreds of species are disappearing every year which could potentially provide new cures for diseases such as HIV.

Forests also contribute to our well-being in other important ways, being highly valued for aesthetic, cultural and spiritual reasons. They provide beautiful landscapes, opportunities to observe wildlife, and a place for outdoor sport and recreation such as walking, running, cycling, camping, canoeing and orienteering. Access to natural areas not only encourages physical activity, which improves fitness and therefore reduces the likelihood of death from causes such as diabetes or heart disease, but can also reduce stress levels and thus improve both mental and physical health. These aspects of forests have usually been ignored by policy makers, but an increasing number of studies are demonstrating that there can be concrete benefits in terms of improved health and happiness for people who have access to green space for recreation.[86] A study in England showed that living close to green space reduced the death rate by 25 per cent,[87] and another study showed that a 10 per cent increase in green space could increase life expectancy by 5 years.[88] These benefits not only increase individual well-being, but also reduce healthcare costs and lost working time, with economic benefits for the rest of society.

Forest protection can have other health benefits, through reducing air pollution from forest fires. Sometimes fires are started deliberately, in order to clear land for agriculture or plantations; and sometimes they start accidentally, though the risk of fires increases significantly when forests become degraded through logging. Fires on peatland are especially damaging, as peat is highly combustible and can burn for years. The health impacts were clearly seen in 1997 and 1998, when much of Southeast Asia was affected by smoke as about 6 million hectares of rainforest and peat were burnt in Indonesia, largely to clear land for oil palm and timber plantations. This released carbon dioxide equivalent to 40 per cent of total annual global emissions from fossil fuels. Around 70 million people were affected by the pollution, and 12 million required treatment. At the height of the fires, simply breathing in the air in Singapore and Kuala Lumpur in Malaysia was said to be equivalent to smoking three packs of cigarettes a day.[89] The total costs to society, including lost tourism revenue, crop damage, healthcare, carbon emissions and loss of forest products, have been estimated as $4–6 billion. During the same period, fires for clearing forest in the Amazon got out of control and destroyed 4 million hectares of forest, resulting in healthcare costs of $3–10 million from smoke inhalation.[90]

Finally, trees have a cooling effect, both by providing shade and through transpiration of water vapour from the leaves. Trees in and around towns and cities can help to reduce the 'urban heat island' effect where buildings and roads absorb and re-radiate heat from the sun. A study of the city of Manchester in the UK found that on a hot summer day with little wind, the town centre could be 12°C hotter than the surrounding woodlands. As temperatures increase due to climate change, this cooling effect will become even more valuable. Increasing the amount of green cover in the city centre by 10 per cent could limit the maximum temperature of a hot (98th percentile) summer day in Manchester city centre to just 1°C above 1961–1990 levels by 2080; but decreasing green cover by 10 per cent could result in maximum temperatures that are 8°C hotter than in 1961–1990.[91]

Valuing the co-benefits

It is hard to put a value on many ecosystem services, especially such intangible benefits as biodiversity and aesthetic pleasure. However, a study by Robert Costanza and colleagues in 1997 valued total global ecosystem services at $33 trillion a year (in 1994 dollars), of which forests provided over $4 trillion, compared to global GDP of $18 trillion at the time.[92] The study was criticised for the methods used to scale regional impacts up to a global total, but although regional studies provide estimates that are lower (per hectare), they are of a similar order of magnitude.[93]

More recently, the TEEB study has compiled illustrative data on the value of ecosystem services in tropical forests by reviewing published studies (Table 4.3). Although this does not claim to provide a complete picture of benefits, the study concludes that the benefits of protecting tropical forests typically outweigh the costs.

Modelling for the TEEB study estimates that the forests and other natural land that are lost *each year* would have provided €50 billion worth of ecosystem services

Table 4.3 Illustrative estimates of the value of ecosystem services in tropical forests

Ecosystem service	$ per hectare per year, 2007 values	
	Average	Maximum
Provisioning services		
Food	75	552
Water	143	411
Raw materials	431	1,418
Genetic resources	483	1,756
Medicinal resources	181	562
Regulating services		
Improvement of air quality	230	449
Climate regulation	1,965	3,218
Regulation of water flows	1,360	5,235
Waste treatment/water purification	177	506
Erosion prevention	694	1,084
Cultural services		
Opportunities for recreation and tourism	381	1,171
Total direct services	**6,120**	**16,362**
Supporting services		
Biodiversity, pollination, soil fertility, etc.	900	n.a
Total	**7,020**	**n.a**

Source: TEEB, 2009

for every year into the future, with a net present value (up to 2050) of €1–3 trillion. This is probably an underestimate, as several important ecosystem services such as pollination, erosion control, rainfall creation and genetic resources were excluded due to lack of data. The value of these services will increase in the future, as environmental pressures such as climate change and water scarcity become more severe, and by 2050 we could be losing €14 trillion each year (7 per cent of predicted global GDP) as a result of cumulative deforestation and habitat loss since 2000: €11 trillion from forests and the rest from other ecosystems. Around €9 trillion of the loss is from carbon storage, with the remaining €5 trillion from the loss of other ecosystem services (soil stability and fertility, flood control, air and water quality, culture, recreation and forest products).[94] Set against this, the estimated annual cost of $17–33 billion in REDD payments to halve the global deforestation rate represents a small fraction of the potential benefits.[95]

Agriculture

The 'Green Revolution' that began in the 1960s changed the face of agriculture. New varieties of rice and wheat were developed with larger seed heads, which more than doubled yields per hectare. These varieties required greater irrigation, and intensive use of synthetic fertilisers, pesticides and herbicides. Mechanisation increased, with tractors and other machinery replacing manual labour. Livestock rearing became more intensive, with animals kept indoors in large 'factory farming' units and fed on grains, beans or oilseed instead of grazing on pasture. These developments enabled global food production to triple since 1960 and undoubtedly made enormous strides towards reducing world hunger.[96]

Yet there has also been an environmental cost. Not only has there been a steep rise in greenhouse gas emissions from agriculture, but soils have become depleted in organic matter; intensive ploughing has led to soil erosion; biocides have reduced wildlife populations and poisoned thousands of farm workers; and fertiliser run-off has polluted streams, rivers, lakes, coastal seas and groundwater supplies. The Green Revolution is now threatening to stall. Although crop yields have increased steadily since the 1960s, the rate of increase is now levelling off for some crops. The diversity of our food supply has declined, and we have become heavily dependent on a small number of staple crop varieties. Rising population, coupled with water scarcity, declining soil fertility, competition for land, and the threat of reduced yields due to climate change are leading to growing concern over food security. A possible taste of things to come has been seen during the last few years (2008–2012), with widespread riots and political instability triggered by steep rises in food prices.

Against this background, climate policy can play a key role in helping to catalyse a shift to more sustainable farming practices. Many techniques for reducing greenhouse gas emissions in agriculture have co-benefits for the environment and for food security, and also help to make agriculture more resilient to future climate change stresses such as increased droughts, floods and temperature rises. Because of their dual role in both reducing climate emissions and adapting to climate change, these practices have been termed 'climate-smart' agriculture in a series of reports by the UN's Food and Agriculture Organization (FAO). The following sections show how these techniques can help to improve soils, reduce pollution, increase biodiversity and boost farm incomes.

Stable and fertile soils

Methods of increasing the amount of carbon stored in soils bring a range of co-benefits for soil stability and fertility.

- *Reducing soil erosion* – e.g. through terracing – stops valuable nutrients from being washed away, cuts the risk of landslides on steep slopes, avoids the creation of erosion gullies which scar the land and make cultivation difficult, and reduces airborne dust and stream sediment.
- *Adding organic matter* such as compost and manure improves the structure and fertility of the soil. The organic matter soaks up water like a sponge, so that heavy rainfall can be absorbed instead of running off the surface and causing soil erosion and possibly flooding. Water can be stored in the soil for longer periods, which can be especially valuable for increasing resilience to droughts and improving food security in water-scarce regions. There will be less need

for additional irrigation, thus avoiding fossil fuel emissions from water pumping and saving money for the farmer. In addition, the soil structure becomes more open and porous, allowing easier penetration of plant roots and aiding the activity of soil organisms. Biochar may also offer benefits for soil structure and fertility, especially in poor tropical soils (see Box 4.2 above).

• *Cover crops* reduce soil erosion, add organic matter, conserve soil moisture, suppress weeds and mop up excess nitrogen in the soil, thus reducing nitrous oxide emissions. Leguminous cover crops can fix nitrogen from the air and add it to the soil when they decompose, reducing the need for expensive and carbon-intensive inorganic fertilisers.

• *'Fertiliser trees'* are leguminous trees that add nitrogen to the soil through their roots and fallen leaves, and are proving of value in improving poor soils in Sub-Saharan Africa. Across the Sahel region, farmers have traditionally retained the African Acacia, *Faidherbia albida*, amongst crops, and trials in Mali, Zambia and Malawi have shown that this can triple yields of maize, and increase yields of sorghum and millet.[97] Shading of crops is avoided because, uniquely, these trees grow during the dry season, shedding their leaves during the rainy season when under-planted crops are growing. As an added benefit, the leaves and pods of the tree are good fodder for animals during the dry season, the bark is medicinal and the wood can be used for fuel and timber. The trees reduce soil erosion and protect crops from dry winds. Experiments with another fertiliser tree, *Gliricidia*, have shown a 5- to 10-fold increase in maize yields when the leaves and twigs are pruned back each year and buried in the soil alongside the maize plants, thus adding nitrogen to the soil.[98]

• *Perennial crops* can also increase soil fertility and carbon levels. One study of a US farm showed that perennial hay meadows retained 40 per cent more carbon and 30 per cent more nitrogen per hectare than annual wheat fields.[99]

• *Low-till agriculture* avoids compaction of the soil with heavy farm machinery, reduces evaporation of water and reduces the loss of beneficial soil organisms such as earthworms, microbes and fungi that are killed during ploughing (though this can be offset by herbicide use).

Taken together, all these benefits for soil structure and fertility can help to improve crop yields and increase resilience to climate change, pests and diseases. Some of these methods could also be used to help restore the 2 billion hectares of land around the world that has become degraded – sometimes even turned into desert – through over-grazing or over-cultivation.[100] Desertification already affects between 100 and 200 million people, and it is thought that a further 2–5 million hectares of land become degraded each year.[101] A combination of methods to combat erosion, add organic matter and increase irrigation can help to restore degraded land (Box 4.4).

Box 4.4 Restoration of degraded land

Terracing and trees in Tanzania

In the Uluguru Mountains of Tanzania, soil is being washed away from degrading farmland and is polluting the rivers downstream. A group of non-governmental organisations (NGOs) have set up the Equitable Payments for Watershed Services programme which encourages farmers to restore their land. Farmers are given tree seeds, and are trained in how to build terraces and plant trees, thus stabilising the soil and returning nutrients at the same time as storing carbon. They are also advised to leave part of their land unused for a year or two to allow the soil to recover fertility naturally.[102]

Zaï pits in Mali

In land that has degraded to leave a hard, dry, barren crust, the soil can be restored by digging a grid of small pits about 30 centimetres in diameter and 10 centimetres deep and adding manure or compost. Breaking through the surface crust means that rain will soak into the pit during the rainy season, instead of just running off the land and causing more erosion, leaving a fertile pit with moist soil ready for planting. The same pits can be used year after year, with more fertiliser being added into the hole each time, gradually building up soil fertility. The use of small individual pits minimises soil erosion and allows precious fertiliser to be targeted precisely where it is needed, around the plant roots.[103]

SALT in the Philippines

After years of intensive cropping, farmers in the Philippines began to suffer from reduced soil fertility and high soil erosion, and yields were falling. In response, the Mindanao Baptist Rural Life Center developed the SALT (sloping agricultural land technology) system. Double hedgerows of mixed leguminous fertiliser shrubs are planted following the contours of the slope, leaving terraces about 4–5 metres wide between the hedges. Every third terrace is planted with permanent crops such as bananas, coffee and citrus trees; the rest are used for rotations of annual crops such as cereals, vegetables and legumes, or for livestock such as goats. Every month, the hedgerow is trimmed back to a height of 1 metre and the prunings are used to mulch the crops, providing a source of nitrogen and suppressing weeds. Stones, rocks and branches can be piled up on the uphill side of the hedges to gradually build up strong terraces that hold the soil in place. The system is labour-intensive at first; but once established, it has been shown to reduce soil erosion by a factor of 50 and increase farm income by a factor of seven compared to traditional farming methods.[104]

Cleaner air and water

Climate-friendly agricultural techniques can reduce air and water pollution by reducing nitrogen emissions from fertilisers and manure, and methane emissions from manure storage and rice farming.

Agriculture is a leading source of nitrogen pollution. When nitrogen fertilisers and manure are applied to the soil, they are gradually broken down by soil bacteria and emit nitrous oxide and ammonia to the air. In developed countries, only about half of the nitrogen added to soil in fertilisers is taken up by plants, and in developing countries the proportion is even lower.[105] It is thought that at least 1–2 per cent of the nitrogen in synthetic fertilisers is lost as nitrous oxide, and around 14 per cent is lost as ammonia, although the rate varies widely depending on the temperature, soil and fertiliser type. Urea and ammonium bicarbonate, widely used in developing countries, emit far more ammonia than anhydrous ammonia which is used in the United States.[106]

Ammonia in the air dissolves in rainwater to form acid rain, and also reacts with other air pollutants to form particles such as ammonium sulphate and ammonium nitrate, which are estimated to form about 8 per cent of particle pollution in Europe.[107] In Europe, ammonia emissions are estimated to cause damage to health and ecosystems worth €15–105 billion per year.[108] Nitrous oxide, as well as being a greenhouse gas, partially breaks down to nitric oxide (NO) which is currently the main source of damage to the ozone layer.[109] It has been estimated that doubling the concentration of nitrous oxide in the air would decrease the ozone layer by 10 per cent; this would increase ultraviolet radiation by 20 per cent, leading to a higher risk of skin cancer and causing damage to ecosystems.[110] Both ammonia and nitrous oxide also contribute to eutrophication of ecosystems when they are re-deposited in water or on soils and plants. Ammonia emissions are thought to adversely affect sensitive plants such as lichens across the whole of Europe, and also affect herbaceous plants in regions with high agricultural emissions such as northern Europe and northern Italy.[111] Methane, as we saw in Chapter 3, is not only a greenhouse gas, but also an air pollutant that contributes to the formation of ground-level ozone, which has damaging effects on health, crops and ecosystems.

As well as being lost to the air, nitrates that are not taken up by crops tend to leach out of the soil and are washed away by rain, or are lost during soil erosion, causing pollution and eutrophication of streams, rivers, lakes and coastal waters.[112] Eutrophication causes problems such as toxic algal blooms, tidal foam, coastal 'dead zones', degradation of coral reefs, loss of biodiversity and possible spread of parasites and diseases. Fertiliser run-off from the US Corn Belt has created a dead zone the size of Ireland in the Gulf of Mexico, for example. In Europe, nitrogen inputs to coastal waters are four times the natural background levels, and nitrogen pollution in freshwater is high enough to cause biodiversity loss in most regions.[113] Nitrates also percolate downwards and contaminate aquifers: 34 per cent of groundwater wells in the European Union are contaminated with nitrates, and 15 per cent exceed regulatory limits, making expensive water treatment necessary.[114] Although nitrates are not directly harmful, there is concern that under certain conditions they can react to form carcinogenic compounds.[115] Part of the nitrogen entering surface water will also eventually be converted to nitrous oxide emissions to the air.

Climate policies that reduce methane emissions or emissions from fertilisers will therefore have the co-benefit of reducing air and water pollution. In particular, reducing the over-application of fertilisers will reduce the amount of 'free' nitrogen that can be released to air or water, as will the use of winter cover crops. The impact on nitrous oxide emissions of shifting from inorganic to organic fertilisers is less clear-cut, as mentioned in 'Relevant climate policies', above, but increasing the organic matter in soils helps to reduce leaching of nitrates from soil to water.

Shifting to a diet lower in animal produce could be the most effective way of cutting agricultural pollution. Meat and dairy production has ten times the eutrophication impact and 100 times the acidification impact of arable crops.[116] The European Nitrogen Assessment (ENA) showed that 85 per cent of the nitrogen in crops harvested or imported into the EU is used as animal feed. In addition, cereal crops convert 30–60 per cent of the applied nitrogen fertiliser into biomass, but only 5–40 per cent is converted for meat, so far more nitrogen is lost to the environment during meat production. A vegetarian Europe could cut the use of nitrogen fertilisers by two-thirds.[117] Intensive livestock production also leads to pollution of water with animal waste, growth hormones, pesticides from feed crop production and antibiotics – this last item is contributing to a growing problem of antibiotic resistance. The resultant pollution of coastal waters is contributing to loss of coral reefs and other damage to marine ecosystems.

However, it should be remembered that the use of fertilisers, pesticides, growth hormones and antibiotics help to increase yields, thus contributing to food security and reducing the pressure for deforestation. This conflict will be explored later in this chapter.

Biodiversity

This chapter has already discussed the major benefits for biodiversity that could result from protecting the carbon stored in forests, but there can also be benefits for biodiversity from some of the methods used to reduce greenhouse gas emissions from agriculture.

Most of the methods for increasing carbon storage in soils will also be beneficial for biodiversity. Reducing soil erosion, reducing soil disturbance through conservation tillage and increasing soil organic matter will boost the population and diversity of soil organisms such as worms, fungi, beetles and microbes: these form an essential part of ecosystems as they are responsible for breaking down decomposing organic matter and thus making nutrients available for other organisms. In addition, these organisms are themselves food for other creatures such as birds, frogs and small mammals. Low-till agriculture also avoids disturbance to ground-nesting birds, which are highly vulnerable in conventional tillage systems: farmland bird populations have declined by 50 per cent in Europe since 1980.[118] However, increased use of herbicide as part of low-till systems is a potential conflict, which we will discuss later.

Conservation tillage usually goes hand in hand with two other practices which help to increase soil carbon as well as providing benefits for biodiversity: cover crops and crop rotation. Maintaining year-round cover crops helps to provide food and shelter for wildlife which would not be available with a system that leaves bare soil during the winter. Crop rotation can also contribute to a more varied farm ecosystem.

Perhaps the most obvious benefit to biodiversity arises from agroforestry and the use of perennial crops. Adding trees and shrubs to farmland instantly provides habitat for a far wider variety of species: perches and nesting sites for birds; bark to shelter insects over the winter; and fruit, flowers, leaves and pollen to feed a variety of insects, birds and mammals. This provides extra habitat for natural pest-controlling predators such as insect-eating birds, bats and frogs, as well as pollinators such as bees. Trees and shrubs planted alongside farmland streams can help to provide a wildlife corridor to enable migration between patches of natural woodland, as well as preventing soil erosion from the stream banks.

Climate policy that reduces pollution also has benefits for biodiversity. As we saw in the previous section, controlling fertiliser application can significantly reduce acidification of soil and water due to ammonia emissions, and eutrophication due to nitrate run-off, both of which have severe impacts on aquatic and coastal ecosystems including coral reefs. Controlling methane emissions will cut ozone formation, which will in turn reduce leaf damage to wild plants.

Cutting meat consumption could also have major benefits for biodiversity. Livestock production is a leading cause of habitat loss – around 26 per cent of the world's land surface is used for grazing, and a third of the 12 per cent that is cropland is used to produce animal feed, bringing the total land used for livestock production to 70 per cent of agricultural land and 30 per cent of total land area (Figure 4.6).[119] Increasing demand for animal products is leading to extensive deforestation to make way for grazing land and feed crops. In the Amazon region in particular, 70 per cent of former forest is now pasture, and much of the rest was cleared for the production of soybeans for cattle feed. This habitat loss is a major threat to biodiversity. Over 300 of the 825 eco-regions defined by the Worldwide Fund for Nature (WWF), as well as 23 of the 35 global biodiversity hot-spots defined by Conservation International and most of the threatened species on the IUCN Red List, are in danger from habitat loss due to livestock farming.[120]

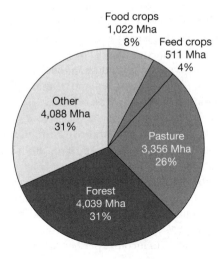

Food crops
1,022 Mha
8%

Feed crops
511 Mha
4%

Other
4,088 Mha
31%

Pasture
3,356 Mha
26%

Forest
4,039 Mha
31%

Figure 4.6 Livestock accounts for 30 per cent of land use (26 per cent pasture; 4 per cent feed crops)

Source: Data from FAOSTAT, 2012

Improving farm incomes

A number of climate policies have the added advantage of improving farm incomes, by cutting costs, boosting yields, enhancing resilience to climate change and diversifying income. These include:

- *Reducing the over-application of fertilisers.* Farmers can save fertiliser costs by applying only as much fertiliser as the crop needs at any one time.
- *Adding organic matter to the soil* or using leguminous cover crops saves fertiliser costs and improves productivity and water retention. Increasing the carbon content of degraded soils by 1 tonne per hectare can increase wheat yield by 20 to 40 kilograms per hectare, for example.[121] In Benin, a cover crop of mucuna (velvet bean) increased maize yields from 1.3 tonnes per hectare to 3–4 tonnes without using any inorganic fertiliser.[122]
- *Conservation tillage.* This can slash fuel and labour costs by reducing the number of tractor passes to cultivate a crop, although there may be increased expenditure on herbicides for weed control. The benefits are greatest where soil erosion is a problem, as erosion reduces soil fertility and in extreme cases can render the land unusable. Zero tillage was estimated to have saved the farming industry in Argentina almost $5 billion since 1991, and saved rice and wheat farmers in India almost $100 per hectare by improving yields and reducing the time and cost of soil preparation.[123] In South America, yields of maize, soya and wheat have been shown to increase by between 10 per cent and 80 per cent with conservation tillage.[124] Manual conservation tillage has even proved profitable in parts of Africa where it can save the cost and time delay

involved in hiring oxen to plough the field, allowing farmers to plant their crops earlier and thus benefit from a longer growing season. However, the increased labour involved in hand-planting and weeding is only worthwhile if there was previously a problem with soil erosion, so that reduced tillage results in higher yields.[125] Leaving crop residues in the field as mulch can also be a problem where crop residues are in demand as animal fodder, fuel or roofing material. Cover crops can help to address both these issues – suppressing weeds and acting as mulch.[126]

- *Agroforestry* can help to diversify and improve farm incomes. Although it can be costly and time-consuming to set up, trees and shrubs will eventually provide additional sources of income for the farmer from fruit, timber and fuel wood, as well as providing habitat for natural predators and pollinating insects and, if leguminous trees are used, a free source of nitrogen fertiliser. Some farmers also grow plants which are natural insecticides, such as the neem tree and fish-poison bean (*tephrosia*), to help control insect pests. In India, a study of 200 farms found that farm incomes per hectare increased by a factor of 10 after planting a variety of trees.[127] As well as serving to reduce soil erosion, trees and perennial crops are often more resistant to floods, droughts, pests and other environmental stresses than annual crops, because their roots are stronger and deeper. In hot countries, trees can protect crops from being scorched by the sun, and provide essential shade for plants such as coffee, cacao and vanilla. In cooler climates, trees and hedges can shelter crops from strong winds. In the United States, for example, planting trees for shelter increased yields of winter wheat by 23 per cent.[128] However, in some climates, the shade cast by trees may reduce crop yields to some extent, so that there is a trade-off with the other benefits provided by trees. Trees can also be beneficial in silvopastoral systems, where animals graze in orchards or timber plantations. The trees provide shade and shelter, and the animals control pests and weeds and provide manure to fertilise the trees. One interesting example is the use of 'masting fodder' trees which drop fruit or nuts that the animals can eat, such as oak forests in Portugal where pigs graze on fallen acorns.

Despite this multitude of potential benefits for farmers, uptake of climate-smart agricultural techniques has still been relatively slow, especially among small-scale farmers in developing countries. A review published in 2011 suggests that there are a number of barriers preventing smallholders from adopting new techniques such as conservation tillage or agroforestry, including lack of information, high initial investment costs (e.g. for building terraces, planting trees or buying a zero-till drill) and lack of access to appropriate equipment, cover-crop seeds or tree seedlings. Even though the potential long-term gains are high, it can take several years for yields to start to increase as soil quality gradually improves. There can even be a decrease in incomes or yields in the short term as farmers try out new techniques and adapt them to suit local conditions, or wait for newly planted trees or shrubs to mature and start producing timber or fruit, or for degraded grazing land to recover when livestock density is reduced.

It is also important to note that not all techniques are suitable for all areas and climates. Cover crops are less suitable where there is a short growing season; conservation tillage may cause problems where the soil is poorly drained; and bund walls are usually good for storing water but can cause waterlogging in wet years.[129] When adopting a new practice such as conservation agriculture or agroforestry, the farmer risks loss of income if the strategy turns out to be inappropriate for local conditions. This highlights the importance of research and demonstration projects and advice schemes to help establish which strategies work best in each area and provide guidance for local farmers.

Conflicts

Several of the policies discussed in this chapter are highly contentious, giving rise to conflicts between different sustainability objectives. Issues include:

- *Problems with forest carbon schemes.* The prospect of including forests in carbon payment schemes such as REDD can lead to land being taken from local communities. Reforestation or afforestation projects can lead to loss of biodiversity if natural grassland or forest is replaced by monoculture plantations.
- *Growing biofuel crops* can reduce the land available for food production, possibly leading to more clearance of forests or to food price increases.
- *More intensive agriculture* based on the use of agrochemicals could help to increase yields, improve food security and reduce pressure for deforestation, but also leads to increased pollution and loss of biodiversity.

Problems with forest carbon schemes

Although protecting and extending forests have tremendous potential for sequestering carbon, protecting biodiversity and benefiting local communities, a number of problems have been identified with forest carbon payment schemes such as REDD.

- *Monitoring, reporting and verification (MRV)* of the amount of carbon saved is very difficult, due to lack of accurate data on forest cover, carbon storage in different types of forest and soil, and emissions from land-use change.
- *Additionality.* Forest carbon payments aim to reward landowners for not cutting down forests, or for planting new ones, but it is hard to tell what would have happened if there were no payments, and therefore whether any 'additional' benefit has been achieved. Using past deforestation rates as the baseline has the drawback of rewarding countries with high deforestation rates and ignoring those which have chosen to keep their forests intact, as well as providing an incentive for countries to increase their current deforestation rates in order to claim more REDD payments in future. There is also a query over whether payments should be allowed for protected areas (forest reserves). Clearly they are not at immediate risk of felling, but if they are not included, there is a risk

that countries could remove the protection in order to qualify for carbon payments.[130]

- *Leakage.* Carbon leakage occurs if protection of one forest simply results in the felling of another one elsewhere, or if protection of forests increases pressure on natural grassland or wetlands for agriculture or development. The risk of leakage is reduced by taking all habitats in a country into account, not just forests, when recording carbon emissions from land-use change.

- *Permanence.* Carbon savings may not be permanent because forest carbon can be released as a result of fire, disease, pests or unauthorised felling. It is not clear who should be responsible if forest carbon is released accidentally. One way of dealing with this is through setting up insurance policies that cover the landowner against accidental loss of the stored carbon, but this could also provide an incentive for deliberate burning of forests in order to obtain insurance pay-outs fraudulently.[131]

- *Land grabs.* There have been cases where governments or private companies have forcefully acquired the rights to forests in anticipation of future payments for carbon storage, dispossessing or evicting local people or denying them access to forests (see Box 4.5 below). This problem is exacerbated by the fact that land ownership is poorly defined in many tropical forests in Latin America, Africa and Indonesia.[132] In Africa, for example, only 2 per cent of forests are owned by communities.[133]

- *Livelihood conflicts.* There is a trade-off between the provision of fuel wood, food and timber for local people, and the carbon storage and biodiversity benefits of forests. Undisturbed forests are richer in both biodiversity and carbon storage, but provide fewer economic benefits for local communities, though there may be revenue from eco-tourism.

- *Monoculture plantations.* New monoculture plantations can be effective at storing carbon, but there can be a loss of biodiversity if they are planted on natural grassland or wetland or if they replace natural forests. There are cases of climate funding being sought for industrial plantations on 'degraded forest land', which raises concern that such funding could be driving further destruction of natural forests.[134] This includes funding for afforestation and reforestation schemes under the Clean Development Mechanism of the Kyoto Protocol.

The danger is that a poorly designed forest carbon payment scheme could achieve very little real reduction in emissions, and could detract from the urgent need to cut fossil fuel emissions. For example, a 2011 report by Greenpeace criticised estimates by McKinsey & Company consultants of potential forest carbon savings in the Democratic Republic of the Congo, claiming that McKinsey underestimated current logging and overestimated future logging to such an extent that logging companies would be able to claim REDD carbon credits while actually continuing business-as-usual.[135] Similar loopholes could allow countries with legally binding carbon reduction targets to claim non-existent carbon reductions, thus reducing the need for them to cut fossil fuel emissions. And if fossil fuel emissions are not cut, forests will die off anyway: one study estimated that 20–40 per cent of the

Amazon rainforest would die even if the global temperature rise were restricted to 2°C, and 85 per cent could die with a 4°C rise.[136] Safeguards to overcome these problems will be discussed in 'The way forward', below.

Box 4.5 Land grabs: carbon cowboys and corruption

The prospect of future payments for stored forest carbon under REDD has already led to land grabs. In Peru, for example, indigenous groups complain that REDD project developers are roaming the jungle and persuading village chiefs to sign away the rights to their forest land, or to the carbon stored in the forests, without a full explanation of the consequences and with little benefit for local people. It is claimed that some land has been sold for as little as 20 cents per hectare.[137] In one incident, an Australian businessman persuaded the president of an indigenous group to sign away the rights to their forest carbon without consulting his own people or other groups using the same forest.[138] Although REDD guidance emphasises the need for 'Free, prior and informed consent', this does not always happen in reality. A review of the World Bank's Forest Carbon Partnership Facility (FCPF) found that indigenous groups are being side-lined, consultation is rushed and inadequate, and governments are handing forest rights to third parties or taking them into state ownership, thus weakening the rights of local communities.[139] As a result, local communities are becoming distrustful of REDD initiatives, which they feel undermine their own rights and often do not deliver on the promised benefits.

Government corruption is also a problem. Critics claim that forestry departments in developing countries are often steeped in corruption, and even that organised criminals are targeting REDD – a real concern in the light of the large-scale fraud uncovered in the European Union Emissions Trading Scheme recently (2009–2010).[140] Some examples are listed below.

- In Indonesia, much of a $6 billion fund for forest restoration was lost during the 1990s due to fraud, corruption and bad management. Although subsequent governments have strengthened forest governance, problems still remain.[141]
- In Liberia, a UK businessman fraudulently obtained 400,000 hectares of forest carbon credits from the Forest Development Authority.[142] A REDD pilot project in Tanzania has also been halted due to suspicions of corruption.[143]

> • In Uganda, it is claimed that 20,000 people were forcibly evicted from
> their smallholdings, and their homes, schools and crops were destroyed
> to make way for commercial timber plantations and a reforestation
> scheme, for which the company has applied for carbon credits.[144]
> Although the people evicted were illegally occupying land on a desig-
> nated forest reserve, most had been there for many years and none have
> received compensation or been offered permanent homes elsewhere.

Bioenergy and biochar

Although bioenergy is seen as a key part of climate policy, first-generation biofuels
(i.e. those derived from purpose-grown sugar, starch and oil crops such as corn,
wheat, soybean, sugar cane, sugar beet, rapeseed and oil palm) are now facing a
backlash for a number of reasons:

• *Energy, water and fertilisers used in production.* Biofuel crops have a high water
demand, and the fossil fuels used for cultivation, harvesting, processing, fertiliser manufacture and irrigation can cancel out a large part of the carbon
savings (see Chapter 5). Water pollution from fertiliser run-off can also be a
problem.
• *Forest clearance.* More than 2 million hectares of rainforest are being cleared
each year in Malaysia and Indonesia to make way for oil palm plantations
(though not all of this is for biofuels – much is for food and cosmetic manufacturing). This is decimating the habitat of endangered species such as
orangutans, Borneo pygmy elephants, Sumatran tigers, and rhinoceroses. It
is estimated that 1,500–5,000 of the remaining 60,000 orangutans in Borneo
and Sumatra are lost each year, and 90 per cent of the Sumatran rainforest
has disappeared since 1975.[145] Oil palm is also wiping out the habitat of lemurs
in Madagascar, and sugar cane plantations threaten the unique wildlife of the
Mabira rainforest in Uganda.[146]
• *Food security.* Rising demand for biofuel crops such as maize, soybean and sugar
cane can contribute to higher food prices, as competition for fertile farmland
and water increases. Around 21 per cent of sugar cane and 11 per cent of grain
and oil crops are already used for biofuels, and this is expected to increase to
33 per cent of sugar, 12 per cent of grain and 16 per cent of oil crops by
2020.[147] It is difficult to separate out the impact of biofuel production from
other factors such as harvest failures, food commodity speculation and growing
demand for food (especially meat) in developing countries, but some estimates
blame biofuels for about 30 per cent of the sharp increase in grain prices
between 2000 and 2007 (Figure 4.7), which led to widespread hardship and
civil unrest such as the 'tortilla riots' in Mexico.[148]
• *Land grabs.* Communities have been evicted from their land in order to make

way for biofuel plantations. In Indonesia, one local NGO is monitoring over 600 land conflicts as a result of oil palm expansion. Asian oil palm companies are now starting to expand into Africa, where 700 million people live on state-owned land with poorly defined tenure rights.[149] Indigenous communities in Honduras and Guatemala have also lost their land to make way for sugar cane and oil palm plantations.[150]

- *Poor working conditions.* In common with other sectors of agriculture, working conditions on biofuel plantations can be very poor, with low wages, limited breaks, informal child labour and payment based on piece-work that results in overwork. However, at the opposite extreme, there are responsible companies that provide good working conditions and also provide healthcare, schools or nurseries for workers.[151]

Tropical forests are particularly at risk of clearance for biofuels, as oil palm and sugar cane thrive in tropical conditions. This has severe biodiversity impacts: research has shown that only around 20 per cent of the species found in natural forests can survive in oil palm plantations, and forest fragments surrounded by plantations contain only half of the species found in continuous forest.[152] Other ecosystems are also at risk: much of the Brazilian Cerrado – the world's most biodiverse savannah – has been destroyed to make way for soy and sugar cane plantations, and sugar cane is also encroaching on natural wetlands in Latin America and Africa. In Europe, expansion of oilseed rape and sugar beet for biofuels onto set-aside land and semi-natural steppe is threatening farmland bird populations and endangered mammals.[153]

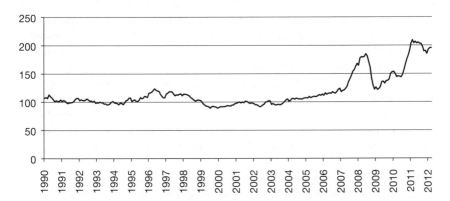

Figure 4.7 The FAO food price index has been rising since 2000, partly due to increased use of farm land for biofuel cultivation

Source: Based on data from the Food and Agriculture Organization of the United Nations (FAO)

Notes: 2002–2004 =100.

Other contributing factors include the rising demand for food due to population growth; increased demand for meat; and commodity speculation. The price crash after the 2008 financial crisis, and the subsequent recovery, can be clearly seen.

These land-use changes can make biofuels more carbon-intensive than fossil fuels. Biodiesel from oil palm planted on cleared rainforest can produce eight times more carbon than fossil diesel, and 20 times more if planted on cleared peat forest.[154] When peat forests are cleared, there can be net annual losses of up to 90 tonnes of CO_2e per hectare,[155] and it has been estimated that it would take 420 years of biofuel production to compensate for the carbon emitted at the planting stage.[156] Oil palm plantations are typically only used for about 25 years.[157]

It is estimated that over half of the oil palm expansion in Malaysia and Indonesia displaces primary forests, and a quarter displaces peat forest.[158] In Sarawak, over a third of peat swamp forests and 10 per cent of rainforests were cleared between 2005 and 2010, and two-thirds of this clearance was for oil palm plantations. Even where plantations are established on existing farmland, this can result in indirect land-use change leading to loss of forests elsewhere. In Brazil, for example, almost half of sugar cane plantations are on rangeland, so rangeland is now expanding into the Amazon rainforest.[159]

Because of these problems, attention is shifting towards biofuel options which do not require high-quality farmland and thus minimise the risk of causing deforestation or competition with food crops. These include biofuels made from food or farm waste, second-generation biofuels made from non-food biomass such as wood, miscanthus grass, switchgrass, crop residues and forestry waste, and trees or perennial grasses planted on degraded land. Biodiesel from oil palm planted on degraded land can save around 10 tonnes of CO_2e per hectare planted per year, for example.[160]

These options can provide co-benefits. Growing trees and perennial grasses on degraded land can help to reduce soil erosion, restore fertility and increase biodiversity, especially when a mix of native species is used, although care must be taken with non-indigenous species such as switchgrass because they can become invasive. Perennial energy crops also require less water, fuel and fertiliser than annual food crops. Biofuels from waste have the advantage of safely disposing of the waste material, which could generate methane emissions and water pollution if landfilled, or black carbon emissions if burnt. There have even been cases of invasive weeds such as water-hyacinth being gathered and used for biofuels.[161] Both first- and second-generation biofuels can also be grown as part of mixed forestry–farming systems; for example, by planting trees on pasture and allowing livestock to graze underneath, or growing sugar cane on part of the pasture and using the residue (bagasse) as cattle feed, so that the total land area does not need to be increased.[162]

Because of these advantages, the IEA expects two-thirds of biofuels produced in 2035 under the climate-friendly 450ppm Scenario to be second-generation.[163] However, these biofuels are currently expensive to produce, because the cellulose and lignin-rich material must first be broken down by hydrolysis to produce sugar, which can then be fermented to produce bio-ethanol. Also, there are often competing uses for the biomass; for example:

- Crop residues could be ploughed into the soil or left in place to reduce soil erosion, improve soil structure and improve fertility.

- Animal, food and garden waste could be used as manure or compost on fields and gardens, although anaerobic digestion does produce solid and liquid by-products that can be used as fertilisers and soil improvers.
- Forestry waste would store carbon for longer if left to decompose naturally in the forest, where it would also add nutrients to the soil and enhance habitat for wildlife. A quarter of woodland species have been shown to depend on forestry waste.[164]
- In developing countries, biomass is also often used for animal fodder, household fuel or building materials.
- Wood and forestry waste is in increasing demand for direct combustion in power stations, partly driven by climate policy targets in developed countries. In fact, this is a more efficient use of the energy contained in the wood than converting the wood to liquid biofuels. However, the growing demand for wood is leading to similar concerns over deforestation, land grabs and food security as for first-generation biofuels, as wood chips and pellets are being exported from developing countries to the EU, United States and South Korea.[165]
- As fossil fuels become more scarce and expensive (see Chapter 5), there may be increasing demand for biomass to substitute for petroleum in materials such as plastics, chemicals and medicines (this will affect both first- and second-generation biofuels). There could also be increased demand for wood as a renewable and less energy-intensive substitute for metal, plastic and brick in buildings, furniture and other products. However, bio-based materials could be used as an energy source at the end of their life as materials.

Despite these competing uses, biofuels can provide genuine benefits in situations where the feedstock would otherwise have gone to waste; for example, by being burnt or landfilled. Promising 'third-generation' options are also being developed, such as biofuels from algae (Box 4.3 above), or from gasification of municipal solid waste, though technical challenges remain to be overcome.

Many of the same issues apply to the production of biomass for conversion to biochar, although it is easy to produce biochar from relatively sustainable sources such as forestry waste, food waste, crop residues, manure, fruit tree prunings or even sewage sludge. Producing biochar from waste could cut the economic and environmental costs of disposing of the waste in landfills or by incineration, although it could also increase the risk of contamination with toxic compounds and could reduce the soil improvement qualities of the biochar. One study estimates that enough biochar could be produced to remove about 12 per cent of current annual carbon emissions from the atmosphere using just forestry waste, food and garden waste, crop residues, manure and biomass crops grown on degraded tropical pastures, thus avoiding the need to clear forests or displace food crops.[166] However, this would divert virtually all of the crop residues and manure that could be used to add organic matter to soils. It would also exclude the use of this biomass for heat, transport and electricity production.

Standards have been developed to try to ensure that biofuels are produced sustainably. The US Renewable Fuel Standards now specify minimum reductions in life-cycle greenhouse gases compared to fossil fuels (including emissions from indirect land-use change) of 20 per cent for corn-ethanol, 60 per cent for cellulosic biofuels and 50 per cent for other biofuels. The EU Renewable Energy Directive specifies a 35 per cent saving, rising to 60 per cent for new facilities in 2018.[167] These standards help to ensure that biofuels are not produced via deforestation of carbon-rich primary forest. In addition, the EU has introduced social and environmental safeguards such as specifying that biofuels cannot be produced from high biodiversity forests or grassland. A number of voluntary standards have been developed, with that proposed by the Roundtable on Sustainable Biofuels (RSB) being amongst the most comprehensive.[168] In Switzerland, biofuels from palm oil, soy and maize are banned altogether, and other biofuels must undergo a rigorous life-cycle assessment before they are approved.[169]

Standards are helpful in protecting the most valuable habitats from encroachment by biofuels, but they have limitations. First, standards often allow biofuel production on land defined as 'marginal' or 'degraded', but this land may be of value as wildlife habitat or be used by local communities and their livestock for foraging and grazing.[170] It can be better to let degraded land recover naturally: one study found that forest degraded through selective logging can recover 84 per cent of its bird species within 30 years, but only 27 per cent of species survived conversion to oil palm. Another analysis concluded that there would be a net loss of biodiversity if biofuels were planted on abandoned farmland that has started to revegetate, and that there would only be a net improvement for plantations on recently abandoned, degraded cropland.[171]

Second, some standards do not take into account the impact of indirect land-use change. This is critical – it has been estimated that the EU Renewable Energy Directive will result in the conversion of 41,000–69,000 square kilometres of natural land to cropland, resulting in net annual emissions of 27–56 million tonnes of carbon dioxide per year, which makes the use of biofuels 81–167 per cent worse than fossil fuels.[172] The IEA has suggested that the only way around the problem of indirect land-use change would be to agree a global cap on greenhouse gas emissions from all sources – fossil fuels and all land-use changes.[173] With these challenges in mind, a strategy for optimising the sustainable use of biofuels will be discussed in 'The way forward' below.

Feeding the world

Feeding the world presents a major challenge. Around 925 million people suffer chronic hunger,[174] and demand for food is expected to increase by 70 per cent as the population reaches 9 billion in 2050. Crop yields are increasing, but more slowly than before, due to higher costs for agrochemicals and energy, expansion of agriculture onto less productive land, and natural limits to techniques such as double-cropping and irrigation.[175] Global food production grew by an average of 2.6 per cent per year between 2000 and 2010, but is predicted to grow by only 1.7 per cent between 2010 and 2020.[176] In the longer term, the FAO predicts that yields could grow by 0.8 per cent per year up to 2050, which would meet 90 per cent of the additional food demand, with an extra 70 million hectares of arable land needed for the remaining 10 per cent.[177]

Yet the increase in yields to date has been based to a large extent on the intensive use of agrochemicals, fossil fuels and irrigation, resulting in pollution, soil degradation, loss of biodiversity and depletion of water resources. The use of mineral fertilisers, for example, tripled between 1961 and 2002, accounting for about 40 per cent of the growth in yield.[178] Overuse of pesticides has led to some species developing resistance, and has wiped out beneficial natural predators. A fifth of pasture is degraded through over-grazing, compaction and soil erosion,[179] and water scarcity is a growing problem. As prices of fuel and agrochemicals increase, small farmers are struggling to afford the inputs on which many have come to depend. These damaging effects are now undermining agricultural productivity and threatening food security.

There is considerable disagreement over the best way forward. One group claims that increased use of inputs (water, energy, agrochemicals), genetic modification and intensive indoor rearing of livestock is the only way of increasing yields and therefore feeding the world without further deforestation. This approach is typified by the FAO report *Livestock's Long Shadow*, which claims that the 'inefficient' techniques practised by small-scale traditional farmers in the developing world are responsible for the bulk of climate emissions and deforestation. The FAO looks forward to a future 'transition from many, small subsistence producers to fewer and larger commercial farmers'.[180] An opposing group points to the adverse impacts of agrochemicals, the animal welfare impacts of factory farming and the unknown consequences of genetic modification. They promote the principles of 'agro-ecology', based on less intensive farming, traditional outdoor grazing and the use of organic and permaculture techniques.

This debate encompasses a number of conflicts relevant to climate policy.

- Artificial fertilisers can increase crop yields from around 2 tonnes per hectare to as much as 8–10 tonnes per hectare,[181] but also create emissions of nitrous oxide from fertiliser application, and carbon dioxide from fertiliser manufacture, as well as increasing air and water pollution.
- Conservation tillage reduces soil erosion, enhances soil carbon and cuts fossil fuel use by farm machinery, but it usually entails intensive application of herbicides for weed control and to break down crop residues, often facilitated by the use of genetically modified herbicide-resistant crops such as RoundupReady soy. These herbicides can cause water pollution and have adverse effects on wildlife.
- Using pesticides and fungicides can help to increase crop yields, improve food security and reduce post-harvest waste of food, but can have adverse impacts on wildlife, including pollinating insects such as bees.
- Eating less meat could lead to significant changes in farming landscapes, livelihoods and associated local culture, possibly including adverse changes to scenic landscapes such as upland sheep and cattle pastures (although many upland grazing areas are unsuitable for arable production). Ploughing the soil to convert pasture to arable land can release stored carbon.

The debate also extends to the issue of how best to adapt to a changing climate. Those in the intensive camp point to the role of genetic modification to produce plant varieties that are better adapted to droughts or shorter growing seasons, and greater use of pesticides to protect against the spread of pests and diseases; while those in the ecology camp look to crop diversity, integrated pest management, restoration of soil organic matter and harnessing traditional knowledge of local crop and livestock varieties to build a more resilient system. In 'The way forward' we will consider how these two opposing views could be reconciled, so that food security can be achieved whilst cutting greenhouse gas emissions and protecting the environment.

The way forward

Climate policy must tackle the growing emissions from agriculture and deforestation, and by doing so can provide important co-benefits including protection for forests, more fertile and stable soils, cleaner water and more sustainable farming. However, badly designed policies can undermine the co-benefits or even make the situation worse. Payments for forest carbon can lead to land grabs, corruption and fraud. Support for biofuels can lead directly or indirectly to clearance of natural habitats, which can wipe out carbon benefits, damage biodiversity and threaten food security. And there is a heated debate over whether agricultural emissions should be reduced by moving to a more intensive or less intensive approach, and what the implications will be for food security. In this section we will try to find strategies to solve these challenges, so that climate policy can deliver real benefits for land use.

Food, fuel, forests: a balanced approach to land use

Natural forests and other habitats are being lost at an alarming rate, as they are cleared for farming, forestry, biofuel plantations, urban development, mining and oil exploration. Vital ecosystem services such as flood protection, soil stabilisation, water supply and biodiversity are being lost because they have no market value. Against this backdrop, efforts to tackle climate change can provide extra motivation and funding for preservation of carbon-rich ecosystems such as forests and wetlands.

However, individual policies such as payments for forest carbon cannot tackle this problem, because all land use is interlinked. Protecting particular forests can simply shift the threat onto other forests or other ecosystems. Protecting all natural ecosystems would limit the land available for food production, causing problems with food security, or leading to an intensified drive to increase crop yields through the use of agrochemicals, which can increase pollution and damage biodiversity. Instead, an integrated and balanced approach will be necessary, which explicitly protects valuable ecosystems while also addressing the drivers of ecosystem loss, such as over-consumption, waste and bad governance, and trying to maximise agricultural yields in a sustainable manner. A potential strategy is outlined below.

- *Expand protected areas* to conserve the most valued forests and other ecosystems, such as tropical peat forests, mangroves, wetlands, natural grasslands and habitats for endangered species. Almost 14 per cent of forests are currently protected in theory, but some of these are 'paper parks' where better enforcement is necessary to prevent encroachment and illegal felling.[182] A target to expand protected areas to 17 per cent of land and inland water areas has been proposed by the Convention on Biological Diversity.[183]
- *Put an economic value on forest carbon*, through payment schemes such as REDD, but with stringent safeguards to prevent undesirable social and environmental side effects (see next section on 'Valuing ecosystems: making REDD work').
- *Address other drivers of deforestation.* Curb rising demand for paper, timber, oil, minerals and farmland by avoiding waste of food and other resources and increasing recycling of paper and timber (see Chapter 6). Increase crop and timber yields sustainably where possible, and reduce meat consumption in developed countries. Cut perverse subsidies for energy, forestry, agriculture or biofuels, such as subsidies for cattle ranching or plantations that encourage clearance of natural forests.[184]
- *Clarify ownership of forests and improve governance*, thus safeguarding the access rights of indigenous communities and tackling illegal logging, fraud and corruption. The World Bank estimates that illegal logging comprises over 10 per cent of the global timber trade,[185] but it has been estimated that as much as 70 per cent of logging is illegal in Indonesia, 50 per cent in Cameroon and 80 per cent in Brazil, causing economic losses of $10–15 billion per year.[186] The United States and the EU now have legislation in place to tackle the import and sale of illegally harvested wood, though the import of illegal wood via third-party processors from China and elsewhere is still a problem.[187]
- *Support certification* schemes to help consumers identify sustainable timber, paper and food.

A similar approach has been modelled by the Netherlands Environmental Assessment Agency. Figure 4.8 shows how 'land-taking' measures such as setting up protected areas can lead to increased land prices and decreased food availability, but combining this with 'land-relieving' measures such as improved agricultural yields, reduced food waste and less meat consumption in developed countries could redress the balance, leading ultimately to less pressure on land and higher food security than with the baseline case. Figure 4.9 shows the benefits of these policies (individually) for biodiversity.

Even with a well-balanced suite of policies, however, great care must be taken over the design of individual measures to avoid the kind of loopholes and undesired side effects that have already sparked vigorous opposition to REDD and biofuels. In the following sections we will summarise important issues to be taken into account when designing climate policy for forest protection, biofuels and agriculture, so as to maximise synergies and avoid conflicts as far as possible.

Change in land prices and food consumption compared to baseline scenario, 2030

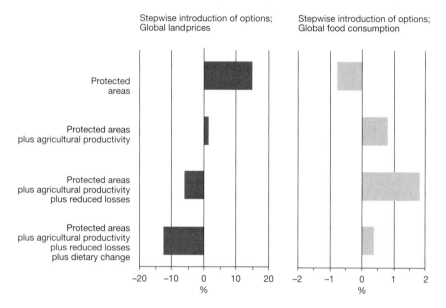

Figure 4.8 Stepwise approach to land-use policy, combining protected areas with increased yields, reduced waste and less meat eating

Source: Netherlands Environmental Assessment Agency, 2010

Valuing ecosystems: making REDD work

A well-designed REDD system could offer large benefits from cutting carbon at the same time as protecting biodiversity and other ecosystem services, but a badly designed system could do more harm than good. Here we look at prospects for funding REDD, and the safeguards and criteria necessary to maximise social and environmental co-benefits and minimise conflicts.

Funding

Crucially, REDD offers the opportunity to tap into a substantial extra source of funding for protecting forests – the carbon market. At present though (2012), all REDD credits are bought as voluntary carbon offsets purchased by individuals and companies. A small amount of forest carbon credits for other tree-planting and sustainable forestry schemes are traded in regulatory carbon markets as part of the Kyoto Protocol Clean Development Mechanism (CDM) and the emission trading schemes in New Zealand and Australia (see Table 4.4 below). Uptake of CDM credits has been low, because the credits are limited to 1 per cent of national baseline greenhouse gas emissions and can only be claimed for up to 5 years.[188] The European Union Emissions Trading Scheme – currently the main global carbon

Change in global biodiversity per option compared to baseline scenario

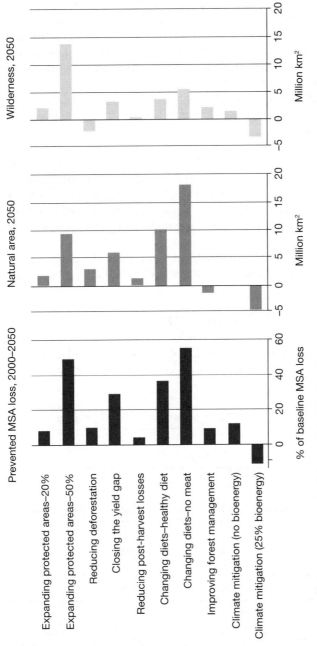

Figure 4.9 Impact of various land-use policies on global biodiversity

Source: Netherlands Environmental Assessment Agency, 2010

Note: 'MSA' is mean species abundance (per cent of original species still present); 'Natural area' denotes areas unused by humans; 'Wilderness area' denotes natural areas with MSA over 80 per cent.

trading forum – does not include forest carbon due to concerns about monitoring and permanence. Other carbon markets are currently in decline following the UN Durban conference on climate change (2011), where action on climate change was effectively kicked into the long grass with a global commitment not expected to come into operation until 2020.[189] Forest carbon credits traded in 2010 were worth $133 million – less than 0.1 per cent of the $142 billion global carbon market,[190] although governments have also committed $8 billion so far in funding for 'REDD-readiness' assessments and pilot schemes.[191]

Including REDD credits in regulatory carbon markets could massively increase the funding available for forest protection. However, there is opposition to this for a number of reasons:

- Flooding the carbon market with cheap forest carbon credits will reduce the urgent need to cut fossil fuel emissions in developed countries.
- Market fluctuations in the price of carbon can undermine the scheme.
- The REDD credits could be vulnerable to fraud, especially as they may be issued in countries with poor governance and poorly defined land tenure legislation.[192]
- Much of the value of REDD credits may be captured by middlemen – verifiers, registrars, agents, consultants and traders – rather than going to forest communities.[193]

In theory, these problems could be tackled by setting tight caps for carbon markets, limiting the proportion of carbon credits that can be supplied by international REDD offsets (to ensure that reduction of domestic fossil fuel emissions remains top priority) and strengthening carbon accounting and social and environmental

Table 4.4 Forest carbon traded in 2010 was mainly in voluntary markets

	Volume traded $(MtCO_2e)$	Value $(\$m)$	Average price $(\$/tCO_2e)$
Voluntary offsets	27.4	126.7	5.63
Chicago Climate Exchange	0.1	0.2	1.18
Total voluntary markets	**27.6**	**126.9**	**5.60**
Kyoto CDM	1.4	6.3	4.49
NSW GGAS	1.1	0.0	*
New Zealand ETS	0.0	0.3	12.95
Total regulated markets	**2.6**	**6.5**	**4.61**
Total primary markets	**30.1**	**133.4**	**5.54**

Source: Ecosystem Marketplace, 2011

Notes: CDM = Clean Development Mechanism. NSW GGAS = New South Wales Greenhouse Gas Reduction Scheme. Chicago Climate Exchange has now ceased trading. The table shows primary markets only; secondary markets from resale of credits are not included.

*Too few data points to disclose average price for 2010.

safeguards. In practice, it should be recognised that existing carbon markets have not yet managed to eliminate problems with over-generous caps. At the time of writing (2012), countries participating in the UN Framework Convention on Climate Change (UN-FCC) have not made a final decision on whether to allow REDD credits to be funded by carbon markets. Without large-scale funding, the supply of REDD credits from forest owners could soon exceed demand from buyers, causing the price to crash. Other funding sources have been suggested, including carbon taxes, redirected fossil fuel subsidies or a financial transaction tax, which could provide up to $650 billion per year.[194]

Verifying carbon savings

It is technically very difficult to accurately measure the change in greenhouse gas emissions as a result of a land-use change project, but there has been considerable progress in developing rigorous standards for monitoring, reporting and verification (MRV), driven by the demands of investors. Over 99 per cent of carbon credits issued in 2010 had third-party verification, with 54 per cent using the Verified Carbon Standard (VCS), originally developed to certify fossil fuel carbon reductions. Other standards include the newly introduced Brasil Mata Viva Standard which was applied to 13 per cent of the market, and smaller specialist standards such as Plan Vivo, aimed at smallholder projects with strong community engagement, and CarbonFix, which is aimed at tree-planting projects.[195] Remote sensing data are helping to improve estimates of current and past forest cover, but it is still challenging to provide accurate MRV at a reasonable cost.

However, some problems persist. It will never be possible to ensure that carbon reductions are genuinely additional to what would have happened anyway. The current approach is to base the analysis on past rates of deforestation, but that penalises countries which have chosen to keep their forests intact. An alternative is to use projected future rates of deforestation, but countries could then receive REDD payments for chopping down more trees in future than they did in the past. Perhaps the solution is to use pro-active investment in natural capital (PINC) alongside REDD (see Box 4.1 above) to reward countries with large standing forests, in recognition of their global biodiversity benefits.[196]

Similarly, it is impossible to ensure that carbon savings are permanent, because of the risk of forests being damaged by fires, pests, diseases or illegal logging. One way of dealing with this is to make small annual payments for 'carbon emissions deferred', rather than large up-front payments for 'permanent carbon savings' that are unenforceable in practice.[197]

Finally, research is needed to provide more accurate assessment of greenhouse gas emissions from land-use change. The Kyoto Protocol only considered emissions from forests – not from peat, wetlands or grassland. If countries were able to include all land-use change in their national greenhouse gas reports, this could help to reduce the problem of carbon leakage where savings in one area are simply displaced elsewhere.

Social and environmental safeguards

If REDD is to deliver co-benefits for biodiversity and local communities, safeguards must address the following issues.

- *Biodiversity.* The project must demonstrate biodiversity benefits – this would prevent logging of old-growth forest or conversion of high-biodiversity natural land to monoculture plantations.
- *Free, prior and informed consent (FPIC)* should be obtained from local communities which own, live on or use the land, including indigenous people who have 'customary use' rights but not legal ownership. Communities should be fully involved in the discussions leading up to implementation of a forest carbon scheme.
- *Forest tenure* should be clarified, preserving fair access to forest resources for local communities, and land grabs and forced evictions should be avoided.
- *Benefit sharing.* Carbon payments should be distributed fairly between land-owners, local communities, land users, local and national governments and scheme operators.

Safeguards to involve local communities from the start of the project can help to avoid problems such as illegal logging, clearance or poaching, which are hard to prevent when funds for maintaining boundary fences and patrolling reserves are in short supply. Projects that educate local communities about the need for conservation, and involve them in drawing up management plans that allow access for sustainable use, are far more likely to succeed than those which attempt to ban access.[198] Community forest management initiatives can ensure that forests provide a secure long-term income source through, for example, tourism, carbon payments or selective timber extraction. A study in Uganda found that poor households living next to a protected area generally experienced improvements to their prosperity over a 10-year period, as their access to free resources such as clean water and building materials was preserved, while households in surrounding areas lost this access as forest cover declined.[199]

Numerous studies have emphasised the importance of strengthening land and resource rights for local communities. This can lead to more sustainable resource management even without extra payments, because it gives people an incentive to look after their resources for the long term or invest in planting more trees.[200] In Niger, for example, a change in the law that gave farmers ownership of the trees on their land (previously all trees were deemed to be the property of the state) contributed to the planting of over 200 million new trees as part of the restoration of 5 million hectares of degraded farmland to productive use.[201] In fact, it has been suggested that clarifying land tenure for local communities and strengthening community organisation is in itself an important co-benefit of forest carbon schemes. However, efforts to secure tenure should proceed with caution to avoid triggering conflict, or excluding resource users from common land that is used for grazing or foraging. It also takes time to build up local capacity to enforce land rights and forest protection rules.[202]

It is worth noting that land grabs are not exclusively a problem of biofuels and forest carbon projects – they are a consequence of growing demand for food, timber, minerals, energy (oil or hydropower) and other resources (see Chapters 5 and 6). In recognition of this, the UN Committee on World Food Security is currently (2012) finalising a set of voluntary guidelines on land tenure, the result of 6 years of consultations between governments and social NGOs, which could help to underpin social safeguards for forest carbon projects. The guidelines aim to protect vulnerable people – such as the rural poor, women and indigenous groups – from losing access to land and natural resources that are essential for their livelihoods; for example, by recognising informal 'customary' rights to land and resources, preventing forced evictions, making land ownership records more transparent and accessible, and providing legal aid to those involved in disputes.[203]

Tremendous progress has been made in developing social and environmental safeguards for REDD. Under the UN-FCCC process, a list of safeguards was agreed at the UN Climate Change Conference at Cancún in 2010, but critics have expressed concern that these are not mandatory and that some were weakened at the subsequent conference in Durban in 2011.[204] Afforestation of natural landscapes and first-time logging of old-growth forest are no longer specifically excluded, for example. However, both the UN-REDD programme and the World Bank's Forest Carbon Partnership Facility – the two main bodies funding REDD pilot programmes – are developing their own detailed safeguards.[205] Investors are also increasingly using the independently developed Climate, Community and Biodiversity (CCB) Standards, which provide a 'Gold' level to reward schemes that go beyond minimum standards and deliver positive benefits for biodiversity and society.[206] In 2010, 60 per cent of all forest carbon projects were certified under this scheme. One example of a successful project is shown in Box 4.6.

Box 4.6 REDD with co-benefits: the Kasigau Corridor Project

This project was started by Mike Korchinsky following a holiday visit to Kenya in 1997. He found that impoverished local people were hunting wildlife and illegally clearing the tropical dryland forest by slash-and-burn to plant maize and produce charcoal. Because there were no permanent water supplies, efforts to cultivate the land were doomed to failure, but that did not prevent more and more of the land being cleared by the desperately poor inhabitants and migrants.

Mike sold his company in the United States and set up a wildlife sanctuary with the aim of providing alternative employment for the local people in order to prevent further deforestation. He established an eco-fashion factory exporting clothes to the United States and Europe, and used the proceeds to employ workers to protect and restore the degraded forest. He also built 18 new classrooms in the local area, paid school and college fees for local children and set up an organic greenhouse and a tree nursery. The project provides citrus trees to local farmers, and even gives elephant dung collected from the reserve to a local women's group for mushroom cultivation.

When REDD was introduced, the project was able to apply for carbon credits for an estimated saving of 3.5 million tonnes of carbon dioxide over 20 years, from 170,000 hectares of protected and restored forest. The additional funds were used to keep the loss-making wildlife reserve afloat, expand the eco-factory and fund more community activities. The project won Gold-level certification under the Climate, Community and Biodiversity Standard for its exceptional co-benefits. As well as the high level of social investment, the area is critical for biodiversity, being part of the Eastern Arc Mountains, a Conservation International hot-spot. It forms a wildlife corridor by linking two large protected areas, Tsavo East and Tsavo West, and contains fragments of cloud forest as well as being home to many rare indigenous plant and animal species including lions, cheetahs, elephants and the highly endangered Grévy's zebra.[207]

Sustainable biofuels

Biofuels can play an important role in meeting climate targets, especially for applications where low-carbon electricity cannot easily be used, such as for air and sea transport and high-temperature industrial processes such as steel making. However, as we saw in the section on 'Conflicts', some biofuels can do more harm than good, mainly because of the land use impacts of biofuel cultivation.

Interestingly, there is a high degree of synergy between carbon savings and other benefits. Changes that are bad for the climate, such as draining peat forest to grow oil palm, are also bad for biodiversity and indigenous communities; whereas changes that are good for the climate, such as growing woody crops on degraded lands, also bring other benefits such as restoration of soil fertility and reduced soil erosion. However, there are also examples of conflicts: the Brazilian Cerrado, for example, which is threatened by biofuel expansion, is low-carbon natural grassland, but is of high importance in terms of biodiversity.

With this in mind, a strategy for optimising sustainable biofuel use might proceed as follows.

1   The carbon impacts of biofuels should be evaluated using life-cycle analysis that includes the impacts of both direct and indirect land-use change. Support should only be provided for biofuels with significant net carbon benefits. This

will automatically eliminate support for the most damaging biofuels: those that result in clearance of rainforest, for example. It will tend to focus support on biofuels from waste, and on second-generation biofuels from woody biomass or algae.

2   Biodiversity and social impacts should be assessed, to prevent expansion of biofuels onto land that has low carbon stocks, but high biodiversity; or that is of high value to local people for grazing, foraging or cultural reasons. Additional support should be offered for biofuels that offer social and bio-diversity benefits, such as mixed native perennial plantings that help to restore degraded land.

3   Strict social safeguards should be applied to prevent land grabs and ensure good working conditions on biofuel plantations.

Only biofuels that satisfy these criteria should be allowed to count towards biofuel targets, or to receive carbon credits or subsidies. It might also be helpful to harmonise some of the 67 separate sets of sustainability criteria that are currently (2012) being developed by various organizations.[208]

National and international targets for biofuel use may have to be scaled down if it appears that they exceed the amount of sustainable biofuel feedstock that is available. Some estimates seem very optimistic. The IEA has estimated that it would only take 10 per cent of agricultural and forestry waste to produce the quantity of second-generation biofuels required in their 450ppm Scenario, leaving plenty of biomass for alternative uses.[209] Critics, however, query how practical it would be to gather and convert even 10 per cent of forest and farm waste, given how widely dispersed and remote much of this is.

Estimates of the land that would be required to meet the projected demand for biofuels vary widely, from 56 to 2,500 million hectares (Mha). One study estimates that between 118 and 508 Mha would be needed to substitute for 10 per cent of current transport fuel demand.[210] The IEA biofuels roadmap estimates that 100 Mha would be required for transport biofuel crops, on top of the amount needed for heat and electricity generation.[211] For comparison, total land area is 13,000 Mha (see Figure 4.6 above), cropland occupies 1,500 Mha and around 36 Mha is currently used for biofuels.

The Worldwide Fund for Nature (WWF) has developed a scenario that would provide almost 100 per cent of global energy from renewable sources by 2050 (see Chapter 5), in which 250 Mha is required for biofuel crops. This is achieved first by cutting transport fuel demand through measures such as improved fuel efficiency and more walking and cycling, and then by maximising production of biofuels from other (non-crop) sources. More than half of the biofuel demand is met from crop residues (35 per cent of residues are used, leaving the rest for soil improvement and other uses) and non-recyclable waste, and the rest is split between wood harvested from sustainably managed forests, energy crops and (after 2030) algal biofuels. After excluding existing cropland, land that would require irrigation, forests, protected areas and a small allowance for future expansion of farmland, they find that there is more than enough land available for biofuel production (Figure 4.10). Most of

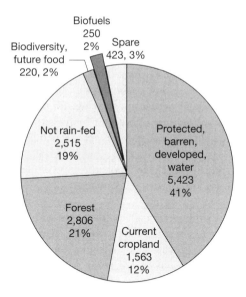

Figure 4.10 Worldwide Fund for Nature (WWF) estimate of land required for biofuels in a
100 per cent renewable energy scenario (million hectares)

Source: Data from WWF *et al.*, 2011

Note: Total land area excludes Antarctica.

this is extensively grazed pasture, which is made available partly through mixed
livestock–energy crop systems, and partly through halving demand for meat in
OECD countries while allowing non-OECD countries to increase their meat intake
by 25 per cent.[212] This would free up much of the 4,000 Mha of pasture and
cropland currently used to support livestock (see Figure 4.6 above).

Sustainable food

If we are to meet the projected demand for 70 per cent more food by 2050 without
increasing deforestation for farmland, we need to increase crop and livestock yields
per hectare of land. The section on 'Conflicts' summarised the heated debate
between proponents of intensive agriculture and those who support less intensive
techniques such as organic farming. This section will try to establish whether a
'middle way' is possible, that can increase yields sustainably at the same time as
cutting greenhouse emissions and pollution, enhancing biodiversity and adapting
to climate change.

Organic farming uses a number of techniques that reduce climate emissions:

•   adding organic matter to the soil, which increases stored carbon (one review
    showed an average of 37 tonnes per hectare for organic systems, compared to
    27 t/ha for non-organic systems);[213]

- using cover crops, which reduces soil erosion and increases soil carbon;
- not using synthetic fertilisers, which cuts emissions of carbon dioxide from fertiliser manufacture, and of nitrous oxide from fertiliser application;
- limiting livestock grazing density, which reduces losses of soil and vegetation carbon from over-grazing and soil erosion.

In addition, organic farming strictly prohibits the use of most herbicides and pesticides. This brings undoubted benefits in reducing water pollution, avoiding pesticide poisoning and improving biodiversity. However, there is considerable debate over the impact of organic farming on crop yields. In developing countries, for example, 20–40 per cent of crops can be lost due to pests and diseases, and another 12 per cent during storage.[214] Pesticides and fungicides could reduce these losses. If yields per hectare are lower, this could drive a need for more farmland, resulting in further deforestation, as well as threatening food security.

A number of studies have addressed the question of whether a completely organic farming system could feed the world. A literature review of 293 examples by the University of Michigan found that yields from organic farming, averaged across all food products and all countries, were 30 per cent higher than those from non-organic farming. In developed countries, yields from organic farming were 8 per cent lower; but in developing countries, yields were 80 per cent higher (Table 4.5).[215] The higher yield in developing countries reflects a number of issues:

- Many farmers in developing countries cannot afford expensive agrochemicals. Fertiliser prices increased by a factor of five between 2005 and 2008, largely due to rising oil prices, and oil prices are expected to continue to rise in the future (see Chapter 5). Therefore organic techniques in these countries can boost yields significantly.
- Organic techniques can improve yields where water is scarce or where torrential rains are common. Adding organic matter to the soil increases water penetration and retention, thus improving resistance to drought and reducing the risk of flooding. As well as boosting yields in developing countries, organic cultivation often gives higher yields in developed countries during drought years.[216]
- Organic farming can be very beneficial in restoring soil fertility in degraded regions.

The study also found that nitrogen-fixing cover crops could supply around 140 million tonnes of nitrogen each year – more than enough to replace the 90–100 million tonnes currently supplied by synthetic fertilisers. As most of the growth in food demand is expected to come from developing countries, the study concluded that organic farming techniques can make a major contribution to food security, poverty reduction and biodiversity, as well as cutting carbon emissions and reducing water pollution, without increasing the land area used for farming.

A similar review of 286 projects in 57 low-income countries found that farmers had increased crop yields by 79 per cent on average by adopting resource-

Table 4.5 Yield ratios of organic versus non-organic farming

Food category	World	Developed countries	Developing countries
Grain products	1.31	0.93	1.57
Starchy roots	1.69	0.89	2.70
Oil crops	1.08	0.99	1.65
Vegetables	1.06	0.88	2.04
All plant foods	**1.33**	**0.91**	**1.74**
All animal foods	**1.29**	**0.97**	**2.69**
All foods	**1.32**	**0.92**	**1.80**

Source: Badgley *et al.*, 2007

conserving techniques such as integrated pest management, integrated nutrient management, conservation tillage, agroforestry, water harvesting, and mixed crop and livestock farming systems. Most of these projects also allowed pesticide and water use to be reduced.[217] The benefits of compost use are illustrated by data from the Tigray Project in Ethiopia (Box 4.7).

Box 4.7 Tigray farmers boost yields with compost

Much of Ethiopia suffers from degradation due to over-grazing, deforestation and soil erosion. Chemical fertilisers are expensive, and poorly suited to the traditional crops. In dry areas of the Tigray region, farmers reported that chemical fertilisers caused problems such as 'burning' of crops, loss of soil biodiversity and increased soil salinity, as chloride salts were building up in the soil and there was not enough water available to flush them out.

In 1996, the Bureau of Agriculture and Rural Development set up the Tigray Project, partly funded by the NGO Third World Network. Farming communities were brought together to discuss the best ways of making agriculture more sustainable and productive. As a result, communities agreed to restrict over-grazing; build ponds, bunds and terraces to conserve water; plant trees and legumes on the bunds; and start making their own compost from farm and household waste and manure.

By 2008, 86 per cent of farmers were using compost in their fields and only 16 per cent were using chemical fertilisers. Analysis of data from over 900 fields showed that compost doubled grain yields compared to untreated fields, and also increased yields compared to chemical fertilisers (Figure 4.11). Fields with compost had better water retention and fewer problems with weeds, pests and diseases. Farmers were able to avoid the cost of chemical fertilisers, and farm incomes improved. Between 2003 and 2006, grain yields in the region doubled, while the use of chemical fertilisers fell by 40 per cent between 1998 and 2005. The system is now being extended throughout Ethiopia, with compost now being added to 16 per cent of farmland.[218]

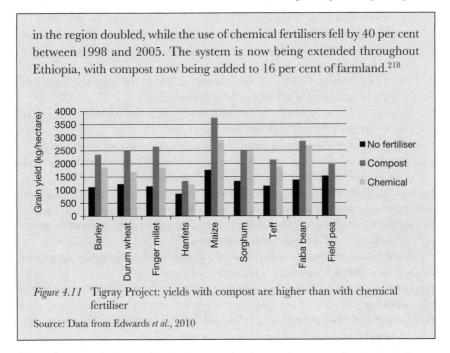

Figure 4.11 Tigray Project: yields with compost are higher than with chemical fertiliser

Source: Data from Edwards *et al.*, 2010

However, when compared to conventional high-input farming, two studies published in 2012 find that yields from organic farming are around 20–25 per cent lower on average, although with best practice and for certain crops there is no yield gap.[219] This indicates that even if organic farming could feed the world in theory, it is possible that yields could be even higher (thus freeing up more land) with judicious use of agrochemicals. Perhaps the best solution is to look for a pragmatic approach, combining ecological farming techniques with limited and efficient use of agrochemicals. For example, there would be far less fertiliser run-off if small amounts of fertiliser were added into planting holes, rather than spraying the whole field.[220] Similarly, genetic modification of plants could be useful where it helps to develop disease resistance or drought tolerance, but the environmental impacts could be negative where development of biocide-tolerant varieties fosters greater use of agrochemicals, or risks creating invasive 'superweeds'. However, it is worth noting that farmers relying wholly on organic techniques can obtain a premium for their produce which would not be available to those following a mixed approach.

A consensus is gradually emerging on a number of 'win-win' techniques that can help to increase yields sustainably at the same time as cutting greenhouse gas emissions and improving adaptation to climate change. These techniques include:

• *development of improved crop varieties* to increase yields and improve resistance to pests, diseases and climate change;

- *conservation agriculture* (reduced tillage with crop rotation and cover crops) to reduce soil erosion, increase carbon storage, increase soil organic matter, improve water retention and save time and money for farmers (but see below re herbicide use);
- *maximising recycling of organic matter* (compost and manure) from the farm back to the soil, to minimise use of synthetic fertilisers;
- *precision application of fertiliser* to reduce nitrous oxide and ammonia emissions, reduce fertiliser run-off and save money for farmers;
- *integrated pest management* to minimise use of synthetic pesticides, by using resistant crop varieties, crop rotation, encouraging natural predators, using pest-repellant plants and burning infected plants;
- *efficient water management,* such as use of drought-resistant crops, rainwater harvesting and storage, and precision irrigation where necessary;
- *agroforestry* to store carbon, stabilise soil, restore soil fertility, improve biodiversity and diversify or improve farm income;
- *mixed farming systems* that combine crops and livestock in order to provide manure for use as an organic fertiliser;
- *harnessing traditional knowledge* of local crop and livestock varieties and farming techniques that are well suited to local conditions;
- *reducing food waste and shifting to a lower-meat diet* in developed countries to give benefits for climate, biodiversity and food security.

Many of these techniques are combined in systems known as 'agro-ecology'[221] or 'sustainable crop intensification'.[222] These approaches aim to provide healthy soil, with a high content of organic matter, good water retention and diverse soil organisms, so that plants will flourish and be more resistant to diseases, pests and extreme weather. By increasing yields while minimising the need for synthetic inputs, agro-ecology can increase the incomes of smallholders in developing countries, thus helping to address food security and combat poverty.

Two conflicts remain to be addressed. The first concerns livestock farming, where the type of production is all-important. Intensive indoor 'factory farming' could reduce the demand for grazing land, thus avoiding deforestation and saving carbon emissions. However, this is a poor solution from the point of view of animal welfare, and the saving in land could be offset to a large extent by the production of nitrous oxide and methane emissions from waste manure and slurry, and the requirement to produce large quantities of feed, which drives deforestation in Brazil. A better option might be to reduce demand for meat and dairy produce in developed countries, where over-consumption is common (see Chapter 8). As well as freeing up enough cereal production to feed over 3 billion people,[223] this would allow the remaining demand for meat to be met by sustainable mixed crop–livestock systems with low grazing density, allowing livestock to play a key role in utilising land that is not suitable for crops and providing manure for crop cultivation. Extensive grazing with no fertiliser input can be climate-neutral, because an increase in soil carbon in regularly cropped pasture can outweigh the effect of methane emissions from the animals and nitrous oxide emissions from manure.[224] Support for low-impact grazing, especially on upland pasture that is unsuitable for arable cultivation, could help to preserve the culture, landscape and local economies of these areas. It is also important to remember that in developing countries, where smallholder farming predominates and under-nutrition is common, livestock can play a vital role in helping to improve household protein intake, diversify farm incomes and reduce poverty.[225]

It is interesting to note that reducing meat consumption would have a far larger effect on greenhouse gas emissions than switching to organic food, as indicated in Figure 4.12 (though this figure underestimates the benefits of organic production because it does not include soil carbon storage). However, in order to maximise the benefits, it will be important to ensure that arable crops grown to replace animal products are produced as sustainably as possible, avoiding unnecessary use of agrochemicals, deforestation, soil erosion or excessive water abstraction.

The second conflict is the use of herbicides for conservation tillage, which can increase farming costs, pollute water supplies and damage wildlife. In Brazil, zero-till systems use 17 per cent more herbicides than conventional agriculture, and the added cost means that farmers sometimes resort to occasional tillage to help remove weeds.[226] Alternatives to herbicide use include the use of year-round cover crops to suppress weeds, and mechanical weeding such as rotary hoeing. However, further research into methods of suppressing weeds without using herbicides would be useful.

It will take considerable effort to make the shift to sustainable food production. One of the first steps should be to eliminate harmful subsidies for agrochemicals, fossil fuels and monoculture plantations, and redirect them to support for sustainable agriculture. In Brazil, for example, there are subsidies per head of cattle

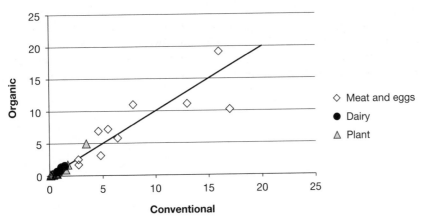

Figure 4.12 Greenhouse gas emissions ($kgCO_2e/kg$ product) from conventional and organic farming systems, excluding changes in soil carbon

Source: Data from FAO, 2011c

Note: Points above the line have higher emissions for organic production.

purchased, but none for improving the quality of the pasture, which could reduce the area needed per animal. This has encouraged expansion of pasture into the Amazon rainforest, which conflicts with polices aimed at preserving forests.[227]

Making low-carbon agriculture techniques eligible for climate change funding, either under schemes such as the Clean Development Mechanism or via the international Green Climate Fund, could help to promote the uptake of these methods, although there has been opposition to this for the same reasons as for forest carbon trading.[228] Another option is the use of payments for ecosystem services (PES) schemes, to compensate farmers for costs incurred in, for example, planting trees or reducing grazing density.[229] However, care must be taken not to encourage 'ransom behaviour' where participants refuse to act sustainably unless paid. It could be more effective to involve communities in determining for themselves the optimum course of action in order to deliver both environmental and livelihood benefits, as with the Tigray Project (Box 4.7 above), and then providing funds to help with the transition.

Considerable investment in research, demonstration and outreach programmes will be needed, especially because less than 1 per cent of agricultural funding so far has been directed at ecological and organic techniques, leaving great potential for yield improvements. It will be important to redirect agricultural investment towards climate-smart activities. Programmes to protect crop and livestock diversity, including seed saving from local varieties and wild relatives, should also be expanded, so that new varieties can be produced that are higher-yielding and more resistant to environmental stress.

Finally, although the focus of this book is on the co-benefits of climate policies, it is worth emphasising that the direct impacts of climate change represent one of the greatest threats to ecosystems, biodiversity, agricultural productivity and food security. Forests, crops and wildlife are threatened by rising temperatures, more storms, floods and droughts, decreased glacial melt-water supply and the spread of pests and diseases. Decreases in agricultural production could be as high as 15–35 per cent in South Asia and parts of Sub-Saharan Africa.[230] Tackling all sources of climate emissions will therefore be crucial in protecting ecosystems, limiting agricultural losses and reducing the need for farmland expansion and deforestation.

5 Secure and safe energy

Adapting to Peak Oil

Key messages

- *Continued dependence on fossil fuels poses risks for energy security.* Production of conventional oil may have already peaked, and the best remaining reserves are concentrated in a small number of countries. Other countries will be increasingly forced to rely on lower-quality unconventional sources such as tar sands, and remote sources in sensitive areas such as deep-water, arctic and rainforest oil. Extraction will be more expensive, energy-intensive and environmentally damaging. As a result, there is a consensus that we face a future of rising oil prices, more frequent supply disruptions and escalating oil-related conflict. Similar trends are seen for gas and coal, with a shift to more environmentally damaging options such as shale gas and lignite as higher-quality conventional reserves become depleted.
- *Climate policy that seeks to reduce reliance on fossil fuels can increase energy security.* Improving energy efficiency, reducing energy demand (especially for transport) and developing sustainable energy sources can deliver co-benefits including lower and more stable prices for energy services in the long term, reduced expenditure on fuel imports, and safer, cleaner energy.
- *The 'Drill, baby, drill!' approach to energy security,* including exploitation of shale gas, will not allow us to achieve the goal of preventing dangerous climate change. Carbon capture and storage (CCS) exacerbates energy security problems because it increases fossil fuel consumption per unit of delivered energy by 25 per cent.
- *A high share of variable renewables such as wind and solar presents a challenge for balancing supply and demand.* This can be tackled by encouraging a diverse mix of renewable sources, and investing in energy storage, electricity grids, control systems and smart demand management.
- *Nuclear power* can improve energy security provided that uranium supplies are secure, but it has other safety and security impacts including

the risk of weapons proliferation and terrorism. Fourth-generation reactors could help to improve these problems in the long term.
* *Other potential conflicts* include the high land, water and energy demand of biofuels, which could be partially mitigated by the development of second-generation biofuels; the visual impacts of wind turbines; and the impacts of large hydropower schemes.

Cheap energy – the end of an era?

The IEA describes energy security as 'uninterrupted physical availability at a price which is affordable, while respecting environment concerns'.[1] But each of these elements of security – physical supply, affordable price and environmental sustainability – is threatened by our high dependence on fossil fuels.

The era of relatively cheap fossil fuel energy that has powered economic growth over the last century may be coming to an end. The richest reserves of oil, gas and coal are being used up, and remaining low-cost reserves are concentrated in a small number of countries, many of which are politically unstable. To satisfy our addiction to fossil fuels, production will eventually be forced to shift to more expensive and more environmentally damaging options such as deep-water oil and tar sands. Global energy demand continues to rise steeply, and the level of investment needed to develop new sources and avert a supply crunch does not appear to be forthcoming. We face a future of rising fuel prices, supply interruptions and increased conflict.

Action on climate change can be instrumental in addressing this problem. Cutting our dependence on fossil fuels can help to ensure a safer, cleaner and more secure energy supply for the future. In addition, renewable sources can provide secure and affordable energy for some of the 1.3 billion people who currently lack access to electricity. This chapter starts by investigating the existing threats to energy security. It then examines how climate policy can help to improve the situation. Finally it considers potential conflicts between climate policy and energy security, and looks for a way forward that delivers benefits for both.

Peak Oil

What is Peak Oil?

The classic measure of how much oil is left in the ground is the 'reserves-to-production ratio' (R/P), which is simply the total proved recoverable reserves divided by current annual production. In 2010, there were 1,300 billion barrels of proved reserves, and production was 82 million barrels per day (of which 69 million was conventional crude oil, the rest being natural gas liquids and a small amount of unconventional oil), giving an R/P ratio of 46 years.[2] This means that if no more oil is discovered, and production stays the same, we will run out of oil in 46 years.

Many people, however, believe that the R/P ratio is not the most useful measure of fossil fuel depletion. For a start, the volume of proved recoverable reserves will change over time, as new discoveries are made, technology makes unrecoverable oil recoverable, and existing reserves are used up. It is also obvious that the demand for oil will change over time – if demand grows rapidly, reserves will be depleted sooner than shown by the R/P ratio. What matters is whether we can produce enough oil to match future demand. If demand exceeds supply, we can expect price rises, shortages, and, if the situation is not resolved, economic disruption and political conflict. This gives rise to concern about Peak Oil – the moment at which world oil production peaks and then starts to decline. If we have not managed to curb the growing global demand for oil by the time Peak Oil occurs, fuel shortages and price rises will be inevitable.

Although the term 'Peak Oil' was coined by geologist Colin Campbell[3] in the year 2000, the concept has been around since 1956, when geophysicist M. King Hubbert published his theory that the production rate of oil from a particular region would tend to follow a bell-shaped curve, increasing to a peak before declining. This is because exploration techniques tend to identify the largest fields first, as these show up more readily from seismic surveys and other geophysical data, and are more likely to be found by exploratory drilling. In addition, oil companies will naturally choose to exploit the largest and most productive fields first. Once these fields are exploited, attention shifts to smaller, deeper or more remote fields. Figure 5.1 shows this schematically – it is a hypothetical case where production

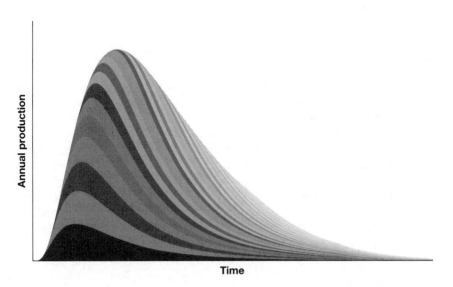

Figure 5.1 Theoretical oil production profile for a region, based on the assumption that each new field is smaller than the last. This leads to a production peak followed by a steady decline

Source: Sorrell *et al.*, 2009

begins with the largest field, and successive fields are then developed that are each smaller than the last. Each layer in the graph represents the production profile of a single field. The figure shows that there will inevitably be a peak in production followed by a steady decline. Further small discoveries will prolong the life of the region, but will not change the date of the peak.

Hubbert correctly predicted that US oil production would peak between 1965 and 1970, and warned that production of other finite resources including gas, coal and minerals would eventually also peak and start to decline.[4] Although the Hubbert curve predicts that the peak will occur when around half of the total amount of oil from a field or region has been produced, studies of real fields and regions show that the peak often occurs when less than half, and sometimes even less than a third of the oil has been produced.[5] There will therefore be a long tail of production at lower levels. Peak Oil is not the same as running out of oil.

How much oil is left in the ground?

Up until the end of 2007, we had extracted just over 1,100 billion barrels of conventional crude oil globally and there were over 1,300 billion barrels of proved reserves. A 2009 review found estimates of 'remaining recoverable reserves', which includes oil yet to be found, ranging from 870 to 3,170 billion barrels – with the lower end of this range being below the current industry estimate of 'proved reserves'.[6] This reflects uncertainty over the size of current reserves, the potential for 'reserve growth' and the prospects for future discoveries. On top of this, there is a large potential contribution from unconventional sources of oil such as tar sands.

RESERVE GROWTH

Proved reserves are defined as reserves which have a 90 per cent probability of being economically recoverable under current economic and political conditions and with existing technology. By definition, this gives a conservative estimate, as only well-quantified reserves can be included. Estimates of proved reserves tend to increase over time – either because more data on a specific field are gathered, allowing more reserves to be classed as 'proved', or because technical advances such as 'enhanced oil recovery' allow more of the oil in a given reserve to be extracted. The estimate will also vary with economic conditions – as the oil price increases, the size of proved reserves will also tend to increase because more oil will be economically recoverable. The ongoing upwards revision of the size of 'proved reserves' for a given area is called 'reserve growth'.

NEW DISCOVERIES

No one knows how much oil will be discovered in the future, but there is a limited range of geological conditions under which oil and gas deposits will form (see Box 5.1). Geologists have identified the regions of the world that meet the conditions for oil and gas formation, and the most promising areas are now fairly well explored; although in a handful of areas, exploration has been limited for technical, political or environmental reasons. After decades of exploration with increasingly advanced techniques, it seems highly unlikely that we will find 'another Middle East', although isolated large fields may be found here and there.

Box 5.1 Why does the Middle East have most of the oil?

Oil and gas deposits only form under a set of specific conditions. They need:

1 A *source rock* rich in organic matter, such as mud laid down in shallow seas containing large amounts of plankton or algae. The organic matter must have accumulated faster than it could decompose. Warm, nutrient-rich, fairly stagnant conditions would be ideal. The main source rocks were laid down during six geological periods, with the richest being the Upper Jurassic and Mid-Cretaceous, providing 25 per cent and 29 per cent respectively of known reserves.[7]

2 A porous *reservoir rock* above the source rock, such as sandstone or limestone, for the oil and gas to migrate into.

3 An *impermeable seal* above the reservoir to stop the oil and gas leaking out. Salt deposits (evaporites) are particularly effective.

4 A *structure to trap* the oil, such as an upwards fold or a fault in the rock layers. Oil migrates upwards and sideways until it collects in a trap. Gas collects on top of the oil.

5 A *period of heating* to turn the organic matter in the source rock into oil and gas. This typically happens when the source rocks get buried deep into the earth; for example, during collisions between tectonic plates. There is an optimum temperature range for the formation of oil: at higher temperatures, the oil is transformed to gas.

The most productive region for oil and gas is known as the 'Tethyan Realm'; it stretches in a wide belt from southern China and Indonesia, through the Middle East, Europe and North Africa to the Caribbean. This region was once part of the Tethys Ocean – a warm, partly enclosed shallow sea that opened and closed several times, splitting the supercontinent of Pangaea (which consisted of all the present continents clumped together) into a northern land mass (Laurasia) and a southern one (Gondwana). Although the Tethyan Realm covers only 17 per cent of the earth's surface, 70 per cent of all known petroleum deposits are found there.

> Within this region, ten countries of the Middle East (Saudi Arabia, Iran, Iraq, Syria, Kuwait, the United Arab Emirates, Bahrain, Oman, Yemen and Qatar) that form only 3 per cent of the earth's land surface contain 60 per cent of the world's proved reserves. Here conditions were exceptional: for 500 million years, the area was part of a warm, shallow marine shelf and experienced almost continuous deposition of sediments, building up a staggering 12-kilometre depth of sedimentary rocks. During the Jurassic and Cretaceous periods, volcanic activity in the nearby ocean caused an upwelling of nutrient-rich currents: perfect conditions for marine life. Thick organic-rich source rocks were laid down, interspersed with carbonate reservoir rocks. As the sea eventually closed up when the African–Arabian and Eurasian plates collided, salt deposits were laid down in the drying sea bed, forming a perfect cap rock. Finally, the collision between the plates was sufficiently gentle to ensure that the oil was trapped in a series of large, gentle folds and salt dome structures – the deformation was not severe enough to destroy the oil, as happened in the Himalayas.[8]

The number and size of new discoveries have been falling since the 1960s (Figure 5.2). In particular, the discovery rate of giant fields (those with more than 500 million barrels of ultimately recoverable reserves) is in decline. Just 500 of the world's 70,000 oil fields are giant fields, but these account for two-thirds of all the oil discovered so far.[9] In fact, just 25 giant fields produce a quarter of the world's oil. But more than half of the world's giant fields were discovered over 50 years ago, and many are past their peak of production. The average size of giant field discoveries has declined from 4.4 billion barrels in the 1960s to 1.2 billion barrels in the 1990s, while the average size of all new discoveries fell from 225 million to under 50 million barrels.[10] Consumption first exceeded the rate of new discoveries in 1980, and it is unlikely that new discoveries of conventional oil will be able to compensate for the depletion of existing reserves. However, reserve growth currently makes up the difference, so that estimates of proved reserves are still growing from year to year.[11]

DOUBTFUL ESTIMATES

The reserve estimates from members of OPEC (Organization of the Petroleum Exporting Countries) are not independently verified, and appear somewhat dubious in several cases. Following OPEC's 1983 decision to allocate production quotas based on the size of each member's reported reserves, the reserve estimates of several members increased sharply over the next 5 years, despite there being no significant new discoveries, and many of these estimates have remained unchanged since then despite continuing production (Figure 5.3, see plate section). Although

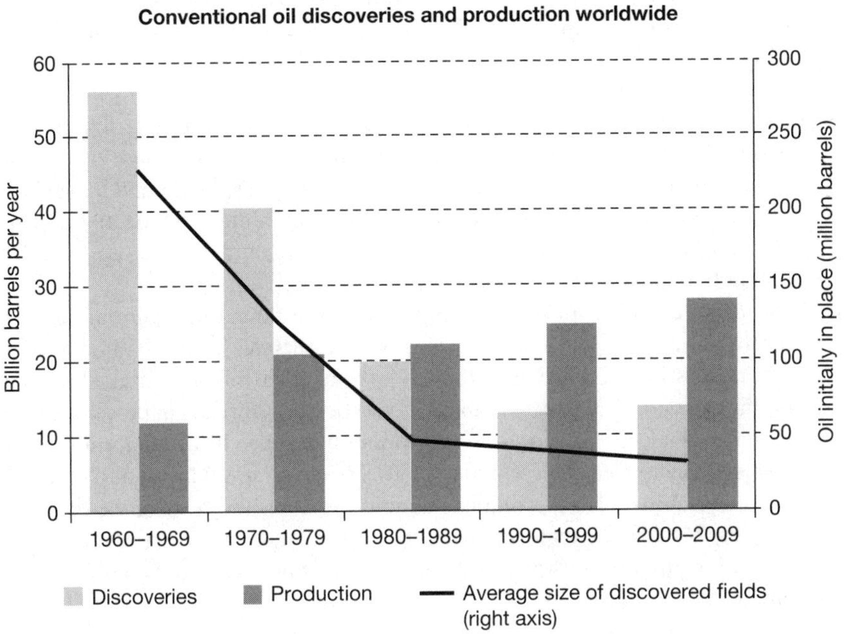

Conventional oil discoveries and production worldwide

Figure 5.2 Discoveries of new oil fields peaked in the 1960s, and new fields are getting smaller
Source: IEA, 2010b. © OECD/IEA, 2010

some of this increase may have been justified owing to previous underestimates and improvements in the recovery rate, some commentators believe that OPEC estimates are around 300 billion barrels too high.[12]

UNCONVENTIONAL OIL

Unconventional sources of oil include tar sands, extra-heavy oil, oil shale and conversion of natural gas or coal to oil.

- *Tar sands* are formed as a result of oil migrating up to the surface, where the lighter fractions degrade, leaving a residue of highly viscous bitumen. For deposits down to a depth of around 50 metres, the rock is excavated by open-cast mining and then heated to recover the oil. For deeper deposits, the oil is extracted *in situ* by pumping steam into the rocks. Either technique requires a lot of water and heating fuel – usually natural gas – and refining the low-quality bitumen or heavy oil is also energy-intensive because there is a high sulphur and nitrogen content. Nevertheless, production is currently profitable if global oil prices are above $65–75 per barrel.[13] Canadian tar sands currently produce over a million barrels of oil per day, and this is expected to increase to over 4 million barrels per day by 2035.[14] Reserves are estimated as at least

170 billion barrels of recoverable oil (assuming a 10 per cent recovery rate) in Canada, with more in Russia, the United States and the Middle East. Total global resources (including yet-to-be-discovered reserves) could be as high as 3,100 billion barrels.[15]

- *Extra-heavy oil* is less thick than bitumen, partly because it is found at greater depths, but it may still require steam injection to aid extraction, as well as extra refining to turn it into a usable fuel. There are thought to be around 500 billion barrels of technically recoverable extra-heavy oil in the Orinoco oil sands of Venezuela, where production is at an early stage, and possibly up to 2,000 billion barrels of ultimately recoverable resources.[16]

- *Oil shale* consists of rocks with a high level of organic material that has not matured into oil. There are vast reserves of oil shale in the United States, China, Russia and elsewhere – thought to total 4,800 billion barrels – but little of this is economically recoverable.[17] It can be burnt directly in power stations to generate electricity, and this was common practice in Estonia until it was displaced by nuclear power in the 1980s. However, it is difficult and costly to produce liquid fuel as this requires mining, crushing and heating the rocks, or heating them *in situ*, then capturing the vapour and converting it to oil. This uses large amounts of water and energy, and generates high greenhouse gas emissions.[18]

- *Techniques for converting coal or natural gas into synthetic oil* also have high energy requirements. Coal has been converted to oil in South Africa since the 1950s, and two coal-to-oil plants are being built in China, but carbon emissions are twice those of conventional oil unless carbon capture and storage (CCS) is used (CO_2 is separated out as part of the process). The process is currently economic at oil prices over \$60–100 per barrel (including CCS).[19]

In short, although there are vast sources of unconventional oil, they are expensive, energy-intensive and environmentally damaging to produce.

When will Peak Oil happen?

It is difficult to predict the year in which oil production will peak. Not only are there wide variations in the estimates of the amount of oil in place, but the date of the peak will depend on the rate at which oil is extracted, which in turn depends on 'above ground' factors such as wars, trade disputes, storms, accidents, terrorism and recessions. The two oil price shocks of the 1970s, for example, each caused consumption to decrease and then to resume growth at a lower rate. Impacts which reduce consumption will tend to prolong existing reserves, causing the peak to shift further into the future.

The Peak Oil debate has raged fiercely ever since Hubbert's first controversial predictions, with a growing number of geologists and scientists warning that a peak is imminent,[20] but with the economists who advise governments tending to believe that technical advances and new discoveries will allow production to continue to meet demand for the foreseeable future.[21] A number of recent studies have predicted a peak within the next few years;[22] these include a thorough review of over 500 studies carried out in 2009 by the UK Energy Research Centre which predicts a peak before 2031, with a 'significant risk' of a peak before 2020.[23] These authors also note that production of oil *per capita* peaked in 1979.

The International Energy Agency (IEA), which advises governments on energy security, previously dismissed fears of Peak Oil, but this has changed in recent years. In the *World Energy Outlook 2010*, the IEA estimates that just 68 million of the 96 million barrels per day that will be produced in 2035 will be conventional crude oil, with the balance being made up by large increases in the production of natural gas liquids and unconventional oil (Figure 5.4, see plate section). Most significantly, the agency predicts that the production of conventional crude oil has reached an undulating plateau at around 68–69 million barrels per day, and 'never regains its all-time peak of 70 million barrels per day reached in 2006'.[24]

This acknowledgement by the IEA that conventional oil production has peaked is tremendously significant, but there is still criticism that their production forecasts are too optimistic. A paper from Uppsala University estimates that total conventional and unconventional oil production in 2030 will be limited to 75 million barrels per day, largely because current investment in developing new fields is inadequate.[25]

Peak Gas

Gas supplies are considered to be more secure than oil supplies – the reserves-to-production (R/P) ratio is currently 59 years, compared to 46 years for oil.[26] The IEA estimates that the remaining recoverable reserves of conventional gas (including fields yet to be found) could be as high as 400 trillion cubic metres – enough for 120 years of production at current rates.[27] But Peak Gas may be closer than we think. The data for gas reserves are even less reliable than for oil reserves, but discoveries of conventional gas peaked in the 1970s and production overtook new discoveries back in 1980.

There are a couple of important differences between gas and oil. First, gas production from each well declines more steeply than for oil, because the pressure in the reservoir rock falls faster, so there is less warning of when a field is going to peak. Second, gas is less dense and so is more expensive to transport than oil. Transport involves either building pipelines or using liquefied natural gas terminals and tankers, both of which require considerable investment. As a result, around 85 per cent of all gas is currently consumed in the same region in which it is produced.[28]

Conventional gas production in Europe, Canada and the United States has already peaked, and the Energy Watch Group[29] believes that the global peak could be as soon as 2025. Regional shortages could occur before then – especially in Europe, where production is expected to decline by 30 per cent by 2035 while demand continues to grow.

As with oil, however, attention is turning to unconventional supplies of gas such as shale gas, tight gas and coal-bed methane:

- *Shale gas* is gas that has been generated within impermeable shale rock. Little will flow out of the rock formation under normal circumstances, but it can be extracted by hydraulic fracturing, or 'fracking', which involves pumping large quantities of fluids into the shale at high pressure in order to fracture it and release the gas.
- *Tight gas* is gas contained within relatively impermeable sandstone or limestone that can be recovered using techniques such as fracturing, acidification or vacuum pumping.
- *Coal-bed methane* is gas that occurs naturally in coal seams. In the past, this was extracted from coal mines for safety reasons and vented to the air, but today it is often recovered for use as a fuel, sometimes by drilling and fracturing coal seams as for shale gas.

These three unconventional sources account for around 13 per cent of current gas production (dominated by US shale gas).[30] The IEA estimates that there are over 200 trillion cubic metres (tcm) of ultimately recoverable shale gas reserves worldwide, plus over 118 tcm of coal-bed methane and 84 tcm of tight gas, although it warns that there is considerable uncertainty over how much of these reserves can be exploited economically. There are also potentially vast reserves of methane hydrates – a mixture of gas and ice found in deep-sea sediments in Arctic regions – but these are extremely difficult to extract and so have not been exploited to date.

Shale gas, in particular, is being hailed as the solution to the world's energy problems, with large reserves identified in the United States, Canada, South America, Europe, South Africa and elsewhere. The United States is abandoning its newly built liquefied natural gas (LNG) import terminals in favour of rapid exploitation of its shale gas reserves.[31] But as with unconventional oil, unconventional gas is associated with a number of environmental problems, which we will discuss later (see 'Safe, clean, sustainable energy'). The European Parliament recently concluded that shale gas reserves in the EU are too small to make a significant contribution to the forecast demand gap of 100 billion cubic metres of gas by 2035; and in view of the environmental damage associated with shale gas production, a better strategy would be to focus on reducing demand and switching to cleaner energy sources such as renewables.[32]

Peak Coal

Coal appears to be more abundant than oil and gas – proved recoverable reserves are sufficient for 118 years of production at current rates.[33] Yet demand for coal is growing faster than for any other fuel, resulting in prices tripling between 2000 and 2010.[34]

Unlike gas and oil, estimates of coal reserves tend to be corrected downwards over time rather than upwards, as once the best coal seams have been exploited, it often transpires that the remaining coal is to be found in thin seams – with complex geology or at great depths, which makes recovery expensive. Estimates of total global coal resources were revised downwards by 55 per cent between 1980

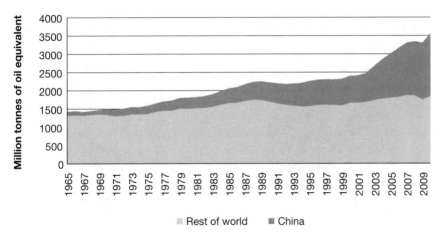

Figure 5.5 The rapidly growing global demand for coal is driven largely by China

Source: Adapted from Mearns, 2010, using data from BP, 2011

and 2005.[35] Unlike oil and gas, there are no 'unconventional' reserves of coal to move to – although production is moving towards lower-quality sub-bituminous coal and lignite as reserves of high-quality bituminous coal and anthracite are exhausted. In the United States, for example, there are 200 years' worth of proved reserves left at current production rates, but much of this is low-quality sub-bituminous coal, lignite or high-sulphur coal. Production of anthracite in the United States peaked in 1950 and bituminous coal peaked in 1990. Productivity per miner has been decreasing since 2000, as the quality of the remaining reserves declines, and although the tonnage mined is growing slowly, production in terms of energy content peaked in 1998. Globally, total coal production could peak between 2020 and 2050.[36]

China produces and consumes almost half of the world's coal, and is responsible for most of the growth in demand over the last decade (see Figure 5.5). Although China has the world's third-largest coal reserves and the reserves-to-production ratio was estimated as 35 years in 2010, reserve estimates have not been updated since 1990, so that 20 years of production should perhaps be subtracted from the reserve estimate.[37] One study estimates that production in China could peak by around 2030.[38]

Market concentration: who's got the power?

Reserves of low-cost oil, gas and coal are increasingly concentrated in a small number of countries. Around 60 per cent of global proved reserves of crude oil are found in the Middle East and 78 per cent are in the 12 OPEC countries.[39] Over half of the world's oil is consumed by OECD countries, but they have only 7 per cent of the proved reserves (Figure 5.6), forcing them to import three-quarters of their oil from the Middle East, North Africa and Venezuela. Developing countries

Share of global proved oil reserves

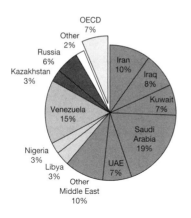

Share of global oil consumption

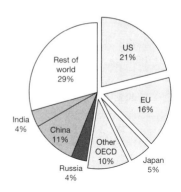

Figure 5.6 OECD countries consume over half the world's oil, but have only 7 per cent of proved reserves

Source: Based on data from BP, 2011

are also vulnerable: China already imports half of its oil, and is expected to import 84 per cent by 2035; India imports three-quarters of its oil and this could grow to 92 per cent by 2035.[40] This raises concerns over energy security, particularly in view of the current political instability in many oil-exporting countries, and the market power of the OPEC cartel which controls prices and limits the production of oil.[41] It is expected that OPEC's share of the oil market will grow from 41 per cent in 2008 to 52 per cent in 2035.[42]

Similarly for gas, over half of all proved reserves are held in just three countries: Russia, Iran and Qatar. Almost half of the world's annual gas production is consumed by OECD countries, but they have only 9 per cent of the proved reserves, and have to import half of their gas. China and India are expected to import over a third of their gas by 2035.[43] The concentration of the gas market is expected to decrease as more gas is supplied as LNG rather than via pipelines, thus reducing the tie-in of customers to particular suppliers.[44] However, there are concerns that the Gas Exporting Countries Forum (GECF), based in Qatar, could seek to use its market power in the same way as OPEC.[45]

The market concentration for coal is similar to that of oil, and is increasing. Three of the four largest producers – China, India and the United States – use almost their entire coal production themselves, so that the rest of the world relies on Australia, Indonesia, South Africa, Colombia and Russia to supply 80 per cent of their coal imports.[46] China became a net importer of coal in 2009, and a significant increase in China's imports could well lead to a global supply crunch.

This dependence on a small number of major producers makes fuel-importing countries vulnerable to supply disruptions and price fluctuations, with the classic example being the 1970s oil shocks (Box 5.2).

Box 5.2 The 1970s oil shocks

In 1973, Arab members of OPEC cut production by 5 per cent and imposed an embargo on oil exports to the United States and other 'hostile' countries, in retaliation for US support for Israel during the Yom Kippur War.[47] They also increased the price of oil by 70 per cent – a move that was influenced by a number of factors as well as the war:

• Oil revenues were declining as the value of the dollar fell following the US withdrawal from the gold standard.
• Costs of importing wheat, petrochemicals and other commodities from the West were rocketing due to inflation as Western countries printed more money to cushion themselves against the impacts of withdrawing from the gold standard.
• Negotiations with Western oil companies to give oil-producing nations a greater share of profits had failed.

Following the embargo, the price of oil immediately rose four-fold, from $3 to $12 per barrel (Figure 5.7). There were shortages and long queues at fuel stations, bans on Sunday driving in some countries, temporary rationing of gasoline sales and a national speed limit of 55 miles per hour in the United States. Although the embargo ended in March 1974, prices remained high and this is generally considered to have been a major factor in triggering the 1974 recession in Western nations that cost OECD countries 7 per cent of GDP.[48]

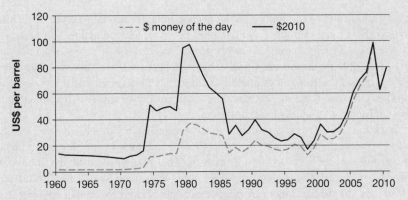

Figure 5.7 Annual average crude oil prices, showing the 1970s oil shocks

Source: Data from BP, 2011, based on Brent crude

Note: These annual averages obscure the high prices of $147 per barrel reached during July 2008. Prices for Brent crude have averaged over $100 per barrel during 2011.

This was followed by a second oil shock in 1979, when supply of oil from Iran temporarily stopped in the political chaos after the Iranian Revolution, again leading to a major recession in Western countries in the early 1980s, with high levels of unemployment and a loss of 3–4 per cent of GDP.[49]

Market concentration leads to growing reliance on a number of crucial 'choke points' on transit routes for oil and gas, which are vulnerable to disruption by political action, terrorism, accidents, freak weather or piracy. Pipelines, such as those carrying oil and gas from Russia to Europe, are particularly vulnerable, but tankers carrying oil or liquefied natural gas can also be at risk. A large part of the world's oil supply passes on supertankers through the Strait of Hormuz, at the southern end of the Persian Gulf, and Iran has threatened to block this vital route if it is threatened by sanctions or military action to force it to halt its nuclear programme.[50] Similarly, about a quarter of all oil carried by sea passes through the Malacca Strait between the Malay Peninsula and Sumatra, en route from the Middle East to China, Japan and other Asian countries. Tankers on this route have been attacked by pirates, and concerns have been raised about possible terrorist attacks, making it necessary to increase the number of security patrols. Piracy off the coast of Somalia is also a problem, and oil tankers have been hijacked. Other choke points include the Bosphorus and the Suez Canal.[51] The Suez Crisis of 1956–1957, for example, cut oil supply by 2 million barrels a day – about 13 per cent of the global total at the time.[52]

Diminishing returns: the EROI problem

Many countries are responding to the decline in conventional reserves by turning to unconventional sources. There are huge reserves, with up to 3 trillion barrels of tar sands oil, 2 trillion barrels of heavy oil and 5 trillion barrels from oil shale waiting to be found. For comparison, 1 trillion barrels of conventional oil has already been extracted and there are an estimated 3 trillion barrels of remaining reserves.[53] The problem, however, is that it is far more costly and environmentally damaging to extract fuel from unconventional sources.

Back in the 1930s, oil production in the United States largely involved sinking a few wells in the deserts of Texas and then sitting back as the oil spurted out of the ground under high pressure. It took the energy equivalent of just one barrel of oil to produce 100 barrels more – the Energy Return on Investment (EROI) was 100:1. By the 1970s, however, production had shifted to smaller, deeper, less productive fields, some of which were offshore. It took more drilling to produce each barrel of oil, and the EROI fell to around 30:1. By the 1990s, many oil fields were past the stage where oil would flow out under its own pressure. It became necessary to pump out the oil, or to use 'enhanced oil recovery' in which fluids are pumped into the rock formations to flush out the remaining oil, reducing the EROI to less than 20:1 for the United States in 2000. Globally, the average EROI for oil production has declined from 35:1 in 1999 to 18:1 in 2006,[54] and currently ranges from 40:1 for Saudi crude to 10:1 for deep-water oil. Including the energy needed for refining and distribution brings the global average down to around 10:1.[55]

With natural gas, the situation is similar – the best reserves are found and exploited first, and the size of the 'pools' of gas steadily falls, meaning that more wells have to be drilled to produce each cubic metre of gas. Further energy is used during compression of natural gas for storage, and distribution to customers. In Canada, it is estimated that the EROI of natural gas production fell from a peak of around 25:1 in 1997 to 10:1 in 2005.[56] For coal, the problem is less acute, but still the global EROI is estimated to have fallen from around 100:1 in 1950 to 80:1 in 2000.[57]

For unconventional sources, the picture is far more serious. Oil from tar sands is estimated to have an EROI of under 4:1; oil from oil shale has an EROI of between 1:1 and 2:1; and converting coal to oil has an EROI of under 3:1.[58] Shale gas might prove to be an exception however – there are little data available at present, but one preliminary estimate suggests that the EROI could be as high as 70–100:1.[59]

The type of fuel used in the extraction process is all-important. Ironically, while shale gas extraction relies mainly on oil (diesel for drills and pumps), tar sands extraction uses gas for steam production. It takes around 1,000 cubic feet of natural gas to extract one barrel of bitumen from tar sands, and another 400 cubic feet to convert it into synthetic crude oil.[60] Fredrik Robelius has estimated that the whole of the proven natural gas reserves of Canada would only be sufficient to convert 29 per cent of the estimated tar sand reserves into crude oil.[61] It is possible to use the tar sands oil itself as the fuel, but as one barrel of oil is needed for each two produced, this dramatically increases the land-use and water-pollution impacts of each barrel produced as well as the emissions of carbon and other air pollutants. However, projects are being developed to gasify the residue that is left over when the tar sands are upgraded to crude oil, using the synthetic gas as a source of steam, electricity and hydrogen for the processing operations.[62] This saves using natural gas, but the capital costs and carbon emissions are high. In theory it would be possible to capture carbon dioxide from the gasification process, to be used for fertiliser manufacture or 'enhanced oil recovery', although this is not commercially viable at present.

In summary, although there are vast reserves of unconventional fossil fuels, extracting and processing them is highly energy-intensive. As reserves of conventional fossil fuels are progressively depleted, and prices rise, extracting these unconventional alternatives will become more expensive and will accelerate the depletion of our remaining high-quality conventional reserves. Alternative approaches that use the unconventional sources themselves to fuel the extraction process can be developed, but this will further increase carbon emissions and air pollution.

Energy crunch: growing demand; inadequate investment

Our demand for energy in all its forms has been growing by an average of 2 per cent per year since 1980. The IEA predicts that, under 'current policies', demand will continue to grow by 1.6 per cent per year, meaning that we will need 51 per cent more energy by 2035 than in 2009. Even under the fairly optimistic assumptions of the IEA's New Policies Scenario, which assume that all future energy-saving policies that had been announced by mid-2011 are fully implemented, demand continues to grow by 1.3 per cent per year and is 40 per cent larger in 2035.[63] Over 90 per cent of this growth in demand is in the rapidly growing economies of non-OECD countries, especially China, although by 2035 these countries will still consume far less energy *per capita* than OECD countries.

Although the New Policies Scenario assumes a significant growth in renewable and nuclear energy, we will still use more coal, oil and gas than in 2009 (see Table 5.1, p. 140). Oil demand grows from 87 million barrels per day (mbd) to 99 mbd in 2035, driven by a doubling of the world's vehicle fleet, again mostly in non-OECD countries.[64] Most of this extra oil production is expected to come from OPEC countries, with non-OPEC countries relying on unconventional sources to offset their declining reserves of conventional oil. The share of unconventional oil will grow from 3 per cent in 2009 to 10 per cent by 2035. It is worth noting that these projections of oil demand have been criticised as being too low: some analysts believe oil demand could be as high as 120 mbd in 2030 and 180 mbd by 2050.[65]

Similarly, demand for gas is expected to increase strongly in all countries, partly because it is seen as a cleaner and lower-carbon fuel than coal and oil. Indeed, the IEA has investigated a 'golden age of gas' scenario, in which demand for gas rises by 44 per cent by 2035, with 35 per cent of this increase coming from unconventional sources (chiefly shale gas).

There is growing concern that supply may not be able to meet this rapidly growing demand. The almost total reliance of the transport sector on oil, and the lack of ready alternatives, mean that oil demand is 'inelastic' – price rises do not significantly reduce demand. This is exacerbated as incomes rise, making rises in the fuel price less significant to consumers. Meanwhile, supply is also inelastic – price rises lead to only a modest increase in production, in the short term at least. This is due to a number of factors that limit oil and gas production (on top of the geological limits discussed above):[66]

- long lead times to develop new production infrastructure (rigs, pipelines, refineries);
- lack of incentive for OPEC countries to expand production, as this could lead to a fall in prices – some commentators believe that OPEC no longer has much 'spare' production capacity in any case;
- mismatch between the increasing demand for lighter oil products and the increasing share of heavy and 'sour' crude oil, as high-quality reserves are depleted and reliance on tar sands and heavy oil increases;
- lack of refinery capacity to deal with the increasing share of heavy oil;

- rising cost of equipment, materials and skilled labour in the exploration and production sector;
- political instability in key areas such as the Middle East and North Africa;
- 'resource nationalism'; i.e. the increasing tendency for oil- and gas-producing countries such as Russia to nationalise the oil industry and restrict access to foreign companies;
- price volatility discouraging investment – the sudden crash in the price of oil after the 2008 financial crisis resulted in cancellation of a significant number of oil and gas projects.

These constraints mean that it will be challenging to increase oil production by 14 million barrels a day (to match the IEA's New Policies Scenario). Production by OPEC countries will have to increase at a faster rate than it did between 1980 and 2008.[67] An analysis of all the currently planned oil and gas 'megaprojects' by Chris Skrebowski predicted a supply crunch by 2015, partly because price volatility (especially the dramatic crash in the oil price during the 2008 financial crisis and subsequent recession) could mean that investors will be looking for a safety margin, perhaps requiring oil prices of above $120 per barrel, to justify the risk of investing in high-cost sources such as deep-water drilling and tar sands.[68] Similarly, an analysis for the World Bank warns that a shortfall of 10 million barrels a day is possible by 2015, as economic growth resumes after the recession, which could push oil prices back up to pre-recession levels of almost $150 per barrel.[69]

Relevant climate policies

Climate policies can play a vital role in improving energy security, by reducing our reliance on fossil fuels. There are four main strategies:

- cutting energy consumption;
- renewable energy;
- nuclear power;
- fuel switching to a lower-carbon and more secure source; e.g. oil to gas, in cases where gas supplies are more readily available.

These policies have already been introduced in Chapter 3, as they all have the additional benefit of reducing air pollution. This section will consider how they can contribute to energy security.

Cutting energy consumption

Any action that saves energy will help to reduce our dependence on fossil fuels, and thus improve energy security. But clearly the situation varies by country. For many countries, the main threat comes from a reliance on imported oil, so it is important to reduce demand in the transport sector, which uses 62 per cent of global oil supplies.[70] There is a strong synergy with climate policy as the transport sector produces 23 per cent of energy-related carbon emissions, and these emissions are predicted to double by 2050.[71]

But supply shortages for gas and coal may also threaten many countries within the next few decades – especially if we start to rely on these fuels to substitute for oil in transport; for example, by switching to electric vehicles while using mainly gas and coal for power generation. So in the long term, action to cut our use of all three fossil fuels will have benefits for energy security as well as for climate change and air pollution. This can be achieved both by improving energy efficiency and by changing behaviour.

Energy efficiency

The potential to cut energy waste is huge. In the US economy, for example, just 13 per cent of the primary energy input is converted to useful work.[72] Even the most efficient countries, including Japan, Austria and the UK, only achieve overall efficiencies of 20 per cent; and in many countries, including Russia, China and India, efficiency is far lower. The UN report *Towards a Green Economy* predicts that energy use in industry could be halved over the next 40 years compared with business-as-usual, and demand for oil-based transport fuel could be reduced by 80 per cent.[73]

Recent work at the University of Cambridge concludes that 95 per cent of global energy use is currently wasted, and that savings of around 73 per cent could be possible.[74] This would require a massive effort to retrofit all buildings up to the Passivhaus standard,[75] using technologies such as triple glazing, super-insulation, air-tight sealing with mechanical ventilation and heat recovery, passive cooling and heat pumps. In the transport sector, fuel use could be cut by 90 per cent by slashing weight – cars would be as little as 300 kilograms, compared to an average of 1,300 kilograms today – and reshaping vehicles to reduce drag.[76] Crucially, though, the study found that we cannot reach climate change targets by energy efficiency improvements alone – we also need to cut material consumption (see Chapter 6), and to change behaviour (see the next section).

As well as cutting fuel bills, energy efficiency can often improve comfort; for example, by insulating homes to make them warmer (see Chapter 7). Energy efficiency is therefore usually seen as a win-win strategy in the fight against climate change. Yet despite the potential benefits, little progress is being achieved. Energy efficiency improved by around 1 per cent per year over the last few decades, but although there were big improvements after the oil price shocks of the 1970s, progress has slowed since then. World energy efficiency actually fell by 0.7 per cent in 2010, and carbon intensity increased by 0.7 per cent.[77] This is partly due to the shift of manufacturing industry to emerging economies such as China, which rely heavily on coal power and often use less efficient technologies. Even in mature economies such as the EU, energy efficiency improvements are slowing down. The EU is not on track to meet its target of a 20 per cent improvement in energy efficiency by 2020, prompting the energy commissioner to consider imposing legal obligations on member states.[78] This is partly because technical efficiency limits are being reached for some industrial processes, and partly because two decades of relatively low energy prices during the 1980s and 1990s led to underinvestment in energy efficiency and established a culture of wasteful consumer behaviour.

It is worth noting that improvements to energy efficiency do not necessarily lead to reductions in overall energy use, due to the 'rebound effect', where the cost savings from efficiency are spent on more goods and services, leading to greater economic growth and more consumption. This is discussed in more detail in Chapter 7.

Energy-saving behaviour

Behaviour change is cheap, simple, has instant results and can have a greater impact on energy consumption than technical improvements.[79] Examples include:

BUILDINGS

- turning off unused lights and appliances;
- turning down the thermostat or air conditioning;
- washing clothes and dishes at lower temperatures;
- minimising appliance use; e.g. air-drying clothes instead of tumble-drying; choosing smaller refrigerators and freezers; not overfilling the kettle.

TRANSPORT

- shifting travel from cars to public transport, walking and cycling;
- shifting travel from air to rail;
- car-sharing and lift-sharing schemes;
- reducing the need to travel through tele-working and videoconferencing;
- better urban design to minimise trip distances between home, school, work and shops;
- voluntary reductions in travel, such as choosing holidays and leisure activities closer to home;
- shifting freight from road and air to rail and water transport;
- buying more local produce to minimise 'food miles';
- optimising freight deliveries; e.g. by trying to eliminate empty return journeys.

Behaviour change is particularly important for energy security because it can have a major impact in the transport sector. Many employers now have green travel plans to try to cut emissions from business travel and commuting. Cutting unnecessary air travel is especially important, as the potential for technical efficiency improvements to aircraft is limited. The Worldwide Fund for Nature (WWF)'s One in Five Challenge, for example, aims to cut business flights by 20 per cent within 5 years. Twelve large employers have signed up, using audio, video and web conferencing to reduce the need for face-to-face meetings.

It might be thought that these simple actions are a far easier way of saving energy than deploying a raft of new energy-efficient technologies or clean energy sources. Yet changing behaviour is notoriously difficult. People in rich countries have become accustomed to abundant supplies of cheap energy, and wasteful habits are hard to break. While buying more efficient appliances ties in with our in-built desire to acquire ever more material goods, energy conservation can be associated with reduced levels of comfort or freedom, and those who advocate it are often branded as killjoys intent on plunging society back into a cave-dwelling dark age. Strategies for overcoming these cultural barriers are explored in Chapter 8.

Renewables

Renewable energy, by definition, is a permanent energy source that will never become depleted. There is enough wind and solar power available on the earth to provide for the world's current and future energy needs many times over.[80] In addition, renewable energy does not tend to suffer from price volatility – although there may be a small effect related to the use of fossil fuels to produce equipment such as solar panels, and for the production of biofuels.

For many countries, renewable energy can reduce dependence on imported fossil fuels. Even if renewable energy is imported from a neighbouring country (e.g as biofuel or via an electricity grid), it can enhance energy security by increasing the diversity of the energy mix. Renewable energy resources tend to be more widely distributed than fossil fuel reserves – most countries can produce wind, solar, wave, tidal, hydropower, geothermal or biomass energy, although some are better suited to particular forms of renewable energy than others. Coastal and upland areas in most regions of the world have high wind speeds, for example; tropical regions have favourable climates for growing biofuels; and volcanically active zones such as Iceland, New Zealand and Indonesia are endowed with geothermal heat.

Renewable sources supplied 20 per cent of the world's electricity and 16 per cent of total final energy in 2009. However, a large share of this – 10 per cent of final energy – is traditional biomass burnt for cooking and heating in developing countries, which is often gathered unsustainably, leading to deforestation. A further 3.4 per cent is from large-scale hydroelectric power. Other renewables – wind, tidal, solar, geothermal and modern biofuels – account for just 2.8 per cent of final energy, but their contribution is growing rapidly.[81]

In 2010, 96 countries had targets to increase their use of renewable energy. The EU is aiming for a 20 per cent share of final energy by 2020; China is aiming for 15 per cent and Germany for 18 per cent, increasing to 60 per cent by 2050. Denmark has just announced a target for a 100 per cent renewable energy system by 2050, with 50 per cent coming from wind power, and Scotland is aiming for 100 per cent renewable electricity as early as 2020. Many countries already have high renewable shares of final energy, such as Iceland (83 per cent), Sweden (50 per cent), Brazil (47 per cent), Norway (46 per cent), New Zealand (37 per cent), Finland (30 per cent) and Austria (29 per cent).[82] Much of this is large-scale hydroelectric power and (for Iceland) geothermal energy, but there is a growing contribution from wind and biomass. There is particular interest in the role that renewables can play in delivering energy to off-grid areas, especially in developing countries (see 'Energy access for all' below).

The vulnerability of the transport sector to oil price rises has driven the uptake of biofuels in the United States, Brazil and other countries (see Chapter 4). Biofuels are unique in that they can substitute directly for liquid transport fuels with little vehicle modification, and they can also replace petroleum as a feedstock for making plastics and other petrochemicals, as well as providing a firm supply of solid, liquid or gaseous fuel to complement intermittent renewables. However, the high requirement for energy, land, water and fertilisers may limit the extent to which biofuels can reduce oil dependence. In the longer term, an alternative option would be the use of renewable electricity to power electric vehicles or to generate hydrogen fuel.

Renewable sources are far less concentrated than fossil fuels, which effectively store millions of years' worth of solar energy, accumulated in plant and animal matter, in a highly condensed form. Despite this, a number of recent (2009–2011) reports have claimed that we could supply almost 100 per cent of our future energy from renewable sources (see Box 5.3).

Box 5.3 Visions of the future

WWF: *The Energy Report*

In this scenario, renewable sources provide 95 per cent of global energy by 2050, cutting greenhouse gas emissions by 80 per cent. Energy efficiency improvements manage to cut demand by 15 per cent from 2005, despite rising industrial output and transport demand that would have doubled energy demand under business-as-usual. Electricity use increases from 20 per cent to 50 per cent of final energy. A small amount of coal is needed for steel making, because the chemical properties are important to reduce the metal ore.

Key technologies include:

- solar, geothermal and heat pumps to provide heat for buildings and industry
- wind, solar, biomass and hydropower for electricity
- smart grids to deal with the intermittent supply of wind and solar power
- more efficient vehicles; more use of public transport, walking and cycling; air and road freight shifted to rail and sea; urban planning to reduce trip distances
- electric cars and trains; fuel cells or hybrid sails (e.g. SkySails) for shipping
- biofuels for aeroplanes, ships, trucks and high-temperature industrial processes

- all new buildings meet Passivhaus standards; all existing buildings are retrofitted with insulation, heat recovery, solar thermal and heat pumps by 2030
- more recycling and better product design in industry
- more efficient biomass and solar stoves in developing countries.

To free up land to produce more biofuels, OECD countries halve their consumption of meat; other countries can increase meat consumption by 25 per cent. Annual capital investments grow from €1 trillion to €3.5 trillion within 25 years, but break-even occurs in 2035, and by 2050 there are net annual savings of €4 trillion (2 per cent of GDP) from reduced fuel use. Break-even would be earlier if the costs of climate change and co-benefits for health were included, or if oil prices rise faster than predicted.[83]

IPCC *Special Report on Renewable Energy*

The Intergovernmental Panel on Climate Change reviewed 164 different scenarios by various research groups, and selected four for detailed analysis, including an 'Advanced Energy Revolution' scenario in which renewable sources provide 77 per cent of primary energy, 95 per cent of electricity and 91 per cent of heat in 2050, with no nuclear power or carbon capture and storage, keeping greenhouse gas concentrations to 450ppm. Electricity is provided largely by wind (24 per cent), concentrated solar power (20 per cent) and photovoltaics (15 per cent). Transport is powered mainly by biofuels (14 per cent) and electric vehicles (57 per cent). Energy efficiency brings energy demand down to half of the IEA's baseline prediction. The cost of this scenario rises from $5 trillion in the first decade to $7 trillion in the second, but averages only 1 per cent of GDP per year.[84]

Jacobson and Delucchi

In this scenario, wind, water and solar power provide all of the world's energy by 2050, using an extra 0.5 per cent of the world's land surface. All heating and cooking is electric; vehicles are electric or use hydrogen fuel cells; and hydrogen is used for air transport and high-temperature industrial applications. Electricity is provided 51 per cent by wind, using 3.8 million large 5MW turbines; 40 per cent by solar, using 90,000 large solar installations and 1.7 billion small rooftop PV systems; and 9 per cent from wave, tidal, geothermal and hydroelectric power. Investment will be around $100 trillion over 20 years, but this will be recovered through sale of energy.[85]

Despite the potential of renewable energy to improve energy security, there is public concern over problems such as the landscape impacts of wind turbines, the safety of hydroelectric dams, biofuel impacts, and the fear that intermittent wind, wave, solar and tidal power cannot 'keep the lights on'. These issues will be discussed in the section on 'Conflicts' below.

Nuclear

Nuclear power can improve energy security relative to gas, and possibly coal, for countries with a high dependence on imported fossil fuels. In the longer term, it could also reduce dependence on oil, as part of a switch to electric vehicles. However, it depends on a ready supply of uranium fuel, which may become more expensive in future years as high-quality reserves become depleted. Other factors may also limit deployment: Chapter 3 dealt with the issue of pollution from waste disposal and accidents, and this chapter will discuss the threat of terrorist attacks or nuclear proliferation, the high water requirement and the high cost of construction.

Fuel switching

In theory, shifting towards lower-carbon, more secure fuels could address both climate change and energy security. In practice, however, the opportunities are limited. The highest-carbon fuel – coal – is generally more secure than both oil and gas, so switching from coal to gas is unlikely to improve energy security. Oil and coal have roughly similar carbon emissions, so switching between the two would have little impact on climate.

That leaves the prospect of switching from oil to gas. This has already happened to a large extent in the power sector, following the 1970s oil shocks, and relatively few oil-fired power stations remain. Most oil is used in the transport sector, where it would, in theory, be possible to switch to gas-powered vehicles. Compressed natural gas is already used to power buses in a number of cities, largely as a way of tackling air pollution (see 'Cleaner vehicles' in Chapter 3). However, this requires a considerable investment in infrastructure, both for converting or manufacturing gas-powered vehicles and for setting up refuelling stations. It might be considered unwise to promote a mass roll-out of gas vehicles to private owners in view of the possible limits on gas supply that could arise within the next two decades. It is possible that a move to electric vehicles where electricity is generated by gas-fired power stations might achieve benefits for both climate and energy security, for countries with abundant gas supplies but limited oil. But electric vehicles would have little climate benefit if the electricity was generated largely from coal-fired power stations.

Finally, it is no longer even certain that switching from coal or oil to gas would achieve a significant reduction in greenhouse gas emissions. Recent work has cast doubt on long-standing estimates of emission factors, claiming that methane emissions from gas production and distribution are higher than previously thought, and that life-cycle greenhouse gas emissions for natural gas are not much lower than for coal and oil, even when the higher efficiency of electricity production from gas is taken into account.[86]

Co-benefits

A strategy based on cutting energy demand and shifting to low-carbon energy can reduce our dependence on fossil fuels, tackling air pollution, energy security and climate change together. By exploiting these synergies, we can deliver energy that is secure, affordable, safe and sustainable, as well as improving access to energy for the billions who live off-grid in developing countries. The following sections look at these benefits in more detail.

Secure energy

Fuel and electricity shortages are rare events in rich nations today, but they are part of daily life for many people in less developed countries. The impacts include:

- blackouts (sudden total loss of electricity);
- brownouts (reduced electrical current causing lights to dim and computers to crash);
- controlled restrictions on gas or electricity supply within certain time slots;
- long queues or rationing to obtain transport fuel;
- steep price increases.

After the oil shocks of the 1970s, many countries took action to reduce the risk of future supply disruptions. Oil-fired power stations were replaced with gas or coal, and there was a shift from oil to gas for home heating. There was also a surge of investment in energy efficiency and renewable energy, as well as intensive exploration for oil in non-OPEC countries. The International Energy Agency (IEA) was formed to co-ordinate a response to future emergencies, and to maintain stockpiles of oil equivalent to 90 days' use. For a while, these measures reduced the reliance of Western nations on OPEC oil supplies. Subsequent supply disruptions such as the Gulf War had less of an effect on the economy – partly because they were shorter and less severe, and because countries had built up emergency stocks of oil, but also because of reduced oil-dependence.

Today, however, many governments are becoming concerned about their increasing reliance on imports. A study in 2011 found that six of the G7 countries (the UK, United States, France, Germany, Italy and Japan) face a high risk of short-term energy supply disruptions, and other major economies including China will face increasing risks in the future.[87]

Potential threats include political and commercial disputes with suppliers; political or financial instability in producing regions (especially in the Middle East and North Africa); war; terrorism; and extreme weather events. Hurricane Katrina, for example, hit oil rigs and refineries in the Gulf of Mexico in 2005, causing the loss of 1.5 million barrels of oil a day from world markets and prompting the IEA to release some of its emergency oil stocks for the first time since 1991.[88] Extreme weather events such as this are likely to become more frequent if we fail to control climate change.

There are two particular problems with energy security today (2012) – the first being a new-found dependence on imported gas, especially in Europe. The 'dash for gas' in the power sector in the UK and other European countries during the 1980s and 1990s appeared to improve energy security at the time,[89] but now that North Sea gas production has peaked, Europe will become ever more dependent on imported gas from Russia, with imports predicted to increase from 61 per cent of supply to 86 per cent by 2035.[90] The fragility of this arrangement was exposed when a dispute between Russia and Ukraine led to the gas pipeline from Russia being cut off for 2 weeks during a spell of very cold weather in January 2009, affecting parts of Europe. This highlights the inflexibility of pipeline supply, which ties customers to particular suppliers. Other countries are also vulnerable, with China expected to import 42 per cent of its gas and India more than 35 per cent by 2035.[91] Europe will have to compete with these growing economies for a share of Russia's gas, and there are concerns that Russia's gas infrastructure may not be up to the task of increasing production. New pipelines will be needed to supply an additional 100 billion cubic metres of gas per year to Europe by 2020, and depletion of the largest reserves will force expansion into small, remote fields in Siberia, where freeze–thaw cycles mean that infrastructure has an unfortunate tendency to sink into the ground.[92]

The second major issue is the reliance of the transport sector on oil, which supplies 93 per cent of transport energy (the rest being from electricity and biofuels).[93] Although vehicle fuel efficiency has improved, the world economy has become ever more dependent on road, sea and air freight transport as a result of globalisation. Many goods now rely on a complex supply network of materials and components sourced from all over the world. Coupled with just-in-time delivery, the whole retail and manufacturing sector is now acutely vulnerable to disruption from transport fuel shortages.

Cutting fossil fuel dependence has obvious benefits for fuel-importing countries, but in the long run, all countries can benefit. In a globalised economy, supply disruptions or price spikes can trigger recessions that affect many countries. Even fuel exporters can be adversely affected by sudden instability in the market.

Fuel scarcity can also trigger conflict – a number of wars and political disputes have already been linked to control of oil and gas resources, including Iraq's invasion of Kuwait in 1990 and, arguably, the US-led incursion into Iraq in 2003 on the pretext of hunting for 'weapons of mass destruction'. Conflict, apart from being disastrous for those caught up in the violence, imposes large economic costs on the participating countries, wreaks environmental damage, and can escalate terrorist activities by groups that identify with the countries that have been invaded. In the long term, nobody wins.

It is self-evident that saving energy will reduce vulnerability to supply disruptions. Countries that can run their economies on smaller amounts of fossil fuel are more likely to be able to meet their energy needs from emergency stocks or from alternative suppliers than those with extravagant fuel needs. At the global level, by limiting demand, we will be able to prolong supplies of conventional fossil fuels long enough to develop sustainable alternatives, thus, it is hoped, avoiding a major supply crunch. However, when it comes to moving to alternative energy sources, the picture is more complicated. There are few situations where a switch to gas can benefit both climate and energy security, with the possible exception of countries with large shale gas reserves, though the climate benefits of shale gas are debatable (see 'Conflicts' below). That leaves the options of renewable energy and nuclear power. Both have the potential to improve energy security and cut carbon emissions, but they have very different characteristics.

Nuclear power relies on adequate supplies of uranium fuel, which generally must be imported, and it also has specific safety and security issues related to the use of radioactive material, which will be discussed under 'Conflicts' below. Renewable energy, in contrast, tends to be indigenous and most countries are able to exploit a variety of renewable sources, although some materials such as rare metals may need to be imported to manufacture the necessary equipment.

Another key difference is that nuclear power stations, together with fossil fuel infrastructure such as coal and gas power stations, oil rigs and refineries, are large, centralised units, which are more vulnerable to disruption by accidents, extreme weather, war or terrorism than small, numerous, widely dispersed renewable sources. It is difficult to imagine a terrorist plotting to bring a country to its knees by attacking a wind farm. The main exception is large hydroelectric schemes, where dams can be at risk from earthquakes or from deliberate sabotage.

Renewable sources that are situated close to end-users, such as rooftop solar panels, can help to increase the robustness of the electricity distribution system and reduce electricity transmission losses and costs. Hydropower plays a particularly valuable role in strengthening the electricity system, as it can provide electricity round the clock but it can also be stopped and started at short notice, making it ideal for meeting demand peaks. Pumped storage schemes are an especially useful way of balancing the grid (see 'Variable renewables' below). The main concern over the role of renewable energy with respect to energy security, however, is the intermittent nature of wind, solar and marine energy. The section on 'Conflicts' will discuss this in detail.

A study for the OECD showed that not all climate policies are equally good for energy security. The study modelled different ways of reducing carbon emissions by 5 per cent for five countries: the Czech Republic, France, Italy, the Netherlands and the United Kingdom.[94] Two indicators of energy security were used – one reflecting energy price volatility (see next section on 'Affordable energy'), and one reflecting the risk of physical energy supply interruptions, which was taken to be equivalent to the share of energy supplied by natural gas pipeline imports, as this was thought to pose the main risk. Strategies based on increasing the efficiency of electricity use, or switching to renewables or nuclear energy for electricity generation, were estimated to improve physical security of supply by between 2 per cent and 38 per cent, depending on the individual circumstances of the country. But switching to biofuels for transport would *increase* physical supply risks by up to 45 per cent, because of the extra use of imported gas for biofuel production. Switching from coal to gas would increase the risk of supply interruptions by between 4 per cent and 87 per cent. This demonstrates the importance of developing integrated policies to deal with climate change and energy security together.

The OECD study looked at modest carbon savings of just 5 per cent. We can get a better indication of the potential scale of energy security co-benefits by looking at the IEA's 450ppm Scenario for 2035, in which global carbon emissions are cut by 29 per cent from 2010 to meet a 450ppm target.[95] Figure 5.8 (see plate section) shows how this is achieved: half of the carbon savings come from improved energy efficiency; a quarter come from switching to renewable energy (including biofuels); and the rest are from nuclear energy and carbon capture and storage. Primary energy demand is 19 per cent lower than under the Current Policies Scenario, and renewable and nuclear energy double their share of primary energy from 19 per cent today to 38 per cent in 2035.[96] The key point is that demand for fossil fuels is greatly reduced. By 2035, fossil fuels provide only 62 per cent of world primary energy, compared with 81 per cent under 'current policies' (Figure 5.9, see plate section, and Table 5.1). Demand for coal is halved, and demand for gas and oil is cut by 25 per cent. This is likely to significantly reduce the danger of supply shocks and price spikes, as well as reducing the need to exploit costly unconventional oil (Figure 5.10, see plate section).

Even more impressive savings are predicted under the IEA's longer-term BLUE Map Scenario, which envisages halving carbon dioxide emissions from energy between 2005 and 2050.[97] Under this scenario, renewable energy provides 40 per cent of world primary energy by 2050. Total primary energy demand is 28 per cent lower in 2050 than under the baseline scenario, demand for oil and gas is halved, and demand for coal is just a quarter of the baseline projection. The reduced need for fossil fuels and the increased diversity of the energy mix will have significant benefits for energy security. More ambitious scenarios such as the '100 per cent renewables' cases presented in Box 5.3 could clearly reduce fossil fuel dependence to almost zero.

Affordable energy

Climate policy can help to make energy more affordable in a number of ways:

- cheaper energy services;
- lower energy prices, especially in the long term;
- improved price stability;
- improved balance of payments for importing countries.

Cheaper energy services

Energy-efficient technologies can deliver energy services to households or industry at a lower cost. For example, it will take less energy to heat and cool a house that

Table 5.1 Predicted energy demand and fuel prices under the three scenarios of the IEA *World Energy Outlook 2011*

	2009	2035			450ppm Scenario compared to	
		Current Policies	New Policies	450ppm Scenario	Current Policies	New Policies
Primary energy demand, mtoe	12,132	*18,302*	*16,961*	*14,870*	−19%	−12%
Compared to 2009		+51%	+40%	+23%		
Oil demand, mtoe	2,539	*4,206*	*3,928*	*3,208*	−24%	−18%
Compared to 2009		+66%	+55%	+26%		
Gas demand, mtoe	3,987	*4,992*	*4,645*	*3,671*	−26%	−21%
Compared to 2009		+25%	+17%	−8%		
Coal demand, mtoe	3,294	*5,419*	*4,101*	*2,316*	−57%	−44%
Compared to 2009		+65%	+24%	−30%		
Nuclear, mtoe	703	*1,054*	*1,212*	*1,664*	+58%	+37%
Compared to 2009		+50%	+72%	+137%		
Renewables, mtoe	1,609	*2,630*	*3,076*	*4,010*	+52%	+30%
Compared to 2009		+63%	+91%	+149%		
Energy mix						
Share of nuclear	6%	6%	7%	11%		
Share of renewables	13%	14%	18%	27%		
Share of fossil	81%	80%	75%	62%		
Import prices						
IEA crude oil, $/barrel	78.0	140.0	120.0	97.0	−31%	−19%
Gas – US, $/Mbtu	4.0	9.0	8.6	7.8	−13%	−9%
Gas – Europe, $/Mbtu	8.0	13.0	12.1	9.4	−28%	−22%
Gas – Japan, $/Mbtu	11.0	15.2	14.3	12.1	−20%	−15%
OECD steam coal, $/tonne	99.0	118.0	110.0	68.0	−43%	−38%
CO_2 emissions	30	43	36	22	−50%	−41%

Source: Data from IEA, 2011a

Notes: mtoe = million tonnes of oil equivalent. Prices in 2010 US dollars.

Current Policies: continuation of existing policies (as of 2011). New Policies: future policies announced by mid-2011 are fully implemented.

is well insulated, and less electricity to light a building that uses low-energy lighting. The financial savings must be offset against the cost of installing the technology, but many energy-efficient technologies have very low payback periods. In the UK, for example, energy-efficient lighting can save a typical household £55 per year, and loft and wall insulation have payback periods of 1–3 years.[98] The potential for cost savings is discussed further in Chapter 7.

Long-term energy prices

We have seen that long-term fossil fuel energy prices are likely to increase, as high-quality reserves diminish, market power becomes more concentrated, and we turn to high-cost unconventional sources. After two decades in which the oil price hovered at around $20 per barrel, the IEA now predicts oil prices of $120 per barrel by 2035 under the New Policies Scenario, or $140 under its Current Policies Scenario.[99] Gas prices are also expected to increase, partly because they generally shadow oil prices, but also because conventional gas reserves are becoming depleted and production costs are rising. In the United States, a glut of shale gas is driving prices down dramatically (Figure 5.11, see plate section), although the IEA expects this trend to reverse within a few years, with US gas prices predicted to double from 2010 levels by 2035 (Table 5.1).

Apart from the obvious impact on household energy and travel costs, energy prices also have an important knock-on effect on the prices of food and other commodities, because of the energy used in production, processing and distribution; and the direct use of oil and gas to make plastics, synthetic fabrics, pharmaceutical products and synthetic fertilisers.[100] The impact on food prices is particularly severe: it takes more than 2 calories of fossil fuel energy to produce a calorie of plant-based food, and 25 calories to produce a calorie of meat.[101]

Climate policy can help to reduce long-term energy prices in two ways:

1   By restraining overall energy demand, we can prolong our reserves of conventional fossil fuels, thus reducing the need to switch to higher-cost alternatives such as tar sands.
2   Developing renewable energy sources will provide a more affordable energy system in the long run.

This last point will perhaps be challenged – many people are accustomed to thinking of renewable energy as being more expensive than fossil fuels, requiring costly government support schemes. Many renewables, however, have been competitive with fossil fuels for some time – large hydropower projects, for example, and onshore wind in favourable locations – and the costs of others are falling rapidly due to technical advances and economies of scale. In remote off-grid areas, renewable sources can easily be the most cost-effective means of supplying energy. However, the high capital costs of many renewables can be a barrier to investment, and this will be discussed in Chapter 7.

A number of studies suggest that climate policy can help to keep energy costs down. The UK energy regulator, Ofgem, foresees increases of up to 25 per cent in UK energy bills by 2020 due to rising fuel prices and the need to replace ageing infrastructure. However, costs would be lower for 'green' scenarios in which the share of renewable sources in electricity generation increases to 30 per cent (compared to 15 per cent for other scenarios), aided by government support such as feed-in tariffs.[102]

At the global level, the IEA predicts that fossil fuel import prices would be significantly lower under the 450ppm Scenario: oil prices in 2035 would be $97

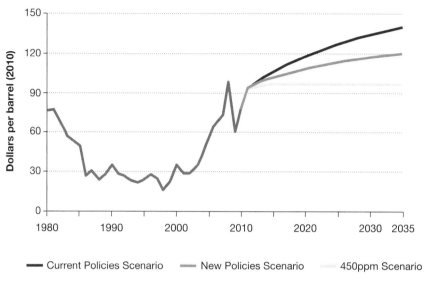

Figure 5.12 Average crude oil import prices for the three scenarios of the IEA *World Energy Outlook 2011*

Source: IEA, 2011a. © OECD/IEA, 2011

Note: Prices are in real 2010 dollars. $120 in real terms (price of oil in central New Policies Scenario in 2035) = $210 in nominal (money of the day) terms.

per barrel, compared to $120 per barrel under the New Policies Scenario by 2035, because lower oil demand means that there is less need to develop high-cost oil reserves (Figure 5.12 and Table 5.1). Similarly, gas prices would be 28 per cent lower in Europe, 20 per cent lower in Japan and 13 per cent lower in the United States, and OECD coal prices would be 43 per cent lower, as a result of a large shift to lower-carbon fuels. In the longer term, the IEA's BLUE Map Scenario predicts that the real oil price could fall to $70 per barrel by 2050, while gas prices would be around a third lower than in the baseline scenario and coal prices around half of the baseline price.[103] Note, however, that prices to consumers would be higher than this as a result of carbon pricing and other government taxes.[104]

Price stability

As well as being on a long-term upwards trend, fossil fuel prices have been highly volatile in recent years (Figure 5.11, see plate section). For oil, this is partly because both demand and supply are inelastic, as described earlier – they do not respond quickly to price changes. This is because there is no short-term alternative to oil in the transport sector, and it takes many years to develop extra production capacity, so even a small reduction in supply can trigger large price rises. The four-fold increase in oil prices in 1973 was triggered by a total global supply reduction of just 9 per cent, and the increase in 1979 (by a factor of 2.5) stemmed from a 4 per cent supply cut.[105] It has been estimated that short-term price rises in response to a supply shock are about ten times larger than they would be if supply and demand were elastic.[106] The problem is expected to worsen in future as global oil production peaks, market power shifts towards less politically stable countries and demand continues to grow.

Gas prices have traditionally been linked to oil prices in supply contracts, and so they are subject to the same instability.[107] This linkage also means that gas prices are unable to adjust in response to changes in gas demand and supply, which may make future gas supply shortfalls more likely. In future, more gas supply contracts may be negotiated separately to oil prices (this is already common in the United States and the UK), but price volatility may well continue as we get closer to peak conventional gas. Having said that, a glut of shale gas production in the United States is currently driving gas prices very low, as discussed above.

Although market prices will naturally vary to some extent in response to changes in supply or demand, sudden price spikes can cause economic chaos, hardship and social unrest. Because the elasticity of demand is low, an oil price rise means that people will spend more on oil and less on other goods and services, affecting other sectors of the economy.[108] The impacts of the 1970s oil price shocks, both of which triggered major recessions with double-digit inflation, have already been mentioned. Even though many countries have reduced oil-dependence to some extent since the 1970s, the sharp rise in oil prices to an all-time high of $147 per barrel in July 2008 is believed to have contributed to the 2008 financial crash and subsequent global recession.[109] In fact, it has been observed that recessions happen whenever world oil spending rises above 4 per cent of global GDP for a sustained period.[110]

The International Monetary Fund (IMF) has investigated the impact of oil scarcity on price volatility and the economy. It predicts that if oil supply grows at a lower rate than before – 0.8 per cent per year instead of 1.8 per cent (which was the average over the period from 1980 to 2005) – then the rapidly rising demand for oil could result in an immediate price spike of 60 per cent, followed by a long-term increase in which prices triple over 20 years, leading to a loss of 3–5 per cent of GDP in oil-importing countries. The impacts could be lessened if alternatives to oil are developed, such as electric vehicles, but could be far worse if supply starts to *decrease* as peak oil is reached – this could lead to a price spike of 200 per cent and a subsequent eight-fold increase in prices over 20 years, with the loss of 10 per cent of GDP for oil-importing nations.[111]

The traditional view of many economists is that rising prices are merely a market signal that will trigger a hunt for alternative energy sources, which will eventually bring prices down again. The problem with this argument, however, is that prices may not start to rise until resources are almost depleted, whereupon they can suddenly increase sharply. This happens because our geological knowledge of a region increases as exploration progresses, helping us to find oil and gas more cheaply. Together with ongoing advances in exploration and production techniques, this helps to keep production costs low, so there may be little warning when depletion finally occurs.[112]

This was illustrated in the United States, where oil extraction costs remained steady or decreased from 1936 right up until the peak of production in 1970, but then quadrupled in the next decade.[113] It is possible that we are currently experiencing this problem on a global scale – despite the significant oil price rise from $30 per barrel in 2003 to over $100 per barrel today (2012), there has been very little increase in oil supply. Not only can a steep price rise cause economic damage, but also this means that there is not enough time to develop alternative energy sources – making a supply shortfall likely, which will reinforce the price rise.

Climate policy can reduce our exposure to fuel price volatility, but the effects depend on the strategy we choose. The OECD modelled how different strategies for cutting carbon emissions by 5 per cent affected the exposure of five EU countries to energy price fluctuations, using an indicator based on market concentration combined with the political stability of the main fossil fuel suppliers. Increasing the efficiency of electricity use, or switching to renewables or nuclear energy for electricity generation, was predicted to improve price stability by between 2 per cent and 4 per cent; but achieving the same carbon savings by improving transport fuel efficiency would improve price stability by between 4 per cent and 8 per cent, because it reduced exposure to highly volatile oil prices. Switching to biofuels for transport would give slightly lower price stability benefits of 3 per cent to 6 per cent, due to the extra energy needed for biofuel production. And switching from coal to gas would have a negative effect on energy security, increasing the exposure to price fluctuations by up to 4 per cent. This demonstrates the importance of an integrated climate and energy security policy.[114]

Balance of payments

Many countries face significant import bills for fossil fuels. Oil is the most traded commodity in the world, accounting for 10 per cent of world exports between 2007 and 2009.[115] India spends around 4 per cent of its GDP on fuel imports; and China, Japan and the EU around 3 per cent. Globally, the oil and gas import bill is predicted to more than double from $1.2 trillion in 2009 to $2.9 trillion in 2035.[116]

Developing countries are particularly at risk from rising oil and gas prices. Few have any indigenous oil and gas reserves – 30 of the 40 on the World Bank/IMF's list of Highly Indebted Poor Countries (HIPC) are net oil importers. In addition, their economies tend to be more energy-intensive, requiring more oil to produce each unit of GDP. Low *per capita* incomes mean that people spend a higher proportion of their income on energy. Energy accounts for more than a quarter of the total import bill for several countries, including Pakistan, India, Madagascar, Ukraine and Sierra Leone. India spends over 45 per cent of its export earnings on energy, and Kenya and Senegal spend more than half. Price fluctuations can be devastating: an oil price rise of $10 per barrel could result in a loss of almost 1.5 per cent of GDP for the world's poorest countries.[117] Policies to cut consumption of imported fossil fuel and develop indigenous renewable energy can reduce these risks, improve the balance of payments and free up money to invest in more socially and economically beneficial goods and services such as health, education or clean technology.

The IEA energy scenarios illustrate the potential benefits of climate policy. Under the 450ppm Scenario, the five major oil-importing regions all have much lower oil import bills compared to the New Policies Scenario (Figure 5.13). Total spending on oil imports by all countries is a fifth lower than in the New Policies Scenario, saving $9.1 trillion over the period from 2010 to 2035. Gas imports are lower in Europe, Japan and China, but are higher in India due to switching from coal to gas. Imports of coal are more than 50 per cent lower, and China regains self-sufficiency in coal.[118]

There is a downside, however – less income (compared to business-as-usual) for the major fossil fuel exporting countries, including the Middle East, North Africa, South Africa, Russia, Kazakhstan, Turkmenistan, Indonesia, Australia, Canada, Venezuela and Colombia. Both the volumes exported and the price per unit would be expected to be lower under an effective climate policy scenario than under business-as-usual. However, prices will still be significantly higher than during the last two decades, which will offset the reduction in volume to some extent. Under the 450ppm Scenario, OPEC's cumulative export revenues from oil and gas over the period from 2011 to 2035 are predicted to be $25 trillion, compared to $32 trillion under the New Policies Scenario, but this is still three times higher in real terms than the revenues during the preceding 25 years.[119]

In the long term, diversification into higher-value-added goods and services can be better for the economy than simply selling off raw commodities as fast as possible (see Chapter 6). Cutting the volume of fossil fuel exports could also be seen as

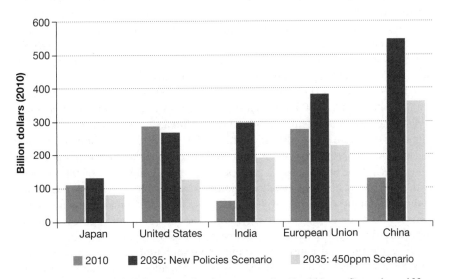

Figure 5.13 Oil import bills for selected regions, comparing the 450ppm Scenario and New Policies Scenario of the IEA *World Energy Outlook 2011*

Source: IEA, 2011a, p. 228. © OECD/IEA, 2011

deferral of income: prolonging the life of the reserves, and storing up a source of revenue for future generations. In addition, many of these countries are well suited to development of their own renewable resources, especially solar heat and power. Several countries of the Middle East now have targets for increasing their use of renewable energy. It is possible that countries in the Middle East and North Africa could eventually export renewable electricity from concentrated solar power plants to Europe.

Safe, clean, sustainable energy

As well as offering a more secure and affordable energy supply in the long term, climate policy can lead us towards safer, cleaner and more environmentally sound energy choices. By cutting our use of fossil fuels, we can not only avoid air pollution, as we saw in Chapter 3, but also reduce the risks of oil spills, coal mining accidents and ecosystem damage from fossil fuel extraction.

Coal mining is an inherently risky and polluting activity. There is the danger of rock falls; build-up of explosive and toxic gases such as methane and hydrogen sulphide; and high levels of dust that causes respiratory diseases such as pneumoconiosis and silicosis as well as posing a major explosion hazard. In addition, mines are often below the water table, so water must be constantly pumped out to prevent flooding. If pumping stops when the mine is closed, the mine fills up with water which then becomes toxic and acidic as heavy metals and minerals such as iron pyrite dissolve out of the rocks. If mine effluent, and rainwater that has percolated through spoil heaps, is not collected and treated, it can pollute local water courses, causing extensive damage to ecosystems. Open-cast mining destroys ecosystems and landscapes, especially through highly destructive techniques such as mountain-top removal, and often leads to soil erosion and blocking of streams with sediment. Deep mining may cause subsidence, damaging buildings and roads, as well as deforestation and landscape damage as land is taken for infrastructure, access roads, spoil heaps and effluent treatment ponds.

There have been over 100,000 coal mining deaths in the United States since 1900, and at least 250,000 in China since 1949.[120] Modern safety techniques can reduce the rate of accidents considerably: in China, the fatality rate per million tonnes of coal mined has fallen from 22 in 1949 to 2 in 2006, although there are still around 2,400 reported deaths per year. For comparison, the rate in the United States is 0.02 for underground and open-cast mining together, or 0.6 for deep anthracite mines.[121] Long-term health impacts can also be reduced, through dust-control techniques.[122] However, the death rates from accidents are still far higher for coal than for renewable energy, with the exception of large hydropower installations in non-OECD countries (see Figure 5.14).

Oil extraction can cause devastating environmental damage through catastrophic accidents involving tankers or oil rigs. The most recent example was the explosion and fire on BP's Deepwater Horizon drilling rig in April 2010, which killed 11 workers and caused extensive oil pollution in the Gulf of Mexico. Around 1,000 miles of shoreline were affected, including salt marshes, beaches, mudflats

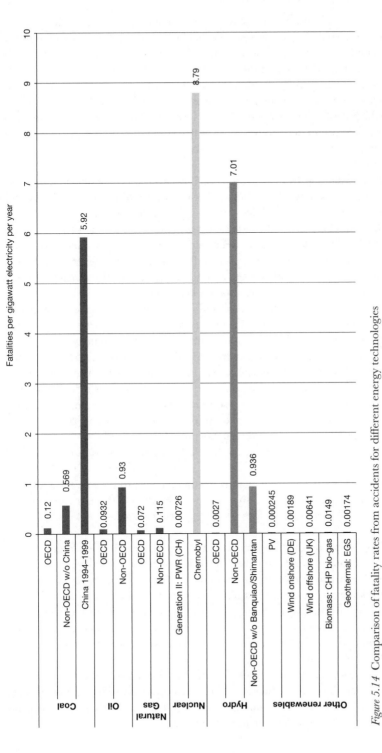

Figure 5.14 Comparison of fatality rates from accidents for different energy technologies

Source: Based on data in IPCC, 2011, Figure 9.15. Data based on the Energy-Related Severe Accident Database (ENSAD) for 1970–2008 for fossil and hydropower; Probabilistic Safety Assessment (PSA) for nuclear power; and a combination of available data, literature surveys and expert judgement for other renewable sources

Note: CH = Switzerland; DE = Germany; UK = United Kingdom; PWR = pressurised water reactor.

and mangroves. Thousands of birds were killed, along with sea turtles and sea mammals, and local livelihoods were affected by damage to fisheries and tourism. Other examples include the capsizing of the Alexander L. Kielland rig in the North Sea in 1980, killing 123 people, and the fire on the Piper Alpha rig in 1988 that killed 167. As shallow fields become depleted, oil will be increasingly produced from deeper water and more extreme environments (such as the Arctic), which will increase the potential for accidents. Of the 14 giant fields discovered between 2000 and 2006, eight are in deep water.[123] As well as accidents, there is concern about the potential for deliberate sabotage: oil tankers could be used as 'floating bombs', for example.[124]

Yet ongoing damage from oil production may be far more significant than these single incidents. Around 200 million gallons of oil leaked from the Deepwater Horizon well, yet it is estimated that over 30 *billion* gallons of toxic waste and crude oil have been discharged into Ecuador's Amazon basin as a result of oil production since 1972, leading to loss of fish in rivers and increases in cancer, skin rashes and other illnesses amongst local people.[125] Another example is the Niger Delta, where gas flaring, which causes acid rain and air pollution, and oil spills have led to severe health problems and ecological devastation in the region.

The shift towards unconventional oil and gas is likely to increase environmental impacts. Open-cast mining of tar sands is particularly destructive, involving stripping the forest and soil cover, causing massive landscape damage and carbon emissions. Large amounts of water are needed to flush the oil out of the sand, producing used water that is heavily contaminated with oil and toxic chemicals, as well as 2 tonnes of oily waste sand for every barrel of oil. This is stored in huge toxic tailings ponds, and local water sources have become polluted. The whole process has been memorably summed up by Matthew Simmons (2007, p. 5): 'It takes vast quantities of scarce and valuable potable water and natural gas to turn unusable oil into heavy low quality oil. In a sense, this exercise is like turning gold into lead.'

Shale gas has higher impacts than conventional gas production, for a number of reasons. These problems can also apply to some extent to the extraction of tight gas and coal-bed methane, in cases where fracturing is used.

- *Drilling.* As gas production from each well declines rapidly, exploiting a shale gas field requires new wells to be drilled almost continuously.[126] Up to six wells are drilled per square kilometre – around ten times more than for conventional gas production – and these require access roads, parking, gas processing and transport infrastructure, drilling fluid storage and treatment facilities and so on, leading to significant land use, landscape damage, wildlife impacts, noise and vibrations.[127] In some cases, mini-earthquakes have been triggered by the drilling operations.
- *Water demand.* Large quantities of water are needed to fracture the shale – around ten times more than is used for conventional drilling. This can cause problems in the increasing number of areas that face water shortages, especially those with a large water demand for agriculture.

- *Water pollution*. Large quantities of drilling fluid, contaminated with drilling chemicals and sometimes by toxic or radioactive elements that have dissolved out of the rock, have to be collected, treated and disposed of safely. There is a significant risk of water pollution if some of the drilling fluids escape into surface water or enter groundwater drinking water supplies – which can happen due to spills, leaks, well blow-outs or escape of fluids from the shale directly into adjacent rock formations. In some cases, as much as 70–90 per cent of the drilling fluids are not recovered.[128] There has been widespread concern over the reluctance of companies to reveal details of which chemicals are being used in fracking fluids, but they are known to include toxic and carcinogenic substances including lead, ethylene glycol, diesel, formaldehyde and benzene.[129] Groundwater supplies in the United States have been found to be contaminated with methane from fracking, creating an explosion hazard,[130] and there have even been instances of local residents being able to set fire to the water coming out of their taps.[131]
- *Air pollution*. Local air pollution from drilling equipment, heavy machinery and gas processing has caused health problems and even animal deaths in Texas, with high levels of hydrocarbon vapour and benzene escaping from gas processing and storage facilities.[132]
- *Methane emissions*. Gas leaks to the air as it is forced to the surface by the injected fluid, or escapes through fractures. Recent work (2011) suggests that more methane is released during shale gas production than with conventional gas, putting shale gas on a par with coal and oil in terms of life-cycle greenhouse gas emissions.[133] Other studies claim that the life-cycle emissions are only a few per cent higher than conventional gas provided that best practice is followed.[134] However, these studies only account for methane recovered through the well, and ignore the potential for methane to reach the surface by escaping into overlying rocks or groundwater.

All these impacts can be reduced by better regulations and improved technology – in the case of shale gas, the IEA has just published a major report on the subject[135] – but they cannot be eliminated. Cutting our use of fossil fuels will be far more effective in reducing impacts – especially as it will reduce the need to exploit more damaging sources such as tar sands. Nevertheless, low-carbon energy sources have their own impacts. Nuclear power leads to groundwater pollution during uranium mining and waste disposal, as well as to occasional serious accidents during operation (see Chapter 3). Renewable energy tends to be inherently safer than fossil fuels and nuclear power, because it does not rely on energy-dense fuels, but for this very reason it requires more infrastructure per unit of power produced, resulting in larger visual impacts. These issues will be discussed in the section on 'Conflicts' below.

Energy access for all

One-fifth of the world's population (1.3 billion people) have no access to electricity. Most of these people live in Sub-Saharan Africa and South Asia, with 84 per cent in rural areas and the rest in slums on the outskirts of large cities, where access to electricity either does not exist or is prohibitively expensive. In these areas, most people rely on candles, kerosene lamps, diesel generators or large batteries that have to be taken to central recharging centres – a time-consuming and costly process. The situation is improving, but slowly – under the IEA's business-as-usual projections, 1 billion people would still lack access in 2030.[136]

In addition, 2.7 billion people rely on traditional biomass (wood, charcoal, dung or straw) for cooking, usually on open fires or inefficient stoves. Again, progress is being made, but population growth will leave this number unchanged by 2030. Another 0.4 billion use coal stoves (mainly in China). As well as causing indoor air pollution (see Chapter 3), firewood collection is time-consuming, often dangerous, and can contribute to local deforestation.

The IEA has estimated that it would cost around $48 billion per year to provide universal access to modern energy by 2030 – just 3 per cent of projected global energy investments. There would be a massive improvement in health and welfare for an increase of just 0.8 per cent in global fossil fuel use and 0.7 per cent in carbon emissions. Renewable energy, including efficient biomass stoves, would play a major role, providing 93 per cent of energy in off-grid areas and 34 per cent in grid-connected areas.[137] Renewable energy can be cheaper than stand-alone fossil fuel alternatives such as kerosene lamps, diesel generators and liquefied petroleum gas (LPG), and can also be cheaper than extending the national electricity grid, especially if the cost of transmission and distribution to remote, rural areas is taken into account.[138] Solar energy is often abundant in developing countries and can be harnessed at a small to medium scale through solar water heating, solar stoves, photovoltaic panels and solar food and crop dryers; or at a larger scale through concentrating solar power stations (for urban supply). Biomass can be used in a number of ways: to make liquid fuels for transport or cooking; in boilers or gasifiers (e.g. as woodchips) to generate heat and power; and converted to bio-gas (for use in heat, power or transport) in anaerobic digesters. Wind and mini-hydropower can operate pumps or mills, including water pumps, and generate electricity. Renewable power from a variety of sources can feed into mini-grids to help power whole villages in off-grid areas.

Clean, affordable energy provides considerable benefits for health, education and local economies: electric light for extra hours of work or study in the evenings; refrigeration for storage of vaccines and medicines in local health centres; power for mobile phones and internet access; and heat and power for local businesses. However, market research and consultation with local people are vital – there have been problems introducing more efficient cook-stoves, for example, because they were not compatible with local needs.[139] Successful schemes must also find innovative ways to make renewable energy affordable, such as pay-as-you-go or micro-credit systems (Box 5.4).

Box 5.4 Low-carbon energy in developing countries

Pay-as-you-go solar charging systems

Eight19, a UK company, is pioneering a pay-as-you-go solar charging system for portable batteries, phones and light-emitting diode (LED) lamps in Kenya. Households pay $1 per week, less than half the cost of similar kerosene-fuelled systems.[140]

Solar power in Bangladesh

In Bangladesh, over 60 per cent of the population has no access to electricity. Grameen Shakti, a branch of the Grameen micro-credit bank, has installed over 150,000 solar home systems, 15,000 efficient cooking stoves and 3,000 bio-gas plants, replacing polluting kerosene lamps and stoves. The company also provides centres where local women and young people are trained as technicians so that they can install, repair and service the systems. Solar systems help local people to set up new businesses such as phone- and TV-charging shops and handicraft businesses.[141]

Efficient cook-stoves in Ghana

Charcoal cooking stoves are used by 31 per cent of households in Ghana. Toyola Energy Limited, founded by two West African entrepreneurs in 2006, makes stoves that are 40 per cent more efficient than the standard design, meaning that householders can save money, avoid harmful indoor air pollution and reduce deforestation at the same time as cutting carbon emissions. Toyola has sold 150,000 stoves so far, saving 150,000 tonnes of carbon dioxide per year. The company's latest strategy is to provide micro-loans to cover the cost of the stoves ($7) and allow householders to pay off the loan over 2 months as they make savings on charcoal costs.[142]

Micro-hydropower in Brazil

A co-operative electricity supply company in Brazil, CRELUZ, found that supply to households at the far reaches of the grid was becoming unreliable as demand for electricity increased. They built six run-of-river mini-hydropower schemes that strengthened the electricity grid, providing reliable power to 80,000 residents with minimal environmental impact.[143]

Bio-gas in China

The Programme for the Development and Promotion of Biogas Utilization in Rural China installed 30 million bio-gas systems between 2001 and 2010, with government financial support for the poorest households. Households with bio-gas digesters save an average of $78 each year in reduced firewood, electricity and fertiliser costs.[144]

Conflicts

Although there are strong synergies between the goals of energy security and climate policy, in terms of reducing reliance on fossil fuels, there are also a number of important potential conflicts.

- *Fossil fuels.* Intensive exploitation of domestic fossil fuels and unconventional sources can improve energy security but increases carbon emissions and other environmental impacts.
- *Carbon capture and storage (CCS)* reduces carbon emissions but increases fuel use, so has a detrimental impact on energy security.
- *Nuclear power* can improve energy security by cutting fossil fuel use, but there may be issues with uranium supply, terrorist activity, nuclear proliferation and high costs.
- *Renewables.* The intermittent nature of some renewable sources such as wind, wave and solar energy poses a challenge for ensuring continuous electricity supply.
- *Biofuels* can improve climate impact and energy security, but place heavy demands on land, water and energy.
- *Other problems with renewable energy* include landscape impacts.

Fossil fuels: 'Drill, baby, drill!'?

As conventional reserves of oil and gas decline, and prices rise, there will be a strong motivation to adopt US politician Sarah Palin's famous mantra of 'Drill, baby, drill!', and attempt to squeeze every last drop of fuel from domestic sources such as tar sands, deep-sea areas and Arctic wilderness. These sources tend to be costly, energy-intensive and environmentally damaging – but they will improve energy security in the short term, and they will be commercially viable if oil and gas prices continue to rise as expected.

The main problems with the 'Drill, baby, drill!' approach are:

- *Lock-in to an unsustainable technology path.* By investing in unsustainable fossil-fuel-based technologies, such as coal-fired power stations, we are wasting time, money and materials needed to develop the alternative infrastructure – renewable energy, super-efficient buildings, electric vehicles and so on – that we will need for a sustainable future.

- *Declining energy return on investment (EROI).* The EROI of most fossil energy sources is gradually declining as the best-quality reserves are used up. We have to put more and more energy in to get each unit of energy out.
- *Carbon limits.* It has been estimated that for a 75 per cent chance of restricting global warming to 2°C, total greenhouse gas emissions from 2000 to 2050 must be less than 1 trillion tonnes of carbon dioxide. This means that we can burn less than half of the remaining proved reserves of fossil fuels.[145] Exploitation of lower-quality fossil fuels will rapidly push us closer to this limit. Tar sands, for example, increase emissions from transport by between 14 and 39 per cent on a well-to-wheels basis.[146]
- *Environmental limits.* Exploitation of remote and unconventional fossil fuels increases environmental impacts and accident risks (see above).
- *Conflict.* This approach fosters a high reliance on fossil fuels, and so tends to go hand in hand with strategic foreign policy interventions in oil-producing countries – leading, in some cases, to war.

Is shale gas the answer?

Shale gas is currently at the centre of a heated debate. It has increased energy security and dramatically reduced gas prices in the United States, but also caused environmental damage. An IEA report published in 2012 sets out *Golden Rules for a Golden Age of Gas*, specifying environmental safeguards which are essential if fracking is to be accepted by the public. One of the main issues is that methane emissions from the fracking process could be high enough to cancel out the carbon advantage of gas over coal for power generation.[147]

However, even if these methane emissions can be reduced, a wider concern is that switching to gas will not keep us within safe climate targets. The 'golden age for gas' scenario, which envisages the share of gas in primary energy rising from 21 per cent to 25 per cent by 2035 (with a third of gas coming from unconventional sources), cuts energy-related carbon dioxide emissions by 1.3 per cent compared to a 'low unconventional gas' scenario, but still leads to a greenhouse gas concentration of 650ppm – well above the safe limit of 350ppm and the policy target of 450ppm. This is partly because the increased availability of gas, and the resulting decrease in gas prices, displaces renewable energy and nuclear power and increases the demand for electricity.[148]

What about CCS?

The other widely touted 'golden bullet' to enable our continued use of fossil fuels is carbon capture and storage (CCS). The problem is that fuel consumption increases by around 25 per cent to provide energy to separate out the carbon dioxide and pump it into underground storage (see Chapter 3). This has a negative impact on energy security as well as causing additional environmental impacts from mining, processing and burning the extra fuel. Other problems with CCS were discussed in Chapter 3.

Nuclear power

Chapter 3 discussed the risk of pollution from radioactive material released during normal operation, accidents or waste disposal. This chapter focuses on safety, security and affordability: the availability of uranium, the impacts of uranium mining, the potential for proliferation or terrorist activity, and problems with cost and time overruns during construction of nuclear power stations.

Peak Uranium?

Nuclear power was once seen as an almost limitless source of energy, but uranium is a finite resource that faces the same depletion issues as fossil fuels. There is plenty of uranium left underground – proved reserves could fuel current reactor require-ments for 62 years.[149] But the best reserves are becoming depleted, and rising demand could lead to a significant risk of short-term supply shortages. A major price spike occurred between 2007 and 2008 (Figure 5.15), due partly to technical difficulties which restricted the supply from certain mines. Longer-term price rises

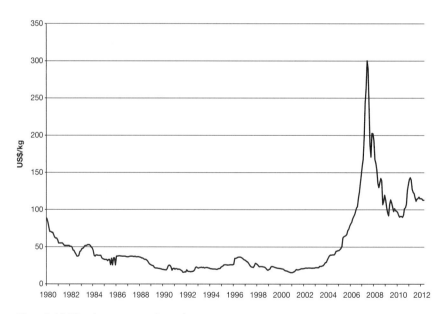

Figure 5.15 Uranium spot market price

Source: Data from IMF, 2012

Note: Most uranium is traded under long-term contracts that are relatively unaffected by short-term fluctuations in the spot market price.

are forecast due to declining ore grades and increasing production costs. Challenges for uranium supply include:

- *Market concentration.* Uranium has been found in 44 countries, but economically recoverable reserves have been almost exhausted in 11 of those.[150] At present, 18 countries produce uranium commercially, and eight of these produce 93 per cent of the world's uranium. Australia, Kazakhstan and Russia hold half of the world's reserves.[151]
- *Declining ore grades and higher production costs.* Although high uranium prices since 2006 have driven increased investment in exploration activities, Figure 5.16 shows that the new reserves being discovered tend to have higher production costs. This is partly due to declining ore grades. Canada is the only country with ore grades above 1 per cent uranium – some deposits have grades as high as 23 per cent. Ore grades elsewhere are below 0.1 per cent, and two-thirds of all deposits have grades under 0.06 per cent. Extracting uranium from low-grade deposits is energy-intensive, and one study estimates that at grades less than 0.02 per cent to 0.01 per cent, it takes more energy to extract the uranium

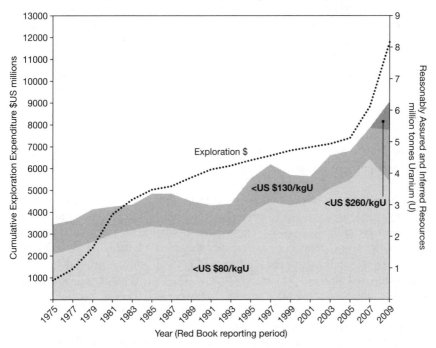

Figure 5.16 Increasing uranium prices drive more exploration – but the cost of production is increasing

Source: World Nuclear Association

than can be recovered in the power station, although this conclusion has been disputed.[152]

- *Loss of secondary sources.* A third of the annual supply of uranium for power stations comes from 'secondary' sources, including decommissioning of nuclear weapons, military stockpiles, reprocessing of spent fuel and re-enriching the 'tailings' waste of previous uranium extraction activities. Some of this supply base may disappear in 2013 when agreements between Russia and the United States for converting weapons-grade uranium into power-station fuel (the Megatons to Megawatts Program) expire. Reprocessing of spent fuel is not currently economically attractive (see below), and re-enrichment of tailings waste may also decline following the end of an agreement between Russia and Europe in 2010.[153]

- *Rising demand.* Despite some countries reconsidering their nuclear programmes following the Fukushima disaster (see Chapter 3), over 60 new nuclear power stations are under construction to add to the 433 currently operating (2012), and 160 more are planned.[154] Uranium demand is expected to double or triple by 2030. Secondary uranium sources could provide only 5–10 per cent of global reactor requirements by 2030, meaning that mining capacity has to increase from 52,000 tonnes of uranium per year to between 90,000 and 135,000 tonnes, requiring considerable new investment.[155]

- *Environmental impacts.* Uranium is extracted either by conventional underground or open-cast mining or by *in-situ* leaching. Both have considerable environmental impacts, which could increase as ore grades decline. Mining creates large quantities of toxic and radioactive waste in the form of mine spoil and tailings, and can lead to local groundwater and surface water pollution. There has been concern over the health and safety of workers, particularly in countries where the voluntary industry safeguards such as efficient mine ventilation and dust control are not rigorously applied. *In-situ* leaching, in which an acidic or alkaline solution (commonly sulphuric acid) is pumped through the deposit in order to dissolve out the uranium, can lead to groundwater contamination, threatening drinking water supplies.[156]

Despite these problems, nuclear power still offers certain benefits for energy security. First, uranium is an extremely dense fuel so that large amounts can be stored on-site at nuclear power stations, thus insulating the operator from short-term market fluctuations in price or availability. Second, the cost of uranium fuel is a relatively small part of the total costs of nuclear power – around 10–15 per cent, compared to 30–40 per cent for coal and 60–85 per cent for gas – which again limits the impact of future price rises on energy costs.[157] It is also worth noting that future reactor designs are expected to be able to extract more energy from each tonne of uranium than current designs.

Two technologies could extend the life of nuclear fuel supplies in the future: fuel reprocessing, and fast breeder reactors. Spent fuel contains unused uranium-235 as well as newly created plutonium-239, both of which can be extracted by reprocessing and incorporated into new fuel, increasing the energy supplied from the nuclear fuel cycle by 25 per cent. Reprocessing plants have been built in the UK, France, India, Pakistan, Russia, Japan and the United States, though only those in France and the UK are currently operating. The problem is that the reprocessed fuel has only half the value of natural uranium, because of contamination with radionuclides that slow down the nuclear reaction and make handling the fuel difficult. Also, reprocessing becomes less economically viable as the rates of fuel burn-up increase in more modern reactors.

Fast neutron reactors, loosely termed 'fast reactors' (because they utilise 'fast neutrons' instead of 'slow neutrons') offer the potential to virtually eliminate problems with fuel supply. They can use uranium-238, which forms 99 per cent of all natural uranium deposits and 95 per cent of spent fuel, instead of just uranium-235 which is only present at concentrations of 0.7 per cent in natural uranium; and they can also consume plutonium and other actinides present in spent fuel. This offers the tantalising possibility of using existing stockpiles of nuclear waste as fuel. Fast reactors can 'breed' plutonium in a blanket of uranium wrapped around the core – when they breed more fissile material than they consume, they are operating as 'fast breeder reactors'. Although this plutonium must be recovered from the blanket through reprocessing, which can be costly and technically challenging, fast breeder reactors could theoretically increase the energy obtained from each tonne of natural uranium by a factor of 60, as well as reducing both the volume and the radioactivity of nuclear waste. To date, however, development of fast reactors has been dogged by persistent technical problems. Sodium-cooled fast reactors have been under development since the 1950s, but without commercial success. Although prototypes were built in the UK, United States, France and Germany, these countries have now largely abandoned their fast reactor programmes as being too expensive. However, other countries are forging ahead, with experimental fast reactors operating in Russia, Japan, India and China, and larger versions under construction in India and Russia. New designs for gas-cooled fast breeder reactors are also under development, but the first prototype is not expected until 2025.

Another possibility for the future is thorium reactors, which use thorium to breed uranium-233. This addresses the supply problem to a certain extent because thorium is approximately three times as abundant as uranium – and unlike uranium, thorium does not require enrichment. A number of prototypes have been tested, and although technical problems remain to be overcome, research and development is going on in Russia, China and India (which has 25 per cent of the world's thorium reserves and is currently building a prototype thorium fast breeder reactor). The International Atomic Energy Authority (IAEA) envisages advanced prototypes being built by 2020, and commercialisation some years after that.

Finally, it might also be possible to extract uranium from coal ash (which averages 0.006 per cent uranium, but can be up to 0.03 per cent in some spoil heaps), or from naturally occurring phosphate deposits, although this would be very expensive.

Proliferation and terrorism

The advantage of nuclear power – generation of large amounts of energy from a small amount of very energy-dense fuel – is also a major drawback. Not only is nuclear energy highly destructive if used in aggression; but also, the relatively small amounts needed to make a weapon make it hard to ensure that it does not fall into the wrong hands.

The very first application of controlled nuclear fission was to breed plutonium for nuclear weapons. The Manhattan Project, set up in 1942, eventually produced the atomic bombs that were dropped on Nagasaki and Hiroshima in 1945. The civil programme for nuclear power began shortly afterwards, with the first commercial power station opening at Sellafield in the UK in 1956, but there has always been an inextricable association between nuclear electricity and weapons development, with concern that civil nuclear power programmes could be used to disguise nuclear arms proliferation by national governments. This is fuelling political tension between Iran and the West, for example. Although 189 countries have signed the Treaty on the Non-Proliferation of Nuclear Weapons to limit the spread of nuclear weapons, including five states with nuclear weapons (the United States, Russia, China, France and the UK), four states with nuclear weapons are not signatories (India, Pakistan, North Korea and Israel).[158] With 60 more countries having notified the IAEA of their interest in civil nuclear programmes, it will become harder to control the risk of arms proliferation.

There is also a risk that radioactive material could be acquired by terrorist groups or rogue states. Fresh fuel poses a comparatively small risk – it has relatively low levels of radioactivity and could not be used directly to make a nuclear bomb without sophisticated equipment to enrich the uranium-235 content from 4 per cent (power-station grade) to 90 per cent (military grade). However, spent fuel is another matter. It is highly radioactive, containing a variety of fission products, and although this in itself would make it very difficult for terrorists to handle it without personal injury, spent fuel canisters could be used directly in 'dirty bombs' that could contaminate wide areas.

The other issue is that spent fuel contains plutonium-239 that was created when the uranium fuel was bombarded with neutrons in the reactor. Plutonium is a particular problem, as it cannot be vitrified in the same way as uranium for long-term storage. The main disposal route for plutonium at present is through reprocessing to separate it from the uranium and other actinides, then incorporation into mixed oxide (MOX) fuel (containing both uranium and plutonium) which can be used in many modern reactors. This in itself creates a security problem, however, as either the separated plutonium or the MOX fuel could, in theory, be used to make nuclear warheads. In the long term, fast reactors could help to address this problem because they are capable of using either spent fuel or separated plutonium as fuel, though the reprocessing step is still necessary.

The risk of material falling into the wrong hands is compounded by the need to transport uranium ore, fuel rods, reprocessing products and high-level waste between mines, fuel enrichment plants, power stations, reprocessing plants and disposal sites. Thirty countries currently (2012) have nuclear power, but the number is growing, and many of the countries about to acquire nuclear power do not have their own uranium mines or fuel-enrichment and reprocessing plants, so that transport of nuclear material between countries is likely to increase as a result.

Security and safety are generally high in most of the rich industrialised countries, but as nuclear power programmes expand in other countries, it will become harder to ensure that watertight security standards are applied everywhere at all times. Large numbers of new reactors are being built in China, India, Russia and South Korea; there are specific plans for reactors in Bangladesh, Belarus, Egypt, Indonesia, Jordan, Kazakhstan, Lithuania, Poland, Turkey, the United Arab Emirates (UAE) and Vietnam; and longer-term proposals for nuclear power in Malaysia, Saudi Arabia and Thailand.[159] Not all of these countries are politically and economically stable, and some are prone to environmental risks such as flooding or earthquakes. Political instability can pose security risks – the break-up of the Soviet Union in 1991, for example, made it harder to keep track of the large amounts of radioactive material that had been held at various locations as part of civil and military nuclear programmes.

Water requirement

Most reactors in operation today require large volumes of water to keep the reactor at a safe temperature and to raise steam for power generation. As a result, these reactors have to be sited next to the sea, lakes or rivers. This leads to two problems. First, the proximity to water means that reactor sites can be at risk of floods and, for coastal reactors in earthquake zones, tsunamis. Second, inland reactors are at risk of disruptions to water supply during dry periods.

Although fossil fuel power stations also need cooling water, there are additional safety concerns for nuclear power stations because of the danger of core meltdown if cooling fails. As a result, nuclear power stations sometimes have to be shut down if water levels in rivers and lakes fall during spells of drought.

Hot weather can also be a problem. It increases the temperature of the water intake, meaning that the water discharged back into lakes and rivers after cooling can be too hot and can have damaging effects on ecosystems. In 2010, for example, output from the Browns Ferry reactors in Alabama had to be halved because of the need to keep the temperature of cooling water discharged into the river below 32°C.[160]

Many inland reactors use re-circulated water, cooled in a cooling tower, rather than a once-through cooling system. This cuts the water intake, although more water is lost through evaporation than with a once-through system. However, cooling towers do not work properly if air temperatures are above 35°C. These systems are also 40 per cent more expensive and 2–5 per cent less efficient than once-through cooling.[161]

Climate change is likely to increase these risks: rising sea levels and more hurricanes could increase the likelihood of flooding for coastal reactors; extreme rainfall events can increase flood risk for inland reactors sited next to rivers and lakes; and more droughts and heat waves could cause problems with water supply and cooling for inland reactors. These trends are already apparent, with a number of reactors in France, Spain and Germany closing during droughts and heat waves in 2003, 2006, 2007 and 2010, and reactors in the United States being shut down due to flooding during 2011.

Cost, construction time and accessibility

Construction of nuclear power stations is capital-intensive and requires long lead times. Cost and time overruns are relatively common – the first two Generation III European Pressurised Reactors being built in Finland and France, for example, are very late and over budget. The Finnish reactor was started in 2005 and was due to be commissioned in 2009, but has now been delayed until 2013,[162] while the budget has increased from an initial estimate of €3.2 billion to over €6 billion. The French reactor is also currently running more than 3 years behind schedule and the budget has increased from €3.3 billion to €6 billion. However, this is partly because these reactors are the first of a kind. Two further EPRs are being built in China and are expected to be cheaper and to have construction times of less than 4 years.

Although nuclear electricity is generally competitive with fossil fuel generation, it is debatable whether the price includes the full costs of decommissioning the plant and disposing safely of the radioactive waste. Taxpayer support is typically used to fund these activities, and to offset liability for any catastrophic accidents (see Chapter 3). There is concern that support for nuclear power can divert public funds away from cheaper, safer and more rapidly deployable options such as energy efficiency and renewable energy technologies.[163]

The high capital costs, complex technology and the risk of proliferation mean that nuclear power is not the best option for providing electricity to the 1.3 billion people who currently lack access, as most of them live in rural areas far from the nearest grid, often in politically unstable countries. However, increasing urbanisation will bring more people within reach of centralised power delivery in future.

Research is proceeding on the next generation of nuclear reactors, termed 'Generation IV' reactors. Six designs are being developed, including three types of fast reactor. These new reactors have the potential to consume far less fuel, produce less waste, use existing reprocessed waste as fuel and have improved safety and anti-proliferation features. There is also the possibility that modular construction using standardised components could reduce cost and construction time, and make decommissioning easier. Plans are in place to construct prototypes of three different Generation IV fast reactors by 2025.[164] In the much longer term, nuclear fusion may become a possibility. The problem with all of these technologies is that they are unlikely to be available soon enough to play a significant role in avoiding a 2°C temperature rise.

Variable renewables: what happens when the wind stops?

There is widespread concern that renewable sources cannot provide a secure and continuous supply of energy because several of the main renewable sources are intermittent or variable. This applies to wind, wave, solar, tidal and run-of-river hydroelectric power.

- Wind power varies with wind speed, which can be seasonal to some extent. Turbines cannot operate when wind speeds are too high or too low.[165] Typically, they operate for about 75 per cent of the time, and over the course of a year they will deliver about 30 per cent of their theoretical maximum capacity.
- Solar power varies on a daily cycle (day/night), on a seasonal cycle (summer/winter) and with daily fluctuations in cloud cover.
- Wave power varies with the speed and height of waves. This in turn depends on wind speeds and directions over the preceding few days. As waves can travel long distances before they reach shore, it is possible to predict fluctuations in wave energy to some extent.
- Tidal energy is intermittent, but it is also completely predictable on a regular daily, monthly and yearly cycle, with tides flowing in and out twice a day.
- Run-of-river hydroelectric power is usually continuous (unless river levels drop dramatically), but the power levels vary slightly depending on the amount of recent rainfall or snowmelt feeding the river.

The first point to make is that fossil and nuclear power stations do not provide continuous electricity. Typically, they operate for about 50–90 per cent of the time, allowing for planned maintenance and unplanned shutdowns due to technical problems.[166] Also, not all renewable sources of energy are intermittent. Sources that are relatively constant include:

- Geothermal power and heat, including heat pumps.
- Hydroelectric power that uses a reservoir, unless water levels fall during a prolonged drought. Hydroelectric power offers the advantage of flexibility – it can be ramped up or down relatively quickly to match sudden peaks or troughs in electricity demand.
- Biofuels which can be stored as solid, liquid or gaseous fuels, although supply can be seasonal or affected by droughts, disease or pests.
- Concentrating solar power (CSP) which can store heat from the sun, usually in molten salts, so that power generation can continue overnight.

Finally, variability is not necessarily a problem provided that supply and demand profiles match well. In cities in hot countries, for example, building-mounted solar photovoltaic panels can provide power that matches the demand for air conditioning, thus reducing the peak daytime load on the grid and improving system reliability. Similarly, winds can be stronger during the autumn and winter, matching the increased demand for light and heat, although there can also be long calm periods of very cold weather.

However, an electricity system that has a large share of intermittent renewable sources will clearly have to cope with much larger variability in supply than one based solely on traditional fossil and nuclear sources. This certainly presents a challenge, but there are four strategies that can help to tackle this problem:[167]

- diversification;
- interconnection;
- storage;
- smart demand and supply management.

Diversification involves using a mix of complementary renewable sources to improve security of supply. A balanced portfolio might include wind power which tends to peak in autumn, winter and spring; solar power which peaks in summer; biomass for back-up power generation; hydroelectric and geothermal power for base-load generation, with hydropower also being used to meet rapid changes in demand; a mix of geothermal heat (including heat pumps), solar hot-water panels and biomass for heating; and liquid biofuels or electric vehicles for transport.

Interconnection between different countries or regions allows the use of renewable sources from a wider geographical area, which increases the chances that the wind will be blowing somewhere within the supply region. In the UK, for example, there were no days between 1970 and 2003 when wind speeds were less than 4 metres per second (the minimum needed for most large turbines) everywhere in the country. Low wind-speed events affecting more than half of the country were present for less than 10 per cent of the time, and high wind-speed events (above 25 metres per second, when most turbines have to be switched off) were even rarer, being present for less than 4 per cent of the time and with each event on average affecting only 0.2 per cent of the UK. Similarly, a 3-year study from 2000 to 2002 found that the longest period where wind generation was less than 1 per cent of capacity was 58 hours in Denmark, 19 hours in Finland and Sweden and just 9 hours in Norway.[168] However, since these studies there has been a 10-day period in January 2009 where wind speeds were almost zero across much of northern Europe.[169]

Combining a large number of renewable sources across a region has the effect of smoothing out variations in supply (Figure 5.17, see plate section). Whereas the power output from a single wind turbine can drop to zero in just a few seconds, the output from a large number of turbines spread across a country the size of Germany would slowly ramp down over around 30 minutes, allowing the grid operator plenty of time to bring back-up capacity on line.[170]

Interconnections between different countries bring even greater flexibility, allowing power to be exported and imported depending on which regions are currently experiencing the best wind, wave or solar conditions. For Denmark, Finland, Sweden and Norway combined, for example, there were no totally calm periods between 2000 and 2002. Good connections to the German and Nordic grids have helped Denmark to achieve a 20 per cent share of electricity from wind power. Taking Europe as an example, a 'super-grid' could provide a renewable portfolio combining wind, wave and tidal energy from northern Europe with hydroelectric power from the Alps and solar power from southern Europe and North Africa. To achieve this, existing grids will need to be extended and strengthened to eliminate 'bottlenecks', and grid management systems will need to be improved. Countries will have to set up systems to trade electricity and co-ordinate the balancing of supply and demand at an international level. This is already practised successfully in the Nordic electricity market, where Denmark sells wind power to Norway and Sweden and buys hydropower in return when the wind is low.

Storage of power is also critical, to help smooth the output from variable sources. Pumped storage is the most widely used option (Figure 5.18). This requires two lakes or reservoirs, one higher than the other. When there is surplus power supply, such as overnight when demand is low, water is pumped up to the upper reservoir, and it can then be released to drive a turbine and generate extra power at peak times. More than 200 pumped storage schemes operate around the world, providing about 100 GW of capacity, with a further 60 GW of capacity being planned. The potential for new schemes is limited by the requirements for two reservoirs in suitable positions, and by environmental objections to new dams. However, capacity could be increased by upgrading existing turbines, or by using the sea as the lower reservoir. In future, the geographical constraints on location could largely be avoided by building networks of underground tunnels to act as the lower reservoir, next to existing lakes.[171]

Other forms of storage are also possible, including batteries, capacitors, flywheels, compressed air, super-conducting magnets and heat storage. These tend to store less power than pumped storage schemes, but they can provide short-term balancing power to smooth out frequency fluctuations.[172] Some of these technologies are not yet commercially viable or technically mature, but they will become more cost-effective as the share of variable renewables increases.[173] Others are mature

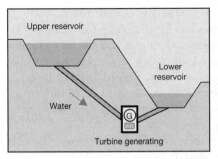 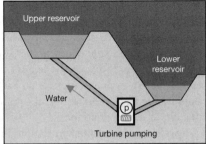

Figure 5.18 Diagram of a pumped storage scheme showing discharge during the day (left) and recharge overnight (right)

Source: Adapted from Inage, 2009. © OECD/IEA, 2009

and are already used to provide back-up capacity to data storage and telecoms facilities.

This brings us to the final part of the strategy – using smarter management systems to help match supply to demand. One obvious approach is to use better weather forecasting techniques, including computer models and satellite data, to predict the variation in renewable supply over the next few hours or days. This will allow more variable renewable sources to be used, provided that there is enough flexible generating capacity to meet the predicted variations in supply. Demand can also be managed to smooth out peaks and troughs, by offering incentives to customers to persuade them to reduce their energy demand at peak times. In the UK, for example, cheaper off-peak tariffs encourage householders to use overnight storage heating, and 'interruptible' contracts with industrial customers offer cheaper rates provided that gas or electricity supply can be withheld if necessary (with compensation).

Smart meters and continuously varying tariffs can take demand management a step further, giving householders even more information on their energy use and allowing them to adjust accordingly to use less energy at peak times. Trials in the United States showed that this could cut peak demand by between 2 per cent and 34 per cent.[174] But the real opportunity for progress lies with smart grids and meters coupled with smart household appliances, which allow automatic demand management. Examples include washing machines and dishwashers that can turn themselves on when there is a surplus of power available (and the price is lower), or refrigerators, air conditioning, heat pumps or water heaters that could briefly turn themselves off when demand exceeds supply.

Smart demand management could be combined with energy storage by using surplus electricity to charge up electric vehicles and other battery-operated devices.[175] It could also be possible for electric vehicles that are not in use to feed energy back into the grid to meet peak demand. In this way, a fleet of electric vehicles could act as a giant battery to store surplus off-peak renewable power. Alternatively, surplus power could be used to produce hydrogen by electrolysis of water, for use as a transport or industrial fuel.

One emerging issue is that some consumers feel uneasy at the apparent intrusion of privacy associated with smart meters and automatic demand management. It will be important to emphasise the opportunities for consumers to save money by using demand management to buy power at cheaper times, and also for customers to be able to manually override the system so that they retain overall control of their energy use. Privacy may be less of an issue in the industrial and commercial sectors, where opportunities for smart demand management could include air conditioning and water heating in commercial buildings, water desalination, metal smelting or ice production.

As well as daily or hourly variations in supply, short-term variations on the scale of minutes or seconds can cause the voltage or frequency of electricity in the national grid to fluctuate. This can cause problems in small grids – it currently limits the instantaneous contribution of wind power on the island of Crete to 40 per cent, for example – though it has not caused problems in the larger grids of

Denmark or Spain.[176] To combat this, electronic control technologies are now being used to enable wind and solar generators to control their output better, thus 'conditioning' the power quality. This carries an extra cost, but it also enables wind farms to help to improve the reliability of the grid by providing services such as fault ride-through (continuing to generate even when the grid voltage dips), providing reactive power (to counteract small-scale fluctuations) and damping power swings.[177]

'Smart grids' will also be critical for enabling a greater share of variable renewables while balancing supply and demand and maintaining power quality. This is a loose term encompassing a variety of advanced technologies which may include:

- control systems to co-ordinate input from different variable renewable sources;
- flexible transmission systems where power can be directed down particular lines;
- high-voltage transmission lines to reduce power losses when transmitting power over long distances (such as from remote wind farms);
- distribution management systems that allow a two-way flow of electricity so that building-mounted solar panels can feed into the grid, and smart appliances or electric vehicles can be remotely managed.

Smart grids can integrate large centralised sources such as wind farms, small distributed sources such as solar panels, as well as storage technologies, electric vehicle charging, smart meters and smart appliances into a smoothly operating electricity network.[178]

Typical electricity grids can already cope with up to a 20–30 per cent share of variable supply with no loss of reliability (Table 5.2).[179] With technical improvements and advanced grid management systems, this could increase to 60 per cent or more by 2050. In theory, therefore, it would be possible to have a 100 per cent renewable power system, if the remaining 40 per cent could be supplied from constant renewable sources such as hydropower, geothermal power, biomass and concentrating solar power with heat storage.

Table 5.2 Maximum share of variable renewable electricity without losing reliability, based on existing grid and storage infrastructure, in various regional electricity systems

Region	Maximum share of variable renewable electricity
Japan	19%
Iberian peninsula (Spain and Portugal)	27%
Mexico	29%
British Isles (Great Britain and Ireland)	31%
New Brunswick area (Canada)	37%
Western interconnection of United States	45%
Nordic market (Norway, Sweden, Finland, Denmark)	48%
Denmark	63%

Source: Based on data in IEA, 2011f

Any electricity system needs back-up and balancing capacity to meet fluctuating demand, using flexible sources such as combined-cycle gas turbines or pumped storage hydropower that can be turned on and off quickly, but more will be required if there is a greater share of intermittent renewable sources. Obviously the existence of this back-up capacity does not cancel out the carbon savings from renewable energy, because it operates only when needed, but it adds to the cost of operating the electricity grid in three ways:

- Short-term balancing services must be purchased by the grid operator more often.
- A larger amount of 'spare' back-up capacity must be retained to ensure that peak demand can always be met (this can generally be provided by retaining some existing capacity that would have been displaced by renewable generation, so new build is not needed).
- Existing fossil power plants lose efficiency because they are turned on and off more frequently in response to fluctuating renewable input.

However, these costs are relatively small – in the UK, for example, it has been estimated that even an 80 per cent share of wind power would add less than 1p/kWh to the cost of wind power – about 10 per cent. The efficiency loss in existing fossil plant was found to be around 1 per cent, but the costs of this were relatively insignificant.[180] The total costs of managing intermittent renewable sources, including back-up and balancing costs and the need for grid extensions, have been estimated as 0.7–3.0 c$/kWh[181] by the IPCC and 0.5–2.5 c$/kWh by the IEA.[182] These costs can be reduced by greater diversification, better interconnection with neighbouring grids, better prediction of wind speeds, and short 'gate closure' times (this means that generation from reserve capacity is not ordered until the last minute, to give variable sources a better chance of being utilised).

Most countries with a high share of renewable electricity rely largely on continuous sources: Norway supplies 99 per cent of its electricity from hydropower, and Iceland provides 75 per cent from hydropower and 25 per cent from geothermal energy. Finland generates 11 per cent of its power from biomass, largely from the forestry industry. But some countries also have a significant contribution from variable sources – chiefly wind. West Denmark has a 30 per cent share from wind; and on occasions, wind has provided 100 per cent of electricity. Ireland has an 11 per cent share from wind, and has achieved instantaneous levels of 50 per cent, while Spain and Portugal generate 15 per cent of their power from wind with instantaneous levels up to 54 per cent and 71 per cent respectively.[183] In Germany, solar energy reached a share of 50 per cent of electricity demand during a recent hot weekend.[184]

In summary, it is possible to supply up to around 60 per cent of electricity from intermittent renewable sources by using a diverse range of renewable technologies, grid interconnections with neighbouring regions, electricity storage mechanisms, flexible balancing capacity and appropriate supply and demand management technologies. However, this will require significant investment in electricity grids, smart supply and demand management systems and extra storage capacity. There will also be an extra operating cost associated with providing additional back-up and balancing capacity.

Finally, it is worth noting that climate change could affect the security of supply of several renewable sources. Biofuels that depend on energy crops may be adversely affected by increased droughts and the spread of pests and diseases, and hydropower could become more intermittent if prolonged droughts cause water levels to fall or if glaciers diminish so that supplies of summer melt-water dry up.[185] Water shortages could also affect the supplies of water needed for concentrating solar power stations. Wind, wave and tidal power are unlikely to be strongly affected, although more extreme weather events could mean that turbines and wave energy devices may have to be designed to withstand stronger wind speeds and greater wave heights.

Biofuels

Although biofuels have an important role to play as a renewable liquid transport fuel and a non-intermittent energy source, they face environmental limits. Chapter 4 looked at the potential impacts of biofuel cultivation on biodiversity, food security and land use. This section will consider the limitations on biofuels that can arise from their demand for water, energy and fertilisers.

Water scarcity is perhaps the greatest threat to maintaining a sustainable supply of biofuels. Cultivating biomass for power generation requires 70 to 400 times more water per unit of power generated than the cooling water requirement of coal or nuclear power stations.[186] A litre of bio-ethanol derived from irrigated corn in the United States typically requires over 500 litres of water, equating to 50 gallons of water for each mile driven.[187] However, the water demand varies considerably depending on the crop, the way it is grown and the local environment and climate. Woody crops, for example, tend to require three to seven times less water, as well as less fertiliser, herbicide and pesticide than food crops.[188] Areas with good rainfall may not need additional irrigation. And in some areas with degraded soil, biofuel crops (especially trees) can improve soil stability and water retention (see Chapter 4).

Energy input also varies considerably between different biofuels. A full 'well-to-wheel' analysis – taking into account the energy used for cultivation (including fertiliser manufacture), harvest, drying, processing, distilling and transporting biofuels – shows that at best they can be almost carbon-neutral, but sometimes they can reduce carbon emissions by as little as 13 per cent compared to petroleum fuels. For sugar cane bio-ethanol in Brazil, for example, the favourable climate and the use of bagasse (sugar cane waste) as a processing fuel means that the ratio of fossil fuel input to energy output is 1:8. Biodiesel from oilseeds has a typical energy ratio of 1:3, and corn bio-ethanol produced through intensive cultivation in the United States has an energy ratio of just 1:1.2.[189] In Chapter 4 we saw that when land-use impacts are also included in the analysis, biofuels planted on cleared forest land can have far worse climate impacts than fossil fuels.

Second-generation biofuels derived from waste use less land, energy and water than biofuels from food crops. However, they do need a secure supply of waste material as feedstock, which can divert biomass from other uses such as using manure, compost or crop residues to improve soil fertility (see Chapter 4). Biofuels from algae also avoid the problem of competition with food crops, although they do require input of nutrients and water, and a recent analysis suggests surprisingly high energy requirements (mainly for pumping water) that may exceed the energy output by a factor of between 2 and 7.[190]

There is also a conflict between energy security and climate policy for 'energy from waste' – burning municipal or industrial waste in an incinerator and recovering heat or power. This is increasingly important as a number of countries have policies to increase the amount of waste incinerated, in order to solve the problem of a shortage of landfill space. The problem is that after removal of recyclable and compostable material such as food, paper, glass and metal, a large part of the remaining waste is often plastic or synthetic textiles – both of which have a very high oil-derived carbon content. Burning this material releases a large amount of fossil carbon to the atmosphere, which may be only partly offset by the avoided combustion of fossil fuels elsewhere in the electricity generation system. Recycling and waste reduction are far better strategies both for climate protection and for conservation of resources (see Chapter 6).

Other impacts of renewables

Although most renewable energy sources offer major benefits for climate, energy security and air pollution, there are some further issues that can restrict their role in providing a safe, secure and affordable energy supply:

*   safety, environmental and social impacts of large hydropower schemes;
*   safety and environmental aspects of geothermal schemes;
*   landscape impacts of wind turbines and power pylons;
*   impacts of tidal energy schemes on estuary habitats of wading birds;
*   reliance on scarce materials such as rare metals (see Chapter 6);
*   high capital costs (see Chapter 7).

Large hydropower schemes are the main source of renewable electricity today, and have a vital role to play in providing a firm electricity supply and instant balancing capability to complement the variable supply from wind, marine and solar power. Yet the potential to increase hydropower capacity is limited by public opposition. Large areas must be flooded to create reservoirs, displacing thousands or even millions of people, destroying wildlife habitats and farmland, and inundating areas of cultural or archaeological significance. It is estimated that 40–80 million people have been displaced by dams, and many have not been provided with adequate compensation or alternative livelihoods.[191] There is also a risk of dam failures – the most devastating example being the Banqiao Reservoir disaster in China, which killed between 26,000 and 171,000 people in 1975 (the upper figure accounting for famine and epidemic after the initial disaster). Dams can also give rise to emissions of methane from rotting vegetation trapped behind the dam, impede the passage of fish or dolphins, and reduce the fertilising effect of regular flooding of farmland downstream of the dam. For these reasons, new hydropower schemes often attract major opposition from local communities and environmental groups, with the long history of campaigning against the Narmada River dams in India being a prominent example. It is worth noting that the risk of dam failures can be reduced by better engineering – safety features were scaled back in the final design of the Banqiao dam. However, it is also possible that climate change could result in greater risks, as failures are often linked to extreme weather events such as heavy rainfall.[192]

Geothermal schemes, which involving drilling into hot rocks hundreds of metres below the surface, have also been linked to safety and environmental concerns including minor earthquakes and groundwater pollution by chemicals in the drilling fluids. Two major projects in the United States and Switzerland were abandoned as a result.[193]

Wind turbines attract opposition in some areas because of their intrusive impact on the landscape – especially in remote and unspoilt areas of natural beauty – and sometimes because of problems with turbine noise. Vocal campaigns in the UK, in particular, have curtailed many onshore wind projects. Offshore wind attracts less opposition, but there have been objections to the siting of associated infrastructure including transformers and power pylons. Studies indicate that acceptance is far greater when local communities have a stake in the project, such as in Denmark, where many wind farms are community-owned and therefore receive a share of the profits. Careful planning guidelines and early engagement with the local community are also important. It is also important to site wind turbines away from major bird migration routes to minimise collisions.

It is worth remembering that the impacts of wind turbines are temporary, localised and reversible. If new energy options become available, the turbines could simply be removed and the landscape returned to its original state. The same is certainly not true of land loss to sea level rise, or the hazard posed by long-term nuclear waste disposal.

The way forward

As cheap supplies of fossil fuel diminish and the remaining high-quality reserves become concentrated in a smaller number of countries, energy security concerns are rising up the agenda. Alternative sources such as deep-water oil, tar sands and shale gas may appear to offer solutions, but also tend to be associated with increasing environmental impacts.

A strong climate policy can offer major co-benefits for energy security, reducing the need to rely on increasingly scarce and expensive fossil fuels. This is particularly relevant for countries that are reliant on imported fossil fuels and have good indigenous sources of renewable energy. Fossil fuel prices are expected to continue to rise and to become more unstable, while the costs of many renewable technologies are expected to decrease. Overall, climate policy that encourages renewable energy and energy efficiency can lead to more secure, safe, accessible and affordable energy in the long term.

However, there are a number of potential conflicts that need to be resolved in order to exploit these synergies. First, concern over energy security could lead us into the trap of rushing to exploit every last drop of fossil fuel, which will lead to increased environmental damage and lock-in to a high-carbon infrastructure. Relying on carbon capture and storage to enable us to continue using fossil fuels will make energy security problems worse, because it increases fossil fuel consumption. Over-reliance on nuclear power could also lead to problems with uranium supplies, safety, security and cost. Renewable energy offers a safer, cleaner alternative, but requires considerable investment to deal with the problem of intermittent supply, as well as problems with the sustainability of biofuels, the landscape impacts of wind turbines and the social impacts of large hydropower schemes.

Modelling studies indicate that even with realistic targets for increasing the share of nuclear power and CCS, it is impossible to reach a 400ppm greenhouse gas target without increasing the share of renewable energy well above the business-as-usual level.[194] Over-reliance on fossil fuels and nuclear power could divert crucial funding away from renewable energy. With this in mind, a policy framework that exploited the synergies and avoided the conflicts between energy security and climate change might include the following elements.

- *Avoid lock-in* to fossil fuel technology by: setting ambitious targets for reductions in fuel use; ending fossil fuel subsidies (see Chapter 7); discouraging new building of coal-fired power stations; and halting exploitation of damaging sources such as tar sands, deep-water oil and arctic oil.
- *Set fiscal policy* to encourage a shift away from fossil fuels; e.g. carbon taxes or caps (see Chapter 7). Governments may be tempted to respond to the rising cost of fossil fuels by cutting fuel taxes, but this is counterproductive – it will simply encourage lock-in to unsustainable technologies as well as depleting tax revenues that could be used to stimulate energy efficiency or alternative energy sources.
- *Cut transport fuel use* to reduce the vulnerability of the transport sector, which is 93 per cent reliant on oil – the least secure fossil fuel.[195] Policy instruments include fuel taxes; investment in public transport and rail freight; promotion of cycling, walking, videoconferencing and tele-working (see Chapter 8); better town planning to minimise urban sprawl; congestion charging; vehicle fuel efficiency standards; and promotion of hybrid, electric or hydrogen fuel-cell vehicles.
- *Enable a greater share of renewables* through fiscal support (e.g. feed-in tariffs) and investment in electricity grids, energy storage and supply and demand management technologies. Aim for a diverse portfolio of renewable technologies, widely distributed, to address the variability of supply.[196]
- *Increase research and development* for key technologies that have both climate and

energy security benefits, including renewables, second-generation biofuels and fourth-generation nuclear reactor designs.

- *Set energy market rules* to enable renewable energy and CHP to meet their full potential; for example, by giving preferred grid access to renewables and CHP,[197] by enabling interconnection and electricity trading over a wide area, and by setting gate closure times to be as short as possible.
- *Set standards* to ensure the sustainability of biofuels (see Chapter 4), the safety of large hydropower and geothermal projects and the acceptability of wind and solar farms. Address public concern over landscape impacts through good planning guidelines to prevent unsuitable developments, early community involvement in the planning process, offering community ownership and sharing the benefits.[198]
- *Improve energy access in developing countries* with technology transfer and financial help to enable households in off-grid regions to install clean local renewable energy sources such as efficient biomass stoves, solar panels, wind pumps for irrigation and bio-gas digesters.
- *Set tougher standards and targets for energy efficiency* in appliances, buildings and vehicles. In the EU, for example, member states are required to present plans for annual efficiency improvements of 1 per cent and to define minimum building efficiency standards for new build and large renovations. More countries could aspire to strict standards such as the Passivhaus zero-energy building standard and Japan's Top Runner Program which uses the current best product as a future standard for vehicles and appliances. Industry needs to know that standards will be continuously tightened, giving strong motivation to try to produce the most efficient products.

6 Less waste

A resource-efficient economy

Key messages

- *Material efficiency is just as important as energy efficiency* for meeting climate targets: over half of all greenhouse gas emissions come from making material goods such as cars, houses, food, clothing and appliances, and these emissions are growing as material consumption increases.
- *We can reduce our consumption of materials by designing out waste from the start of the production process*: creating products that are durable and upgradeable; choosing low-impact materials; cutting waste; minimising packaging; and increasing recycling, reuse and repair.
- *There is a wide range of co-benefits.* We can avoid resource-scarcity impacts, including supply constraints, price spikes and conflict; reduce the environmental and social costs of resource extraction, such as air and water pollution, deforestation, biodiversity loss and displacement of local communities; avoid waste disposal problems, such as odour, litter, water pollution, methane emissions and lack of space for landfill sites; save money and make businesses more competitive.
- *Improvements in resource efficiency may be outweighed by increased consumption.* So we also need measures to ensure that total consumption stays within sustainable limits. Ultimately this means that we need to find ways for people to live well while consuming less. There are important implications for employment, lifestyles, international trade, development and economic systems.

Materials: a hidden source of carbon emissions

Saving energy is a major tool of climate policy, but few people realise that only around 40 per cent of global greenhouse gas emissions come from direct use of energy in buildings and vehicles. The remaining 60 per cent is linked to our ever-growing consumption of material goods: construction of buildings, roads and other infrastructure; manufacture, distribution and disposal of the food, clothing,

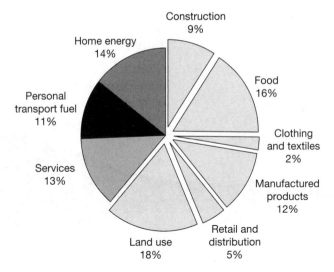

Figure 6.1 Global greenhouse gas emissions, 2001, showing around 60 per cent were from production of material goods, food and infrastructure

Source: Based on data from Hertwich and Peters, 2009; land use from EDGAR database (EC/JRC-PBL, 2011)[1]

Note: Land use differs from Figure 2.1 because these data are for a different year (2001 compared to 2008).

furniture and other products that we use; and forest clearance for production of food, paper and timber (Figure 6.1).

The high climate impact of material goods arises for a variety of reasons. For metals, extraction and processing is very energy-intensive due to the need to mine, crush and melt the ore. Producing a tonne of steel typically generates almost 2 tonnes of carbon dioxide, and a tonne of aluminium generates over 8 tonnes.[2] For concrete, in addition to the energy needed to extract limestone and sand, large amounts of carbon dioxide are released from the limestone (calcium carbonate) as it is converted to lime (calcium hydroxide). Production of glass, bricks, paper, chemicals and plastics is also very energy-intensive. The industrial sector accounts for 77 per cent of our direct use of coal, 40 per cent of total electricity use and 35 per cent of direct gas use.[3] Further greenhouse gas emissions arise from the agriculture and forestry sectors for production of food, fibre and timber.[4] Finally, emissions also arise from disposing of used goods and materials: methane from landfill sites and carbon dioxide from waste incinerators account for 2 per cent of global greenhouse gas emissions.

Trends in material use

Over the last century, annual global consumption of materials (including fossil fuels) has soared from 7 billion tonnes to 60 billion tonnes (Figure 6.2) and the area used for agriculture has doubled from 15 per cent to 30 per cent of the earth's land surface.[5] Under a business-as-usual scenario, consumption of materials would

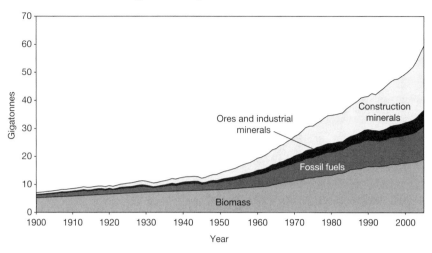

Figure 6.2 Global resource extraction is escalating

Source: Data from Institute of Social Ecology (Krausmann *et al.*, 2009)

Note: Includes resources extracted and used. Excludes unused resources (see Table 6.7, p. 212, for definition).

increase to 100 billion tonnes by 2030.[6]

This growth in our use of materials is strongly linked to our exploitation of fossil fuels – partly because large amounts of energy are needed to produce material goods, and partly because economic growth, driven by cheap fossil energy, has raised the demand for goods. In pre-industrial times, consumption was limited by reliance on renewable sources of energy – sun, wind and biomass – and consumption *per capita* was roughly constant. But during the Industrial Revolution, with the gradual introduction of coal as an industrial energy source, *per capita* consumption of both fossil fuels and other materials began to rise.

Consumption escalated with the rapid uptake of oil-based transport and electric power after the Second World War, with *per capita* consumption of materials growing on average by 1.6 per cent each year from 1945 to 1973, compared with just 0.2 per cent from 1900 to 1945 (see Figure 6.3). Yet after the oil price shocks of the 1970s, the rate fell to just 0.6 per cent. Intensive development of energy efficiency technologies allowed most industrialised countries to stabilise their *per capita* material and energy use, despite continued growth in GDP.[7]

Since 2000, however, growing demand from consumers in the developing world has pushed average material use *per capita* up by 3.7 per cent each year – more than double the post-war growth rate. In 2005, the average US citizen consumed 28 tonnes of materials and the average EU citizen used 15 tonnes, but the average Indian used just 4 tonnes. If all countries catch up to current Western levels of consumption, total material consumption will reach 140 billion tonnes by 2050.[8] Demand for steel, concrete, aluminium, paper and plastic – which together account for 56 per cent of industrial carbon emissions – is likely to more than double under business-as-usual.[9]

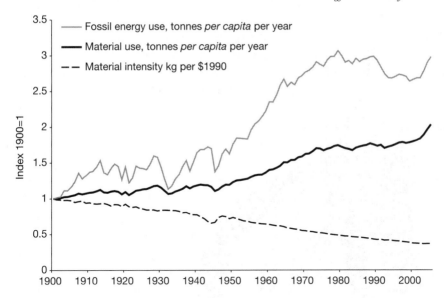

Figure 6.3 Global material and fossil fuel use *per capita* stabilised after the 1970s oil shocks but is now rising again

Source: Data from Institute of Social Ecology (Krausmann *et al.*, 2009)

Carbon omissions

The carbon impacts of the goods that Western consumers buy are partly hidden, because many of these goods are now made in other countries. Around 26 per cent of global carbon emissions from fossil fuel burning are, in effect, exported from the country of origin,[10] with the main flows being from China and other emerging markets to the West (Figure 6.4). In 2004, the main carbon importers were the United States, which imported almost 700 million tonnes (Mt) of CO_2, followed by Japan with 284 Mt and the UK with 253 Mt. The main exporter was China with 1,147 Mt, followed by Russia with 286 Mt and the Middle East with 177 Mt.[11]

National greenhouse gas inventories based on UN-FCCC guidelines ignore the emissions embodied in imported goods. If these hidden imported emissions (which have been termed 'carbon omissions')[12] were taken into account, the emissions attributed to Western European countries would increase by around 38 per cent. For the developed countries that signed the Kyoto Protocol, the inclusion of imports reveals that emissions have actually increased by 7 per cent between 1990 and 2008, compared to the fall of 2 per cent currently reported. Most of this increase has occurred since the year 2000. At the same time, China's reported emissions would fall by a fifth if exported carbon were taken into account.[13]

The material efficiency challenge

To meet climate targets, emissions from the industrial sector must be halved, despite the predicted doubling or tripling of demand for materials.[14] The main approach

Figure 6.4 The largest flows of carbon emissions embodied in traded goods, in million tonnes of CO_2 per year

Source: Davis and Caldeira, 2010; by permission of *Proceedings of the National Academy of Sciences* of the United States

to date has been to focus on improving industrial energy efficiency, but this will not be enough. Many production processes are now approaching their thermo-dynamic efficiency limits, and further improvements will be small. A 2010 study estimated that even with the most optimistic assumptions, energy efficiency could achieve only a 20–40 per cent cut in emissions for the five main industrial mate-rials.[15] And for some metals, the decreasing quality of ore reserves means that more energy will be needed in the future to extract and refine each tonne of metal.

Because of this, it is important to focus not just on energy efficiency, but also material efficiency. By providing the same goods and services but using fewer materials, we can cut the amount of energy needed to extract, process and transport the materials, assemble them into manufactured goods and eventually dispose of them. We can do this by reducing waste, increasing reuse and recycling, increasing product lifespan and using eco-design to make products that are lighter, more durable and more recyclable.

The IPCC estimated that the industry sector can save almost as much carbon through material efficiency as through energy efficiency.[16] In the UK, for example, it has been estimated that material efficiency could save up to 28 per cent of emissions from the goods and service sectors, with knock-on savings in the raw material, transport and energy sectors.[17] In Europe, one study estimated that cutting emissions from material production could save as much as 800 million tonnes of carbon dioxide – 16 per cent of total emissions from all sectors – and this ignores the potential of key strategies such as eco-design.[18] And in the EU-27, more ambitious recycling targets could save 244 million tonnes of CO_2e, which is almost a third of the target for reducing greenhouse gas emissions by 20 per cent by 2020.[19]

The scale of the challenge is daunting. Continuous improvements in material efficiency over the last few decades have been far outweighed by the impacts of pop-ulation growth and increasing affluence (Figure 6.5). This is partly due to the rebound

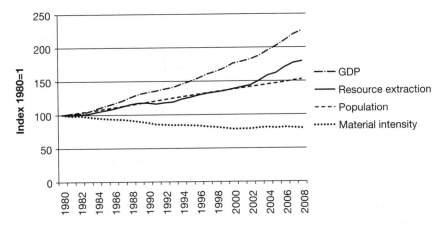

Figure 6.5 Global resource extraction has grown as rising population and affluence have outweighed improvements in material intensity

Source: SERI, 2011

effect, which will be explored in Chapter 7. Ambitious and carefully designed policies will be needed to reduce our total use of materials to a sustainable level.

The good news is that policies to reduce material use have a very wide range of co-benefits. As well as cutting carbon emissions, we can avoid much of the environmental and social damage associated with material extraction and manufacturing, including habitat destruction, air and water pollution, mining accidents and the displacement of indigenous communities. In a world of increasingly scarce resources, we can cushion our economies against supply shortages and price fluctuations and avoid resource-related conflict and instability. There are financial benefits too – lower costs for industry, savings for householders and fewer costly and unpopular landfill sites and incinerators.

This chapter will look in more detail at the techniques that can reduce material use, and illustrate the potential co-benefits. It will also examine conflicts between climate policy, material efficiency and other objectives – especially the economic and social implications of reducing material use – and evaluate policy approaches that can maximise the benefits and minimise the conflicts.

Relevant climate policies

Cutting material use can reduce climate impacts in two ways – first, by avoiding the emissions associated with making material goods; and second, by avoiding emissions from waste disposal.

A wide range of measures can be used to reduce material use. These can be broadly categorised under the waste hierarchy of 'reduce, reuse, recycle, recover', which is intended to convey the message that the preferred option is to reduce the amount of waste generated in the first place, with reuse being the next best option, followed by recycling and composting, then energy recovery (burning waste to generate power), leaving landfill as a last resort. Energy recovery and composting do not cut material use, although the benefits of compost for improving soil fertility and avoiding synthetic fertiliser production were discussed in Chapter 4. Of the other options, the emphasis in most countries to date has been firmly on recycling, with relatively little effort expended on reuse or waste prevention. We will therefore address the hierarchy in reverse order, beginning with recycling as the most well-established and familiar strategy and then moving on to more radical approaches:

- recycling – using the materials in end-of-life goods or waste to make new products;
- reuse, repair and re-manufacture – reusing goods or their components;
- resource-efficient production – making goods with less material and less waste;
- cutting consumer waste – buying less, making goods last longer and wasting less;
- dematerialising the economy – providing the same level of welfare with fewer goods.

Collectively, these techniques are referred to by a variety of terms such as 'resource efficiency', 'eco-efficiency', 'sustainable consumption' and 'zero waste'. Energy efficiency is, of course, integral to this approach, but as that has been discussed in Chapter 5, the focus here is mainly on methods of achieving material efficiency and waste reduction.

Recycling

For almost all materials, recycling uses far less energy than extracting and processing raw materials from scratch. Making a tonne of aluminium from scrap, for example, takes just 5 per cent of the energy used to make it from bauxite ore. This is partly because it takes less energy to melt aluminium scrap (with a melting point of 660°C) than to melt bauxite (which melts at 900°C), and partly because the energy needed to mine, prepare and transport bauxite ore to the processing plant is avoided.[20] Energy savings of 60–70 per cent can be achieved for steel, 70–85 per cent for copper, 60–80 per cent for lead, 60–75 per cent for zinc, 64 per cent for paper, 37 per cent for glass and 80–88 per cent for plastics.[21] Table 6.1 shows typical greenhouse gas savings for recycling one tonne of various materials, compared to landfill or to incineration with energy recovery.

In 2005, it is estimated that the EU-27 saved over 150 million tonnes of carbon dioxide emissions by recycling 37 per cent of its municipal waste.[22] However, there is considerable scope for improvement to recycling rates in many countries. In 2008, for example, 46 per cent of municipal waste was recycled in the EU-27, but rates in individual countries ranged from 70 per cent in Austria to under 1 per cent in Bulgaria.[23] Electronic waste is a particular problem, with only 15 per cent being recycled globally despite the high content of valuable metals,[24] and less than 1 per cent of the rare metals being recovered.[25] It is estimated that only 2.5 per cent of mobile phones in OECD countries are recycled in such a way as to allow recovery of critical metals such as platinum and palladium.[26]

Further savings could be achieved by increasing the quality of recycling systems as well as the quantity, and this has been identified as a priority in a review of the European Commission's waste strategy.[27] The key is to collect recyclables in separate streams, to avoid contamination.[28] For example, paper and textiles cannot be recycled effectively if they are contaminated with glass. Plastic bottles can be

Table 6.1 Greenhouse gas savings from recycling

	Compared to landfill	Compared to incineration
	tonnes CO_2e saved per tonne of material recycled	
Paper and cardboard	1.40	0.62
Glass packaging	0.60	0.43
Plastic packaging	1.00	1.00
Steel packaging	1.34	0.79
Aluminium packaging	7.00	7.00

Source: Parfitt, 2006; based on data in WRAP (2006), which is a review of 55 separate life-cycle analysis studies

Note: It is assumed that steel (but not aluminium) is recovered from the ash after incineration. The savings include avoided methane from decay of paper in landfill sites.

recycled into new food-grade bottles if they are not mixed with other plastic waste, and aluminium cans should be kept separate from aluminium foil if they are to be re-melted for new drinks containers. Food and garden waste are particularly important as they are responsible for most of the methane generated from landfill sites. If collected separately they can be composted, or food waste can be sent to anaerobic digestion plants which convert waste to bio-gas and fertilisers.

Typical separate collection schemes require householders to place dry recyclables in a box, which is then sorted manually by collection staff into different containers on the collection vehicle. Unfortunately there has been a trend in some countries (such as the UK) towards co-mingled collection, in which mixed dry recyclables are collected in large bins or bags, compacted on board the collection vehicle and then sorted at a Materials Recovery Facility (MRF). Although this can reduce the cost of collection, it results in low-value recovered materials and a high degree of waste – typically 12–15 per cent of the material collected is wasted at English MRFs,[29] compared to less than 1 per cent for separate collection. Many paper mills will not accept paper from MRFs, and glass from MRFs (which is crushed and not sorted by colour) tends to be used for low-grade applications such as road aggregate, with little environmental benefit.

Reuse and re-manufacture

Conventional recycling offers major benefits compared to the use of virgin materials, but it is actually 'downcycling' – it loses much of the value and embodied energy of the original product. Reuse and re-manufacture, in contrast, preserve this value. It is generally far more efficient to reuse a glass bottle or a steel girder, for example, than to melt them down to make new ones. Similarly, it is far more efficient to upgrade an old computer, keeping most of the components intact, than to shred it and recover only low-value plastic and metal fragments.

Reuse includes such obvious examples as returning bottles to be refilled, reusing shopping bags, reusing pallets, refilling laser toner cartridges, reusing scrap paper and donating or buying goods through the charity and second-hand sector. A less obvious example is the use of modular carpet tiles, pioneered by Interface, which allows worn or damaged tiles to be replaced without replacing the whole carpet. The importance of reuse is emphasised by the finding that it will not be possible to reduce carbon emissions from industry to sustainable levels unless half of the steel and aluminium that is currently recycled is diverted to reuse; for example, by

Table 6.2 Reuse generally saves more emissions than recycling

	Television	Cotton t-shirt	Sofa	Office chair
	tonnes CO_2e per tonne item (negative values are savings)			
Reuse	−8.19	−12.8	−1.45	−2.96
Recycling	−0.07	−0.84	0.00	−0.91
Landfill	0.05	0.20	0.75	0.10

Source: WRAP, 2011b

finding ways to demolish buildings without damaging steel girders.[30] Table 6.2 shows that reuse generally saves far more emissions than recycling.

Re-manufacturing is a type of reuse which transforms old products to new ones of equal or better quality, so that they can be resold complete with warranty. This involves recovering used products, taking them apart, testing the components and then replacing, repairing or refurbishing any worn or damaged components before reassembly and testing. Re-manufacturing works well with products which are high-value, complex and durable, but are not considered to be 'status' items: examples include electric pumps and motors; car, truck and aircraft engines; machine tools; computer equipment; photocopiers and vending machines. It works best in cases where manufacturers sell direct to customers, as this makes collection of old products easier. Firms involved in re-manufacturing include Caterpillar, Sony and Xerox (see Box 6.1).

Box 6.1 Re-manufacturing case studies

Xerox

Xerox, which makes photocopiers and printers, began to collect used equipment for re-manufacturing in the early 1990s. Equipment is designed for easy disassembly and high durability, with standardised and coded parts and with strictly controlled use of chemicals. Over 80 per cent of the parts can be reused. After disassembly and re-manufacture, there is a rigorous testing programme to ensure that the new product meets the same quality standards as one made from scratch, so that the re-manufacturing process is invisible to the customer. As well as the re-manufacturing programme, Xerox has a Waste-Free Factory initiative and 93 per cent of non-hazardous materials are recycled.[31]

Caterpillar

Caterpillar makes earth-moving equipment and is one of the world's largest re-manufacturing companies, with a turnover of over $1 billion. Around 70 per cent of a Caterpillar machine can be reused directly, and 16 per cent recycled, giving an overall material recovery rate of 86 per cent. Re-manufactured machines offer cost savings to customers. In 2008, its re-manufacturing business continued to grow despite the recession, whereas sales of new equipment declined.

Sony

Sony has always used a re-manufacturing system for repairs to its PS2 Playstation, and now offers a 'by return' repair or upgrade service to customers. Across Europe, the company has recovered 6.8 million components in 3 years.

Careful product design is essential to facilitate upgrade and re-manufacture. Components can be standardised so that upgraded parts will fit into old casings, and products can be made so that components can be easily removed and replaced without damage. One aspect of this is the emergence of 'design for disassembly' to make it quicker and easier to take products apart and reuse or recycle the components. This includes simple strategies such as replacing screws with snap-fit fastenings, and coding parts to identify them, together with more novel techniques such as the use of soluble glues, soluble or pressure-sensitive fasteners and screws made from shape memory alloys which lose their thread when heated.

In industry, a more far-reaching vision of the potential of reuse and recycling has evolved into the concept of 'industrial ecology'. This is inspired by the observation that in nature there is no such thing as waste – dead and decaying organisms are broken down by fungi and bacteria to provide nutrients for the next generation of life. Industrial ecology treats the flow of materials and energy through the manufacturing sector as if it were an ecosystem. By taking a step back and looking at the whole process, we can design circular 'closed loop manufacturing' or zero-waste systems. Reuse and recycling on-site is maximised, and any waste which does arise is recovered, treated and used as input for a different process which can be on-site, off-site or even in a different sector.

This identification of waste streams as valued raw materials for other processes is sometimes called 'industrial symbiosis'. In the UK, a national industrial symbiosis hub has been set up which has helped to identify hundreds of new uses for process waste. Examples range from the use of polystyrene packaging from an IKEA store to make beads for filling soft toys at a nearby toy factory, to the use of assorted food waste and lime dust to make a slow-release lime-based fertiliser.[32] Industrial symbiosis is also used in China and South Korea, with a more centrally planned approach using a network of eco-industrial parks, and in Denmark, Australia and the United States.[33]

Resource-efficient production

Rather than trying to clean up pollution and waste at the end of the production process – for example, by installing end-of-pipe effluent treatment systems – it is far more efficient to try to 'design out waste' from the start of the process – avoiding pollution and saving materials, energy and money. Techniques that can be used range from simple housekeeping measures to more radical process and product redesign.[34]

Good housekeeping

Simple changes to operating practices are often low-cost and quick to implement. Typical improvements might include:

- employee training, incentives, bonuses or suggestion schemes aimed at identifying opportunities to reduce waste and improve efficiency;

- improved maintenance procedures to keep equipment at peak efficiency and prevent breakdowns, leaks and spills;
- better material-handling procedures to prevent loss due to mishandling, contamination, expiry of time-sensitive materials or inadequate storage conditions;
- waste segregation to reduce the volume of hazardous waste by not mixing it with non-hazardous waste.

Process design

Efficiency can often be improved and waste reduced by changes to the production process, modifications to equipment or the use of automated systems. For example, one electroplating company managed to reduce metal losses to zero with process improvements which included installing a spectrometer to allow more precise control of the concentration of metal plating solutions, improving surface pre-treatment so that pre-plating with copper was no longer necessary for nickel plating, and using a new design of evaporator to fully recover copper and nickel from effluent. This cut energy, chemical and water use and effluent generation and saved £1.5 million in 5 years.[35]

Choice of materials

Improvements can also be made by switching to materials which are less energy-intensive to produce; create less pollution, less waste or less hazardous waste during production; or make the product more environmentally friendly. For example, Goodyear replaced the traditional tyre fillers of carbon black and silica, which are energy-intensive to manufacture and create a waste disposal problem, with a filler based on corn starch. The new filler used less energy during manufacture, came from a renewable source, created less waste and also produced tyres which were lighter, with lower rolling resistance, improved braking performance and lower fuel consumption.[36]

Product eco-design

Changes to the manufacturing process, as described above, can reduce the 'cradle-to-gate' impacts of a product, from the extraction of raw materials to the moment when the product leaves the factory gate. Eco-design, on the other hand, aims to reduce the impacts of the product over its entire life-cycle, including end-use and disposal – these are the 'cradle-to-grave' impacts. Better still, products and processes can be designed so that any waste that does occur can be recycled back into the process – this is known as 'closed loop manufacture' or the 'cradle-to-cradle' approach.

Eco-design often begins by looking at the potential for products to be made using less material, thus cutting the energy needed to manufacture the material and transport the finished goods. This is known as light-weighting, right-weighting or lean production. Typically it can be applied to packaging, structural steel frames, buildings, electrical goods, household goods, furniture and vehicles. The Container Lite initiative in the UK, for example, found that most drinks containers can be made 10–20 per cent lighter,[37] even though the average weight of metal drinks cans has already been cut by 77 per cent since the 1960s.[38] Houses can be made using 10 per cent less material using modular off-site construction methods which reduce waste.[39] In the case of vehicles, light-weighting has the added benefit of increasing fuel efficiency while the vehicle is in use. Light-weighting has been facilitated by the development of materials such as carbon fibre composites with a high strength-to-weight ratio.

Eco-design also looks at the impact of products during use. One obvious example is that products which use energy, such as electrical appliances, can be made more efficient. Other examples are perhaps more surprising. Life-cycle analysis carried out by retailer M&S showed that over 50 per cent of the lifetime carbon impacts of a cotton t-shirt were from the energy used to heat water for washing it; this led to initiatives to design clothes, washing machines and washing powders that allowed washing at lower temperatures.

Finally, eco-design takes into account the end-of-life phase. There is considerable potential to reduce material impacts by designing products to be easily repaired, upgraded or re-manufactured. As an example, a team of students has designed a laptop that can be taken apart for recycling in just ten steps, with no tools, and separated into different materials such as plastics and metals. By comparison, it took a team of three engineers 45 minutes and 121 steps to dismantle a conventional laptop.[40] Eco-design also includes the selection of materials which are more easily recycled or composted after use. There can be conflicts here as lightweight composite materials are harder to recycle, and excessive light-weighting can make products less durable or packaging less protective, resulting in more waste.

Cutting consumer waste

Opportunities for cutting waste by consumers include:

*   reducing packaging; e.g. through cutting out unnecessary layers of packaging, making containers thinner, allowing customers to refill containers (e.g. shampoo bottles) and discouraging the use of disposable shopping bags;
*   allowing consumers to opt out of receiving junk mail through schemes such as the Mail Preference Service in the UK. Eradication of junk mail could save 550,000 tonnes of paper use in the UK, which is 4.4 per cent of the UK's consumption of paper and cardboard;[41]
*   encouraging consumers to avoid food waste, through education and awareness campaigns;
*   lifetime optimisation – not throwing away goods which are still usable.

Table 6.3 shows the estimated greenhouse gas savings from avoiding various types of consumer waste, showing that avoiding the generation of waste typically saves far more emissions than recycling.

% of urban
population

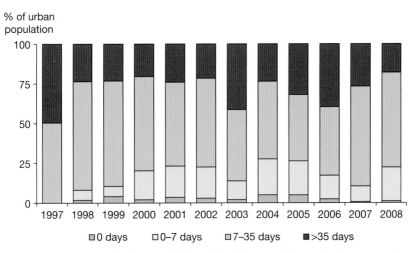

Figure 3.1 Percentage of urban population of the European Economic Area exposed to
particle (PM_{10}) concentrations exceeding the daily limit value set by the EU
Air Quality Directive

Source: EEA, 2010a

Note: Most EU member states have failed to achieve the target of exceeding the limit on fewer
than 35 days per year by 2005.

Exceedance of the 180 µg/m3 ozone information threshold
Interpolated around urban and rural stations

Reference period: summer 2003 (April – August)

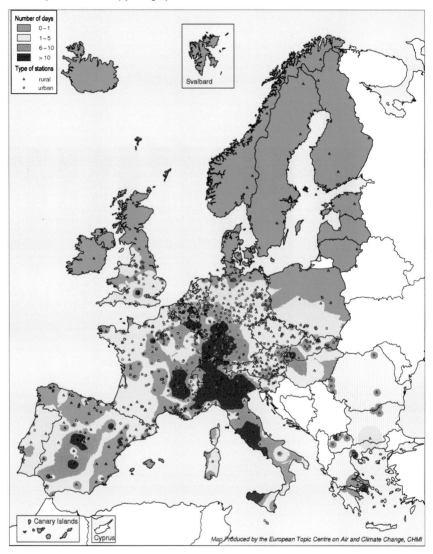

Figure 3.2 Ozone exceedances in Europe, summer 2003

Source: EEA, 2003

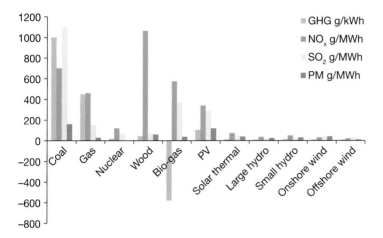

Figure 3.5 Life-cycle air pollutant emissions from different electricity generation
technologies

Source: Data from Pehnt, 2006 (renewables); WEC, 2004 (coal: UK plant with flue gas desulphurisation and selective catalytic reduction; gas: low NO_x burners; nuclear: mid-range estimate).

Note: Assumes advanced pollution control for coal and gas, but not for wood and bio-gas.

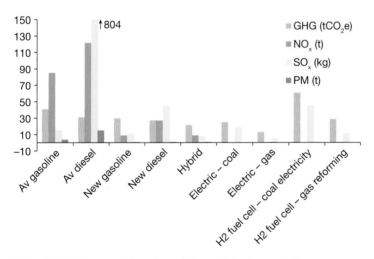

Figure 3.7 Lifetime emissions from different light-duty vehicles

Source: Based on IEA, 2010a, Table 17.3

Note: 'Av' = global average Euro II; 'New' = Euro V. Lifetime emissions include fuel or electricity
production and assume vehicles drive 15,000 km/year for 10 years.

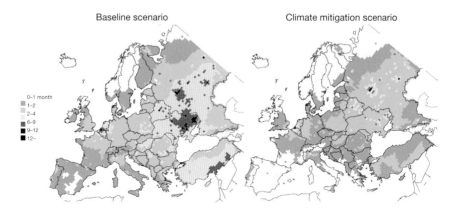

Figure 3.8 Loss in statistical life expectancy (in months) across Europe due to fine
particle pollution in the baseline scenario and the climate mitigation scenario
in 2050

Source: ClimateCost study (Holland *et al.*, 2011b)

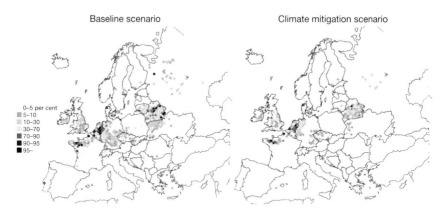

Figure 3.9 Area exceeding critical level for acidification in Europe in the baseline
scenario and the climate mitigation scenario in 2050

Source: ClimateCost study (Holland *et al.*, 2011b)

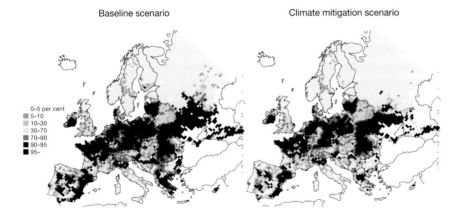

Baseline scenario   Climate mitigation scenario

0–5 per cent
5–10
10–30
30–70
70–90
90–95
95–

Figure 3.10 Area exceeding critical level for nitrogen deposition in Europe in the baseline scenario and the climate mitigation scenario in 2050

Source: ClimateCost study (Holland *et al.*, 2011b)

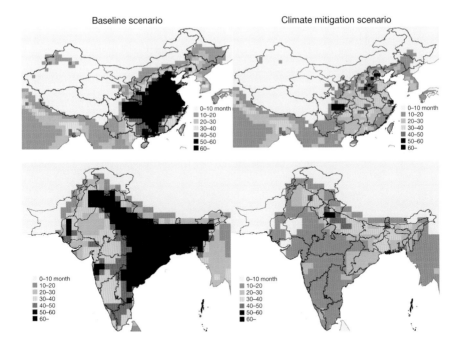

Baseline scenario   Climate mitigation scenario

0–10 month
10–20
20–30
30–40
40–50
50–60
60–

Figure 3.12 Loss in statistical life expectancy (in months) across China (top) and India (bottom) due to fine particle pollution in the baseline scenario and the climate mitigation scenario in 2050

Source: ClimateCost study (Holland *et al.*, 2011b)

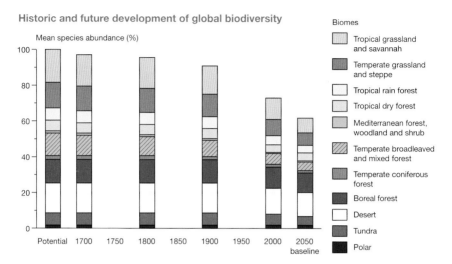

Figure 4.4 Biodiversity is declining

Source: PBL Netherlands Environmental Assessment Agency (Bakkes *et al.*, 2008)

Note: Mean species abundance is an estimate of the average abundance of original species compared to their abundance in an undisturbed environment.

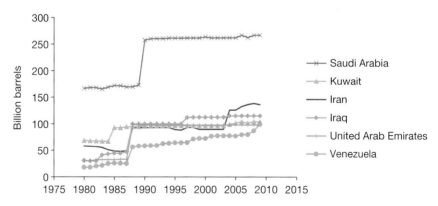

Figure 5.3 OPEC reserve estimates from several countries, showing sudden unexplained increases during the 1980s

Source: Based on data from US Energy Information Administration, 2012

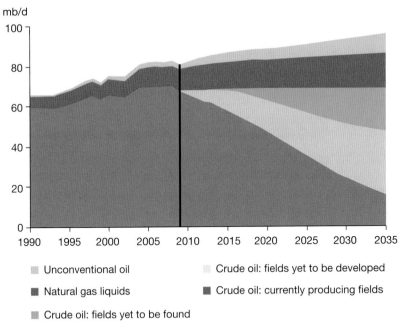

mb/d

- Unconventional oil
- Natural gas liquids
- Crude oil: fields yet to be found
- Crude oil: fields yet to be developed
- Crude oil: currently producing fields

Figure 5.4 Predicted world oil production in the IEA *World Energy Outlook* 2010 (New Policies Scenario), showing a peak in conventional crude oil production in 2006

Source: IEA, 2010b. © OECD/IEA 2010

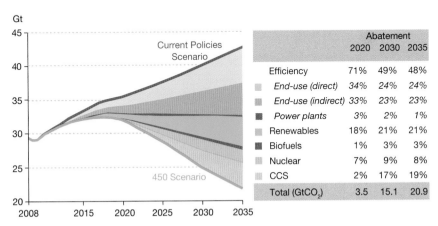

Gt

	Abatement		
	2020	2030	2035
Efficiency	71%	49%	48%
End-use (direct)	34%	24%	24%
End-use (indirect)	33%	23%	23%
Power plants	3%	2%	1%
Renewables	18%	21%	21%
Biofuels	1%	3%	3%
Nuclear	7%	9%	8%
CCS	2%	17%	19%
Total (GtCO$_2$)	3.5	15.1	20.9

Figure 5.8 Carbon savings under the IEA's 450ppm Scenario (GtCO$_2$)

Source: IEA, 2010b. © OECD/IEA 2010

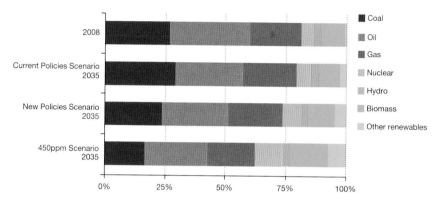

Figure 5.9 Share of world primary energy under the three scenarios of the IEA *World Energy Outlook 2010*

Source: IEA, 2010b. © OECD/IEA 2010

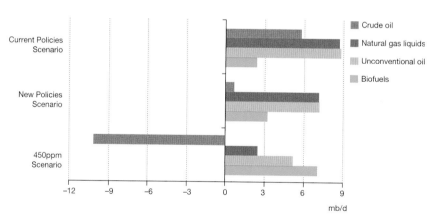

Figure 5.10 Change in oil supply from 2008 to 2035 under the three scenarios of the IEA *World Energy Outlook 2010*

Source: IEA, 2010b. © OECD/IEA 2010

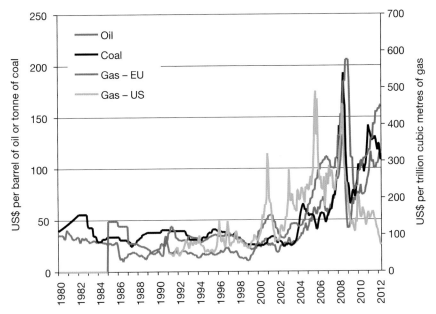

Figure 5.11 Monthly average oil, coal and gas prices, showing recent volatility

Source: Data from IMF, 2012

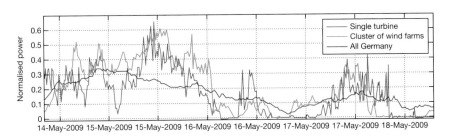

Figure 5.17 Smoothing of wind power in Germany: normalised power produced by a
single turbine, a cluster of wind farms and all Germany

Source: Fraunhofer Institute, 2011

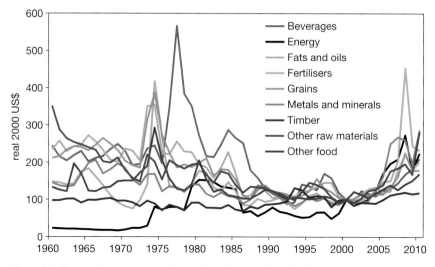

Figure 6.6 Commodity prices are rising after a long decline (indexed relative to the year 2000)

Source: Based on data from World Bank, 2011b

Note: Beverages = tea, coffee, cocoa.

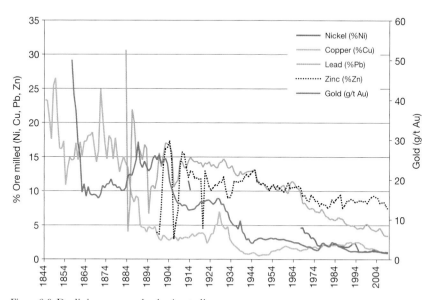

Figure 6.8 Declining ore grades in Australia

Source: Prior *et al.*, 2010

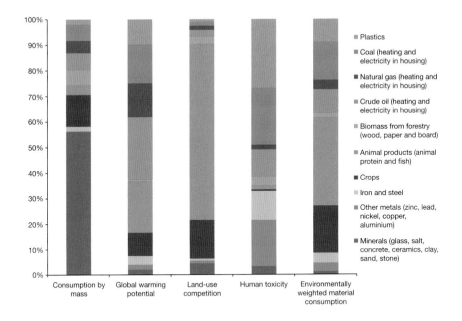

Figure 6.9 Relative environmental impact of different materials for the EU-27 plus
Turkey, 2000 (total of the 10 material groups set to 100%)

Source: UNEP, 2010a, adapted from van der Voet *et al.*, 2005

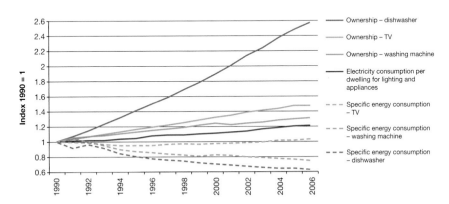

Figure 7.1 Growing appliance ownership outweighs efficiency improvements in the EU

Source: EEA, 2010c

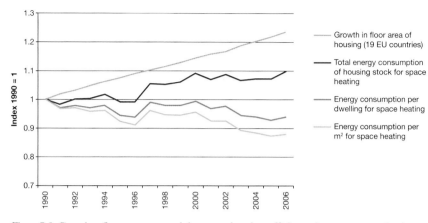

Figure 7.2 Growing floor area outweighs space heating efficiency improvements in the EU

Source: EEA, 2010c

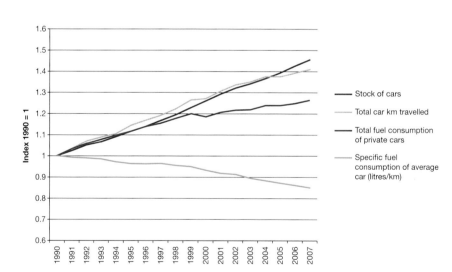

Figure 7.3 Growing car use outweighs fuel efficiency improvements in the EU

Source: EEA, 2010c

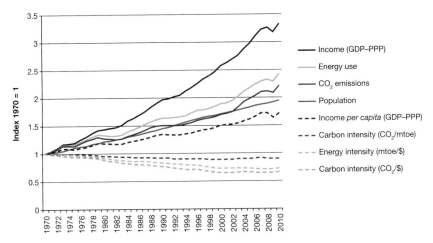

Figure 7.4 Although global carbon and energy intensity decreased from 1970 to 2010, total carbon emissions increased due to growth in population and GDP per person

Source: GDP and population from United Nations online statistics; energy use and carbon emissions from BP, 2011

Note: GDP measured with purchasing power parity (PPP) in constant 2005 US\$. Energy use is primary energy consumption in million tonnes of oil equivalent (mtoe).

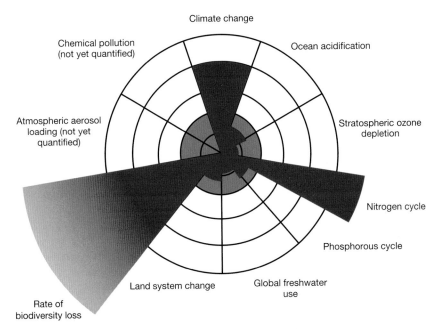

Figure 7.5 We have already exceeded three of the nine 'planetary boundaries'

Source: Adapted from Rockström *et al.*, 2009

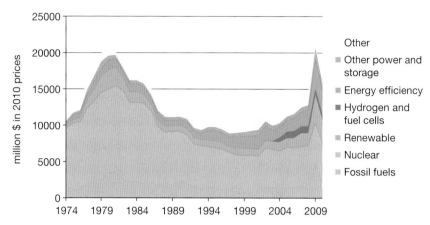

Figure 7.7 R&D budgets of IEA countries

Source: Based on data from IEA Data Services, 2012

Note: The peak in 2009 was caused by US stimulus spending.

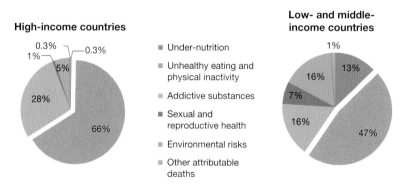

Figure 8.2 Estimated attributable deaths by risk factor, 2004

Source: Data from World Health Organization, 2009a

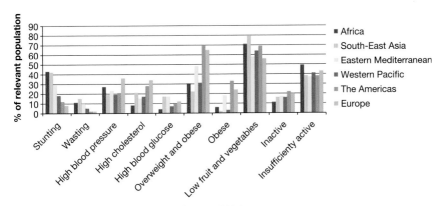

Figure 8.3 Prevalence of health risk factors, 2004

Source: Data from World Health Organization 2009a

Note: Stunting and wasting refer to children under 5; physical activity refers to those over 15 years old; and the remaining categories refer to those over 30.

	Air	Water	Biodiversity	Energy security	Resources and waste	Cost
Energy efficiency	☺☺	☺☺	☺	☺☺	☺	☺☺
Material efficiency	☺	☺☺	☺	☺	☺☺	☺☺
1st-generation biofuels	–/X	XX	X	☺☺	X	–
2nd-generation biofuels	–	–	–	☺☺	☺	?
Renewable energy	☺☺	☺	–	☺☺	–/?	?
Fuel switching (coal to gas)	☺	–	–	X	–	–
Nuclear	☺☺	X	–	☺	X	–
CCS	X	X	–	X	X	X
Forest protection	–	☺	☺☺	–	–	X
Sustainable agriculture	☺	☺	☺	☺	☺	–
Geo-engineering	XX	X	XX	X	–	X
Green lifestyle	☺☺	☺☺	☺☺	☺☺	☺☺	☺

Figure 9.3 Summary of climate policies and their co-benefits and conflicts

Notes: The ratings refer to the potential change if the deployment of this technology or policy was significantly increased compared to the current situation. ☺☺ = large improvement; ☺ = improvement; – = no significant change; X = deterioration; XX = large deterioration; ? = variable or uncertain.

'Water' refers to both water demand and water pollution.

'Geo-engineering' refers to sulphate aerosols and iron fertilisation of the oceans.

Table 6.3 Greenhouse gas savings from recycling or avoiding 1 tonne of material compared to the average UK disposal mix of landfill and incineration (tonnes CO_2e)

Material	Tonnes saved by recycling	Tonnes saved by avoidance
Aluminium	9.00*	6.00
Food waste	0.45	4.48
Glass	0.32	0.69
Paper/card	1.32	3.90
Plastic bottles	1.25	2.80
Mixed plastics	0.60	2.80
Steel	1.28	2.70
Textiles	1.75	19.5
White goods	3.31	–
WEEE (small items)	0.88	–

Source: WRAP, 2011d, p. 31

*Note that recycling aluminium appears to save more carbon than avoidance because the figure for recycling is compared to virgin aluminium, whereas the figure for avoidance is based on the average composition of aluminium which includes a large proportion of recycled material.

Dematerialising the economy

The methods described above can all cut material use by helping us to make goods with less raw material and therefore fewer carbon emissions. The problem is that despite continual improvements in material efficiency, our total material use is still increasing, because the demand for goods is escalating due to population growth and rising affluence (see Figure 6.5, p. 177). Dematerialisation can help to address this problem by providing the same level of consumer service, but with fewer material goods. Examples of how this can be achieved include:

- Repair – extending the lifetime of goods through repair and maintenance.
- Durability – manufacturing goods with longer lifetimes, as opposed to the present paradigm of planned obsolescence.
- Efficient use of existing infrastructure; e.g. meeting housing needs through retrofitting and refurbishment rather than demolition and rebuild. This can be facilitated by designing buildings with moveable partitions or easily replaceable panelling and windows.
- Leasing – sometimes described as 'sale of service', 'product-service' or 'goods to service'. Suppliers provide an ongoing service, such as provision of photocopying equipment or company cars, as opposed to the one-off sale of a product. This gives them a motive to supply more durable goods which can be easily upgraded, repaired or re-manufactured.
- Hiring or sharing – by hiring rather than buying infrequently used goods (e.g. lawnmowers, tools, second cars, 'special occasion' clothing), customers reduce the total number of goods demanded by society.
- Reduced consumption – consumers choose to buy fewer material goods, or buy goods with lower impacts (e.g. vegetarian food instead of meat). This is sometimes referred to as frugality or voluntary simplicity.

Although dematerialisation requires little in the way of new technology, it faces a far greater barrier – the need for a major cultural change in which we move away from a throwaway society to focus more on reuse, repair, maintenance and even, more controversially, a certain degree of abstinence from the joys of shopping. This raises many interesting issues, such as the impact of reduced consumption on economic growth and employment, which will be discussed in Chapters 7 and 8.

Co-benefits

The co-benefits from cutting our use of materials and reducing waste include:

- conserving scarce resources and avoiding resource-related conflict and instability;
- reducing the impacts of extracting and processing resources, such as pollution and landscape damage;
- reducing the costs and impacts of waste disposal in landfill sites and incinerators;
- saving money for householders, by improving the durability and lifetime of goods and reducing waste of food and other goods;
- making businesses more competitive by saving money on materials, water, energy and waste disposal.

Peak everything: conserving valuable resources

Peak Oil, as discussed in Chapter 5, is a familiar concept, but the security of other resources is now attracting concern – both non-renewable, such as metals and minerals, and renewable, such as fish stocks, fertile land and fresh water. 'Material security' and 'food security' are now competing for attention with 'energy security'.

Back in the 1970s, there was a flurry of interest in resource scarcity driven by the oil price shocks and influential publications such as *The Limits to Growth*.[42] But during the 1980s, interest gradually waned as new reserves were discovered, commodity prices continued their long-term decline, and the focus of the environmental movement turned to the more pressing issue of climate change.

Since the early 2000s, however, scarcity issues have reappeared on the agenda. The surge in demand for raw materials, driven by growth in developing countries, has created price spikes and temporary shortages for several key commodities (Figure 6.6, see plate section). Although prices fell after the financial crisis of 2008, they are already rising again.

Care should be taken in interpreting commodity price trends. Economists have traditionally argued that the long-term decline in commodity prices proves that scarcity (for metals and minerals) and environmental limits (for food and timber) are not a problem. If demand for copper, for example, outstrips supply, the price will rise and this will stimulate companies to increase their prospecting activity, ultimately leading to increased production and a resumption of the downwards price trend. Yet, as discussed in Chapter 5, there is a counter-argument that the long-term decline in prices is largely a reflection of the decrease in production costs as a result of technical developments such as automation, improved geological exploration techniques and the exploitation of cheap fossil energy.[43] Producers may not know that a resource is becoming depleted until it is almost exhausted, whereupon a sudden steep price rise can result.[44]

It certainly seems that supply constraints are now beginning to bite for many commodities. Figure 6.6 shows that the long-term decline in prices has given way to a steep increase since the early 2000s. The impacts have been felt around the globe. Food prices have been a particular focus for concern, as mentioned in Chapter 4, with increases driven by rising demand in developing countries, extreme weather events such as the 2010 drought in Russia which cut wheat production, competition with biofuels for land, and commodity price speculation. Higher energy prices (see Chapter 5) have also fuelled the price surge for energy-intensive materials such as metals, and for food that depends on input of synthetic fertilisers.

Future food, timber and biofuel supplies are vulnerable to increasing constraints on two other vital commodities – water and phosphorous. Phosphorous, largely derived from phosphate rock, is vital for plant growth and is an essential component of the fertilisers on which agriculture around the world depends. Many commentators predict that supplies will peak by 2030, and long-term depletion may occur within 50–100 years.[45] Water shortages are also hitting the headlines: almost 3 billion people already live in areas where water is in short supply and this could increase to 4 billion by 2030.[46] The main use of water is for agriculture, and there is a growing demand from power generation (see Chapter 5), but other water-intensive sectors include paper, steel and textiles – all of which tend to result in contamination of the water after it has been used.[47]

Other biological resources are also under threat. Approximately 60 per cent (15 out of 24) of the ecosystem services examined during the Millennium Ecosystem Assessment are being degraded or used unsustainably, including fresh water, natural fisheries, air and water purification, climate regulation and natural hazard regulation.[48] Biodiversity is declining rapidly (see Chapter 4). These ecosystem pressures are directly related to the endless quest for raw materials to fuel our economies: food, energy, timber, paper, minerals and metals. The amount of land used by humans is increasing, leaving less space for other species.

This can be expressed in the measure of HANPP – human appropriation of net primary production – which is an estimate of the proportion of the world's annual production of biomass by plants that is taken by humans through harvesting or land-use change, and therefore is not available to other organisms (see Table 6.7, p. 212). It is estimated that HANPP is currently around 24 per cent – meaning that humans take or destroy around a quarter of the world's production of biomass for their own use.[49] A striking indication of our dominance of the planet is the estimate that humans and domestic animals form 98 per cent of the weight of all vertebrates on earth.[50]

Another measure is the ecological footprint: the area of land or sea needed to provide the food, energy and other materials which we consume, and absorb the greenhouse gases we create. Since 1961, our ecological footprint has increased from

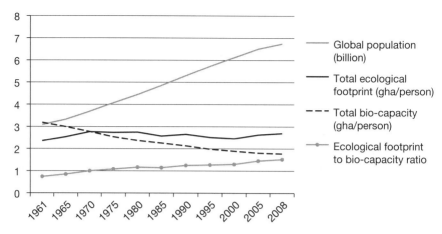

Figure 6.7 Our ecological footprint exceeds the available bio-capacity

Source: Based on data from Global Footprint Network, 2010

Note: gha = global average hectare.

2.4 to 2.7 global average hectares per person, while the available bio-capacity has fallen from 3.2 to 1.8 hectares per person (Figure 6.7). Our footprint first began to exceed the available bio-capacity in the late 1970s; and in 2011, we exceeded the earth's carrying capacity by 35 per cent. Earth Overshoot Day – the day of the year by which it is estimated that we have used up that year's allocation of renewable resources – fell on 27 September in 2011.[51]

On top of Peak Oil, Peak Phosphorous and ecological overshoot, we are now faced with the possibility of Peak Minerals. In most cases, the concept of Peak Minerals does not refer to an absolute shortage – most metals and minerals are fairly abundant within the earth's crust. Instead, scarcity issues are related to three distinct problems:

1   uneven distribution of reserves;
2   declining quality of reserves and increasing cost of extraction;
3   increasing concern over the environmental and social impacts of extraction.

To make extraction of a mineral viable, it must be present in a reasonably concentrated form. But high-quality ore deposits are unevenly distributed across the earth, with production being concentrated in a limited number of countries. For example, China supplies 97 per cent of the world's rare earth metals;[52] Brazil supplies 92 per cent of niobium; and South Africa supplies 79 per cent of the world's platinum.[53] This means that importing countries are vulnerable to price rises and supply restrictions. China has already begun to restrict exports of critical raw materials including phosphorous and rare earth elements, triggering complaints from the United States and the EU to the World Trade Organization (WTO).[54] Countries such as Russia, Pakistan, India and Ukraine have also restricted exports of commodities such as scrap metal, iron ore, coke and wood at various times.[55] As well as the issue of geographical distribution, concerns have been raised over the concentration of production in the hands of a relatively small number of very large multinational corporations.[56]

As high-quality reserves are exploited, production must shift to lower-quality reserves characterised by lower ore grades, greater depths and more remote or inaccessible locations – all of which will increase costs. Figure 6.8 (see plate section) shows how ore grades have gradually declined in Australia since the 1840s – a pattern being repeated around the world. More energy is needed to extract lower grades of ore, because more rock must be recovered and processed to extract the same amount of metal. As a result, costs, carbon emissions and other environmental impacts will all be greater. This will be exacerbated as the prices of fossil fuels increase in the future (see Chapter 5).

Although scarcity is not generally related to an absolute shortage of minerals, there are cases where reserves appear quite small compared to current production rates. Figures from the OECD indicate that known reserves of copper, lead, nickel, silver, tin and zinc will be exhausted within 20–40 years at current production rates.[57] Of course, new discoveries are likely to be made as prices rise, but continued steep increases in demand could cause supply shortages. Even if production grows at 5 per cent per year, known reserves of iron and aluminium would be exhausted within 50 years.

A study for the UN International Resource Panel concluded that the following metals have supply risks:[58]

* resource appears small relative to annual extraction: gold, copper, zinc, platinum-group metals (platinum, palladium, rhodium);
* resource appears small and most uses are non-substitutable: indium, rhenium, europium, hafnium, erbium;
* principal source countries are politically problematic: cobalt, tantalum;
* very large energy requirements: aluminium, titanium;
* limited by high toxicity: lead, mercury, arsenic, cadmium.

Demand for many metals has escalated as a result of technical developments and new applications in the electronics industry. For example, a sudden rush for tantalum to make mobile phones caused supply shortages in Europe in 2000, and ruthenium prices increased almost ten-fold in 2006 due to its increasing use in advanced hard disk technology.[59] Several rare metals are essential for new 'green' technologies, such as gallium and indium for solar panels and rare earth elements for permanent magnets used in wind turbines. Reserves of indium and tantalum could be exhausted by 2030 according to one estimate.[60] Few of these metals are readily substitutable by other materials – there is no known substitute for the use of europium in colour liquid-crystal display (LCD) screens, for example, or for erbium in fibre-optic cables.[61]

The problem is exacerbated by the fact that many metals are mainly produced as by-products – indium, for example, is a by-product of zinc mining and its supply is restricted when zinc prices are low; and tellurium is a by-product of copper mining. The situation is even worse for rhenium and hafnium, which are by-products of by-products.

Governments around the world are now looking at resource scarcity with a new sense of urgency. Some key initiatives are listed in Box 6.2.

Box 6.2 International action to address resource scarcity

Europe

The EU-27 is highly dependent on imported metals: it imports over 80 per cent of its iron ore, bauxite and tin, almost half of its copper, and all of its cobalt, indium, magnesium, rare earth elements and platinum group metals. In total, 14 metals and minerals are identified as having a high supply risk over the next 10–15 years.[62] Illegal export of scrap metal from Europe is leading to a shortage of secondary raw materials, which are used for 40–60 per cent of European metal production.[63]

The European Commission aims to increase material efficiency through a number of initiatives including:

- 2005 Thematic Strategy on the Sustainable Use of Natural Resources;
- 2005 Thematic Strategy on the Prevention and Recycling of Waste;
- 2008 Action Plan on Sustainable Consumption and Production (SCP);
- 2008 Raw Materials Initiative;
- 2011 Flagship Initiative for a Resource-Efficient Europe (part of the Europe 2020 Strategy).

Several European countries have set national targets for improving resource productivity (see 'The way forward' below). The French government has identified short- to medium-term supply risks for 17 rare metals,[64] and a report for the UK government warns that the UK faces supply risks for indium, lithium, phosphorous, rare earth elements, fish and aggregates.[65]

United Nations Environment Programme

The United Nations International Resource Panel was launched in 2007 to try to find ways of decoupling economic growth from resource use, building on the UN's 10-Year Framework of Programmes on Sustainable Consumption and Production (otherwise known as the Marrakech Process). Regional strategies have been developed for Latin America and Africa, with the aim of using North–South co-operation to allow developing countries to leapfrog to sustainability.

Other UNEP initiatives include:

- The World Resources Forum, launched by the International Resource Panel, which met in 2009 and 2011;
- Resource-Efficient and Cleaner Production;
- The Green Economy Initiative.

The United States

The US National Research Council reported in 2008 that five metals which were critical to the US economy were at high risk of supply disruption: indium, manganese, niobium, the rare earth elements and the platinum group metals. The US Department of Energy is preparing a report on a strategy for ensuring supplies of rare earth elements and a report for the Pentagon on the dependence of the military on scarce metals. The US Geological Survey is also co-ordinating the Global Mineral Resource Assessment Project (GMRAP) which aims to assess the extent and location of the world's undiscovered mineral resources.

China

Although China has significant reserves of many important raw materials, rapid economic growth has been accompanied by increasing levels of environmental damage related to the extraction and use of resources. In 2009, China enacted a Circular Economy Promotion Law which aims to reduce the environmental impacts of production through an industrial ecology approach.

Japan

Japan has limited resources of its own, and is a heavy importer of raw materials. Stockpiles of seven rare metals are maintained[66] and the government has a strategy to 'support key resource acquisition projects by promoting active diplomacy' including the use of overseas aid.[67] In 2000, Japan launched an initiative to establish a 'Sound Material-Cycle Society'. The country later took this initiative to the international community as the 3R Initiative (Reduce, Reuse, Recycle) which was adopted at the G8 conference in 2004 and now has an Asia-wide forum.[68]

Resource efficiency can help countries to minimise the threats posed by increasing scarcity. It can make national economies less reliant on imports from other countries, and help to avoid the risk of supply disruptions and sudden price fluctuations which can damage industry. An efficient economy which maximises the recycling and reuse of materials can insulate itself from these risks to a large extent, thus making its industries more competitive and its economy more stable.

A 2010 UK report illustrates the benefits of 13 'quick win' strategies for resource efficiency, including production strategies such as lean production, waste reduction, recycling and material substitution; and consumption strategies such as lifetime optimisation, reuse and refurbishment, dietary changes, reducing food waste and moving from goods to services.[69] The study estimates that the following benefits could be achieved by 2020:

- greenhouse gas savings equivalent to 10 per cent of the UK's target reduction, plus additional savings in other countries due to the reduction of imports;
- savings of over 38 million tonnes in annual use of the materials included in the study (15 per cent of the current consumption of 260 million tonnes);
- reduction of 10–25 per cent in reliance on rare earth metals, cobalt and lithium;
- cut of 6 per cent in water abstraction associated with UK consumption;
- reduction of 5–7 per cent in the UK's ecological footprint.

Strategies that focus on consumers were found to have the greatest potential impacts – especially lifetime optimisation and a shift from goods to services.

Resource wars: reducing conflict

Competition for resources is a natural feature of all ecosystems and all species, and humans are no exception. Although natural resources are seldom the sole cause of conflict, they are often a contributory factor. According to UNEP, at least 18 violent conflicts have been fuelled by the exploitation of natural resources since 1990 (see Table 6.4); and over the last 60 years, at least 40 per cent of all intrastate conflicts have a link to natural resources.[70]

The UNEP report identifies three ways in which resources are linked to conflict. First, and most obviously, conflict can be triggered by the desire to control scarce or valuable resources. The invasion of Sierra Leone by a Liberian-sponsored group in 1991, for example, was motivated by the desire of Liberian warlord Charles Taylor to seize control of lucrative diamond mines close to the Liberian border. This war, which left over 200,000 dead and more than 2 million people displaced, continued until the UN Security Council imposed sanctions on diamond exports in 2001. Taylor then switched to using Liberian timber as his main source of revenue, and this persisted until 2003 when sanctions on timber exports finally caused Taylor's regime to collapse. Similarly, the conflict in the Democratic Republic of the Congo between 1998 and 2002 has been linked to the increase in demand for (and price of) tantalum for mobile phones.[71] Other conflicts, including

Table 6.4 Recent civil wars and unrest fuelled by natural resources

Country	Duration	Resources
Afghanistan	1978–2001	Gems, timber, opium
Angola	1975–2002	Oil, diamonds
Burma	1949–	Timber, tin, gems, opium
Cambodia	1978–1997	Timber, gems
Colombia	1984–	Oil, gold, cocoa, timber, emeralds
Congo, Dem. Rep. of	1996–2008	Copper, coltan, diamonds, gold, cobalt, timber, tin
Congo, Rep. of	1997–	Oil
Côte d'Ivoire	2002–2007	Diamonds, cocoa, cotton
Indonesia: Aceh	1975–2006	Timber, natural gas
Indonesia: W. Papua	1969–	Copper, gold, timber
Liberia	1989–2003	Timber, diamonds, iron, palm oil, cocoa, coffee, rubber, gold
Nepal	1996–2007	Yarsagumba (caterpillar fungus used in Chinese medicine)
PNG: Bougainville	1989–1998	Copper, gold
Peru	1980–1995	Coca
Senegal: Casamance	1982–	Timber, cashew nuts
Sierra Leone	1991–2000	Diamonds, cocoa, coffee
Somalia	1991–	Fish, charcoal
Sudan	1983–2005	Oil

Source: UNEP, 2009b

those in Darfur (in the Sudan) and the Middle East, have been triggered by the scarcity of vital resources such as fertile land and clean water, or by environmental degradation related to resource extraction (such as Shell's polluting activities in the Niger Delta).

Second, once conflict is under way, it can be prolonged by the use of revenue from exploitation of natural resources. The prolonged civil war in Angola in the 1990s has been described as the ultimate resource war, as the rebel group UNITA was funded mainly by illegally extracted 'conflict diamonds' and the government was funded by oil revenues. The outcome of the conflict was affected to a large extent by the price of oil relative to that of diamonds. In Cambodia, forest cover was reduced from 73 per cent in 1969 to 30 per cent in 1975 as timber was used as the principal revenue source for the Khmer Rouge anti-government forces. Bizarrely, the Cambodian government also raised funds by selling certificates of origin to the loggers at $35 per cubic metre, thus enabling timber exports to continue. This only stopped when Thailand, under threat of withdrawal of US aid, closed its border to Cambodian timber, leading to the eventual disintegration of the Khmer Rouge in 1998.

Finally, unfair allocation of access to natural resources can prevent efforts at peace building and reconciliation. Conflicts associated with natural resources are twice as likely to relapse into conflict within the first 5 years after resolution.

In a bitter vicious circle, conflict itself causes further environmental damage and further pressure on resources. High-profile examples include deliberate acts of aggression such as the burning of Kuwait's oil wells during the 1991 Gulf War, but a huge amount of collateral damage also occurs through the destruction of ecosystems and release of pollution as a result of military activities, and through the over-exploitation of resources by people displaced by war who are forced into a subsistence existence or pushed onto marginal land. In Afghanistan, for example, the UN found in 2003 that over half of the natural pistachio woodlands had been felled to provide fuel wood, income or space to grow food for people whose livelihoods had been damaged by conflict. This has led to complete deforestation in some areas, and resulted in soil erosion and damage to water supplies, forcing many farmers to migrate to towns and cities.

As resource constraints grow, conflict is likely to increase. The UNEP report warns that:

> As the global population continues to rise, and the demand for resources continues to grow, there is significant potential for conflicts over natural resources to intensify in the coming decades. In addition, the potential consequences of climate change for water availability, food security, prevalence of disease, coastal boundaries, and population distribution may aggravate existing tensions and generate new conflicts.
>
> (UNEP, 2009b, p. 5)

The report concludes (2009b, p. 5) that 'the effective governance of natural resources and the environment should be viewed as an investment in conflict prevention'. It argues for better legal protection for natural resources during conflicts, for sanctions to prevent trade in 'conflict resources' that sustains violent conflict, and for peace negotiations to take into account the need for sustainable management of natural resources.[72] The Millennium Ecosystem Assessment echoes these concerns, warning that the impacts of the degradation of ecosystems are being borne mainly by the poor and are contributing to inequality, poverty and conflict.[73]

Resource wars have horrific impacts in the countries directly involved, which tend to be mainly in the developing world, but resource scarcity can also lead to economic conflict and trade disputes that affect developed countries, particularly as many key mineral resources are located in regions which are politically or economically unstable. A recent example is the restriction of exports of rare earth elements by China (see 'Peak everything: conserving valuable resources' above).

Further tensions may arise as countries with large raw material requirements begin to buy up agricultural land and mineral rights in other regions. The World Bank reports that the 2008 boom in commodity prices stimulated interest in land purchases on an enormous scale. Investors are a mix of agribusiness firms, financial speculators in developed countries such as the UK and the United States, and countries such as China, Russia, North Africa and the Gulf States which are resource-hungry or dependent on food imports. Around 56 million hectares of agricultural land were bought up in large-scale international deals in 2009 alone – almost 4 per cent of the global cultivated area of around 1,500 million hectares. Around 37 per cent of the land purchases were for food crops, 21 per cent for industrial cash crops and 21 per cent for biofuels, with the remainder being for game reserves, livestock and plantation forestry. Two-thirds of the area bought is in Sub-Saharan Africa, with half the remainder being in East and South Asia and the rest in Europe, Central Asia, Latin America and the Caribbean. The report warns that investors tend to target countries with weak land governance, leading to adverse social consequences such as displacement of local people, loss of land rights and poor environmental safeguards.[74]

But what role can resource efficiency play in reducing international conflict? Here we enter a minefield of complex arguments concerning economic justice, trade and development. As developed countries decrease their reliance on imports of raw resources, they may be more able to select resources which are not associated with fuelling conflict. At the same time, developing nations will gain the environmental space they need to increase their own levels of consumption. But there is also a danger that developing nations may lose a vital source of export income. These issues will be explored in more detail in 'Developing countries: escaping the resource curse' below.

Clean planet: reducing environmental and social impacts

Few shoppers stop to think about the environmental cost of the products they buy, yet the extraction of raw materials and manufacture of finished goods is responsible for a wide range of impacts such as deforestation, pollution and industrial accidents. Three examples illustrate the environmental and safety risks of metal production:

1   *The Hungarian 'red sludge' disaster.* In October 2010, the dam surrounding the tailings pond of an aluminium plant burst, flooding the surrounding area with highly caustic red mud which killed ten people, injured over 100 and poisoned rivers leading to the Danube. Aluminium production creates 2 tonnes of caustic red mud for every tonne of metal produced.[75]

2   *The Chilean copper mine accident.* In August 2010, 33 miners were trapped when the roof collapsed and were finally freed in a dramatic televised rescue in October 2010 after 69 days underground. Despite the happy outcome to this particular incident, an average of 34 people have died every year in mining accidents in Chile since 2000, with a high of 43 in 2008.[76] As copper prices rise, smaller firms with lower safety standards enter the market, often reopening mines which were formerly closed as uneconomic. As a result, there is a link between the price of copper and the number of accidents in Chilean mines.[77]

3   *The town of La Oroya in Peru* has been listed as one of the top ten most polluted places on earth, due to emissions from a metal smelting plant. Lead levels in the blood of local children are three times the World Health Organization limit, and acid rain has destroyed vegetation in the area.[78] Protests over pollution of rivers from copper mining in Peru are common.[79]

Extraction and processing of raw resources are often dangerous for workers. Globally, it is estimated that around 2 million people per year die from work-related illness, of which 900,000 deaths are from exposure to hazardous chemicals, and there are also over 300,000 fatal accidents at work. Mining and quarrying are particularly high-risk occupations, averaging 18 deaths per 100,000 workers in the United States, for example, compared to a rate of 4 deaths per 100,000 workers for all sectors. Agriculture, fishing and forestry are even worse, averaging 30 deaths per 100,000 workers in the United States.[80]

Resource extraction can also have social costs. Indigenous people and other local communities often lose the use of their land due to encroachment by other users. Aerial photographs of uncontacted tribes in the Amazon rainforest taken in 2011 prompted international concern over the threat that deforestation, mining and oil exploration pose to their way of life – impacts that have already severely affected many other tribes in the area.[81] Many other communities around the world, especially in Sub-Saharan Africa, are threatened by 'land grabs' for biofuels, agriculture, minerals or oil (see Chapter 4).

Typical impacts arising from resource use are summarised in Table 6.5.[82]

Table 6.5 Summary of main environmental and social impacts arising from resource extraction and processing

Resource	Activity	Main impacts
Minerals and construction aggregates	Quarrying	Large-scale landscape devastation; waste generation; transport pollution
Metals and minerals	Mining	Landscape damage; waste generation; ground and surface water pollution; accidents; health impacts
Fossil fuels	Oil and gas extraction	Landscape damage; biodiversity loss; displacement of local communities; air pollution; water pollution; greenhouse gases; accidents
Metals	Smelting	Air pollution; greenhouse gases
Timber, paper	Forestry	Biodiversity loss; deforestation; displacement of local communities
Paper	Pulp and paper manufacture	Water pollution; water abstraction; greenhouse gases
Food and fibre from crops and livestock	Agriculture	Water abstraction; water pollution; eutrophication; acidification; biodiversity loss; greenhouse gases; landscape damage; deforestation; displacement of local communities; health impacts from use of pesticides
Fish	Industrial-scale fishing	Biodiversity loss; loss of livelihood for local communities
All	Manufacturing industry	Air pollution; water pollution; water abstraction; waste generation; accidents

Figure 6.9 (see plate section) shows the relative impact of different resources on global warming, land use and human toxicity. Fossil fuels are a major source of both global warming and toxic emissions; food production is the main pressure on land use and is a major contributor to global warming; and both plastics and metals are a major source of toxic emissions.

The outsourcing of production to the emerging economies of Asia exacerbates these problems, as health, safety and environmental standards tend to be lower in developing countries. Developing countries also tend to contain more valuable ecosystems – tropical forests, for example, are far more biodiverse than temperate forests – as well as being home to a number of indigenous communities; so sourcing raw resources from these areas can have greater impacts.

Impacts are also increasing as production shifts to lower-quality or more inaccessible resources. Lower ore grades mean that more rock must be processed to recover each tonne of mineral, which increases the amount of tailings waste, landscape damage, water pollution from mining, air pollution from smelting, the area of land affected and habitat lost, and the risk of accidents per tonne of product. Figure 6.10 shows how the energy needed to extract metals increases exponentially as the ore grade declines, and Table 6.6 shows the increased use of energy, water, chemicals and limestone associated with extracting lower grades of copper ore. In addition, production is now expanding into remote and vulnerable ecosystems such as the Arctic – a move which, ironically, is made easier by the onset of global

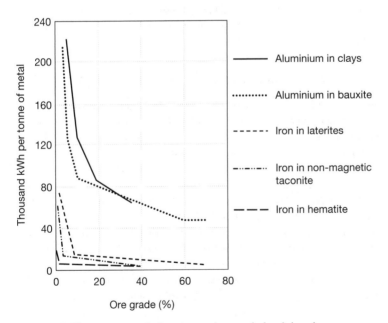

Figure 6.10 The energy needed to extract iron and aluminium increases exponentially as ore grades decline

Source: Page and Creasey, 1975

Table 6.6 Flows related to the production of 1 tonne of copper (by pyrometallurgy)

Copper concentration in ore	2.8%	0.99%	0.44%
Ore (tonnes)	40	125.2	272
Used water (cubic metres)	132	418	902
Steel (kg)	34	90	191
Lime (kg)	50	149	340
Primary energy for mining (GJ)	2.9	8.57	21.3
Primary energy for processing (GJ)	11.2	19.62	47.5
Total primary energy (GJ)	14.1	28.19	68.8
Chemicals for flotation	4.1	11.8	25.8

Source: Massman *et al.*, 2009

warming – and the Amazon rainforest, which now supplies almost half of Europe's imports of iron ore.[83]

By moving towards a circular economy, in which materials are continuously reused and recycled, we can continue to provide the goods that we need but with a much lower environmental impact, cutting the adverse impacts associated with resource extraction activities such as mining, quarrying, metal smelting and forestry. A study published in 2011 found that if all EU countries were just to adopt current 'best practice' recycling rates, the EU would free up 30 million hectares of productive land (largely from recycling paper and wood) and save carbon emissions equivalent to 25 per cent of the 2020 target, as well as significantly reducing pollution and resource depletion.[84] A serious move towards a zero-waste economy could achieve far more than this.

Less waste: reducing disposal problems

Waste prevention, together with more reuse and recycling, can cut the total amount of residual waste generated by householders and industry. This will reduce the costs to taxpayers of collecting, treating and disposing of the waste, as well as reducing the environmental impacts of landfill sites and incinerators.

Waste is closely related to affluence. Figure 6.11 shows that the total amount of municipal waste generated in OECD countries has grown by 78 per cent since 1980, and that waste *per capita* is increasing. In Europe, for example, about a third of the material used each year ends up as waste.[85]

Landfill sites are associated with a range of problems: they are unsightly; take up valuable land space; and generate noise, traffic, odour, pests, litter, groundwater pollution and methane emissions. In developed countries, better management has reduced many of these impacts. For example, impermeable liners reduce the risk of groundwater pollution; prompt covering with topsoil reduces litter; and a significant proportion of methane emissions can be recovered for energy generation. Despite this, there is still a shortage of suitable areas for new landfill sites in many countries. In the UK, for example, current capacity is predicted to run out within 10 years,[86] and local authorities only have landfill space for a quarter of the

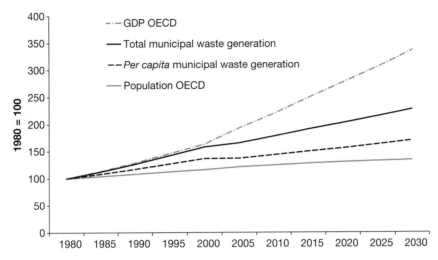

Figure 6.11 Historic and predicted municipal waste generation in OECD countries, showing how waste increases with affluence

Source: OECD, 2008b

additional tonne of waste that will be generated by each household by 2020 if current trends continue.[87]

Incinerators also generate intense public opposition due to concerns over toxic emissions such as dioxins, even though pollution-control technologies have reduced these emissions significantly in developed countries. Public opposition has restricted the building of new incinerators in the United States, Germany, the UK and other European countries.[88] As well as emissions to air, incinerators produce toxic ash and flue gas cleaning residues which present an additional hazardous waste-disposal problem.

In developing countries, the problems are even worse. Waste disposal is often poorly managed or even unmanaged, with shallow uncovered landfill sites or uncontrolled dumping and burning of waste leading to severe public health and safety problems including the spread of diseases such as dysentery and typhoid.[89] In addition, there is a growing problem related to export of waste (especially waste electrical equipment) from Western countries to developing countries, where it is often processed under conditions where environmental and safety standards are poor or non-existent.[90] Marine litter from poorly controlled disposal on land or at sea is also a growing hazard: over a million birds and 100,000 marine mammals and sea turtles die each year from ingesting plastic particles or becoming entangled in waste material such as plastic bags.[91] The problem is particularly acute in the areas known as the Great Pacific Garbage Patch and the North Atlantic Garbage Patch, where millions of square kilometres are covered with mats of floating debris and sludge.

For industrial waste, some of what is generated is toxic, flammable, explosive, corrosive or infectious. The hazardous component of industrial waste is growing as the use of chemicals and rare metals in manufacturing increases.[92] Again, the problem is particularly severe in developing countries where much of the world's production now takes place. Around 70 per cent of industrial waste in developing countries is discharged directly to surface water without treatment. There have also been instances of Western companies illegally dumping toxic waste in developing countries, such as the Trafigura case in which thousands of people in West Africa were hospitalised.[93]

Biodegradable waste – which includes food and garden waste, agricultural waste, wood and paper – is a particular concern because it produces methane when landfilled. Composting avoids methane emissions as well as producing compost that can be used for agriculture. Anaerobic digestion of food or farm waste is a good alternative, producing bio-gas for energy and solid or liquid fertilisers.

A good waste-management strategy will reduce the waste sent for disposal (through prevention of waste at source as well as more recycling and composting) and also improve the disposal facilities themselves (through better controls on landfill sites and incinerators). In the EU-27, this approach has cut greenhouse gas emissions from landfills and incinerators by over 30 per cent since 1990 – the highest reduction rate of all sectors – as well as reducing water and air pollution from disposal sites.[94] Yet there is considerable variation between member states, and much potential for improvement. The Belgian region of Flanders shows what can be achieved with a strong, integrated strategy (see Box 6.3). Outside Europe, the global potential is even more significant. Under a green economy scenario modelled by the United Nations, global recycling rates would triple by 2050 compared to business-as-usual, and the amount of waste sent to landfill would fall by 87 per cent.[95]

Box 6.3 Flanders moves towards Zero Waste

The Belgian region of Flanders is densely populated and lacks space for landfill sites. Public opposition to landfill sites, with concern over soil and water pollution and greenhouse gas emissions, led to the introduction in 1998 of a ban on landfilling or incinerating recyclable materials or unsorted waste. Two years later, a ban on sending combustible residual waste (unrecyclable waste) to landfill was introduced, diverting this waste to incineration with energy recovery.

The landfill and incineration bans were part of an integrated range of measures including:

- separate collection of a wide range of recyclable materials;
- promotion of home composting and subsidies for householders to buy chickens (to recycle food scraps into food);

- charges for waste disposal (pay-as-you-throw) for householders – including recyclables;
- high taxes on landfill and incineration;
- provision of a wide range of container parks where people can bring recyclable materials;
- a network of reuse centres which collect, refurbish and sell items such as furniture and electrical appliances;
- limits on the building of new incinerators to discourage incineration of recyclable waste.

The landfill ban was introduced gradually, with plenty of warning for local councils and industries and with clear and strict enforcement measures. Various exemptions were granted in the first few years of the scheme, to allow time for companies to adapt and for new waste-processing capacity to be established, but it was made clear that these exemptions would be gradually withdrawn.

The scheme was highly successful in diverting waste from landfill and increasing material recovery rates. The proportion of waste recycled increased from 27 per cent to 45 per cent, and the waste sent to biological treatment (composting or anaerobic digestion) increased from 20 per cent to 23 per cent, while the proportion landfilled fell from 25 per cent to just 3 per cent in 2007. An impressive 71 per cent of household waste is now separately collected for recycling or composting.[96]

Saving money: don't waste our waste

Waste can be viewed as a failure of the industrial system – a sign of inefficiency. By throwing away waste, we are losing valuable resources. In the words of Karl-Heinz Florenz, member of the European Parliament, 'We can no longer afford to waste our waste'.[97]

Cutting waste can save money for both businesses and households. One of the simplest strategies is 'lifetime optimisation' – in other words, not throwing away things that are still usable. In the UK, it is estimated that a third of clothes, appliances, household goods, cars and other products thrown away could still be used.[98] As well as saving resources and carbon emissions, this would save money for consumers. Households in the UK could each save £1,800 per year by using products to the end of their useful life.

Food waste is a particular concern. In the UK, around 20 per cent of food purchased by households is wasted, costing the average family £680 per year. More than half of this waste, which adds up to £12 billion per year, is avoidable.[99] Cutting food waste and optimising product lifetimes together could save UK households a total of £47 billion per year.[100] In the United States, around 40 per cent of food is wasted, worth $48 billion each year.[101] Global hunger could be eliminated with just the food that is wasted in the United States.[102]

Savings in water use will also become increasingly important for households, as water scarcity increases and costs rise. In the UK it has been estimated that 4 million households may already suffer from 'water poverty' – defined as being when households spend more than 3 per cent of their income on water – and this is likely to increase as water bills are estimated to rise by 5 per cent a year.[103]

Similar savings are possible in the service sector. It has been estimated that the UK service sector could save almost £3 billion per year simply by using goods such as carpets, clothes and computers until they became obsolete.[104] However, lifetime optimisation will require a change in consumer attitudes – an issue which will be explored further in Chapter 8.

Resource efficiency offers plenty of 'quick wins'– almost instant savings with very little up-front investment. A UK study estimated that simple low-cost or no-cost resource efficiency actions targeted at energy, waste and water could save businesses up to £23 billion per year, and actions with a payback of over a year could save a further £33 billion – giving a total of £55 billion of possible savings and potentially increasing the gross profits of the industrial sector by 15 per cent.[105] Some £40 billion of these savings are from avoided waste, with a further £11 billion from energy and £5 billion from water. These resource cost savings can help to make businesses more competitive, while saving 90 million tonnes of carbon dioxide, which is 13 per cent of annual UK emissions. Similar results apply in other countries – a study in Germany estimated that resource efficiency could save 20–30 per cent of industrial production costs, and generate a million jobs.[106] In South Korea, Extended Producer Responsibility (encouraging producers to take back end-of-life goods) is expected to save $1.6 billion.[107]

The most obvious waste-related savings for businesses arise from avoided waste disposal fees. For example, if the UK Waste and Resources Action Programme (WRAP) achieves its target of diverting 8 million tonnes of business waste from landfill, it will have saved businesses £1 billion in waste disposal fees as well as cutting carbon emissions by 5 million tonnes.[108] Yet far greater savings can be made from preventing waste at source. On average, diverting a tonne of waste from landfill (for example, by recycling or composting) in the UK is estimated to save £43 and 0.32 tonnes of carbon dioxide, whereas preventing a tonne of waste from arising in the first place saves £593 and 0.99 tonnes of carbon dioxide. The biggest savings come from lean manufacturing and reduction of waste during production – these account for three-quarters of the long-term material efficiency savings, with the remainder coming mainly from material substitution, recycling and sustainable building.[109]

An earlier UK study noted that waste prevention has hidden benefits that extend beyond the direct costs to the company of waste treatment and disposal. By cutting waste, companies can also save raw material costs, energy and water costs, packaging costs and time and effort. These 'hidden' costs are typically 5–20 times greater than the direct avoided waste disposal costs. The largest hidden benefit was found to be raw material savings, which accounted for 60 per cent of the total savings from avoided waste.[110]

Much of the waste that companies throw away has a significant economic value that could be recovered through recycling. Waste plastic, for example, is typically worth around €300 per tonne – more than coal, wheat or iron ore.[111] Yet in 2004, around half of the recyclable materials in the EU-27 were either landfilled or burnt. It is estimated that these materials would be worth €5.25 billion if recycled, and this would save 148 million tonnes of CO_2e. In the UK in 2008, 24 million tonnes of recyclables were wasted, equivalent to 19 million tonnes of CO_2 and worth £650 million.[112]

Often, the waste from one industrial process can be a valuable input material for other processes. The UK National Industrial Symbiosis Programme (NISP) was set up in 2005 to help businesses find markets for their waste, and currently has over 13,000 members from multinationals to individual entrepreneurs. By December 2011, the programme claimed to have achieved numerous benefits, many of which will continue to accrue for years to come:[113]

- over 39 million tonnes of waste diverted from landfill;
- over 2 million tonnes of hazardous waste eliminated;
- over 53 million tonnes of virgin material saved;
- over 35 million tonnes of CO_2 savings;
- over 64 million tonnes of water saved;
- additional sales for industry worth £900 million;
- over £860 million of cost savings to industry;
- over 8,000 jobs created and saved.

Examples of some successful projects are given in Box 6.4. Similar industrial symbiosis programmes are now becoming established in other countries, including China, Brazil, Mexico and the United States. It has been estimated that industrial symbiosis could save €1.4 billion a year in the EU and generate sales worth €1.6 billion.[114]

Box 6.4 Industrial Symbiosis case studies

Cow-powered cement

John Pointon and Sons Ltd is the largest single-site animal renderer in the UK, converting animal waste to harmless by-products. The NISP programme found that energy-rich meat and bone meal which had previously been sent to landfill could instead be used to fuel a number of nearby cement kilns. Burning the meat and bone meal also generated calcium salts which could be used as a raw material replacement. In the first year, the project saved 277,000 tonnes of carbon dioxide, diverted 150,000 tonnes of waste from landfill and created ten jobs. The operation is now expanding.

From foam to homes

Sekisui Alveo makes foam products for the automotive sector, such as door panels and instrument panels, and needed an alternative use for foam waste that had to be compacted and landfilled. The NISP team put the company in contact with Globally Greener Solutions Ltd, which had devised a method of converting foam waste into an inert material called GLOWASOL, which can then be used to make a variety of products including panels for building low-cost, sustainable housing. In the first year of operation, Sekisui Alveo saved £25,000 on waste disposal costs and 400 tonnes of material were diverted from landfill, saving 400 tonnes of raw material extraction and 6,000 tonnes of carbon dioxide.

Hazardous waste to high-quality wheels

Manufacturing company DENSO was paying £30,000 per year to dispose of a hazardous potassium aluminium fluoride-based material which was a by-product from its production of air conditioning units and engine cooling systems for the automotive industry. The NISP register was able to identify a use for the material as input to a metal foundry, Mil-Ver Metals, which used it to produce aluminium ingots which were then sold on to make high-quality alloy wheels.[115]

A growing proportion of valuable resources such as metals are now contained 'above ground', in the form of used and end-of-life consumer goods, infrastructure, stockpiles, scrap materials and even in landfill sites. These stocks of material in society can be considered as a separate resource base, giving rise to the concept of 'urban mines'.[116] In fact, precious metals such as gold, platinum and rhodium are often present in higher concentrations in end-of-life goods than they are in natural ore deposits.[117]

For many countries, a large part of the potentially available resource base is imported in the form of finished products, such as rare metals in electronic goods. There is a major opportunity to exploit this free stock of resources through reuse, re-manufacture and recycling of imported products (Box 6.5).[118] Often, however, materials are lost either through disposal or through export of end-of-life goods to other countries for reuse or recycling. In Germany, for example, the platinum contained in catalytic converters in scrap cars which are exported to other countries is equivalent to a third of the annual platinum use in Germany.[119] In 2006, the EU-25 exported 10 per cent of its metal waste and a staggering 71 per cent of its plastic waste to non-EU countries, mainly to Asia.[120] Import imbalances tend to create a low-cost transport option that encourages this export of waste – containers that arrive in Europe full of goods and materials can easily be filled with waste for the return journey. Illegal export of end-of-life goods, particularly vehicles and waste electrical and electronic equipment, also drains valuable raw materials from EU countries.[121] Those responsible often claim that the goods are for resale, but in fact they end up being recycled in countries with poor waste-treatment facilities. Controls on shipments of waste to other countries are currently being tightened.

Trade of waste does not in itself present a problem – in fact, export to countries with lower labour costs can allow lower-quality waste (such as co-mingled materials) to be more effectively recycled. Also, if China is making goods for the rest of the world, then it makes sense to make them with recycled material if possible. But problems arise if the waste is inadequately recycled, meaning that valuable materials are lost to the world economy – for example, if just the steel or aluminium is recovered from cars or appliances, but not the rare metals, or if the goods made from the waste end up in countries with no recycling facilities.[122] There can also be severe health and safety problems (see Chapter 7). Better recycling systems, such as separate collection of different waste fractions, can allow more waste to be reused in the country of origin, which reduces the need to export waste and retains valuable materials for national use.

Both the establishment of industrial symbiosis networks and the recovery of valuable resources from 'urban mines' come with an additional benefit for business – a steady supply of locally sourced input material which is not vulnerable to the same supply risks and price fluctuations that apply to virgin raw materials sourced from a small number of global suppliers. Europe is particularly vulnerable to supply risks, with the world's highest net imports of materials per person. Some 14 per cent of EU jobs (30 million) depend on imported minerals.[123]

Box 6.5 Urban mining case studies

ecoATM – cash for used phones

In California, ecoATM has launched street kiosks which give consumers cash for used electronic goods such as mobile phones. Consumers in the United States buy about 500 million new electronic gadgets every year, and most of the old goods end up in landfill sites. ecoATM has trialled the use of eCycling stations – automated kiosks which identify electronic goods and then offer consumers a buy-back price. Around half of the phones are sent to secondary markets and the rest are recycled.[124]

Umicore turn scrap into gold

Umicore Precious Metals Refining is one of the world's largest recyclers of precious metals. They accept a variety of scrap, including circuit boards from electronic goods, cables and used catalysts from vehicles and industry. Electronic scrap is either mechanically shredded or manually dismantled and 17 different metals are recovered. Manual dismantling gives better recovery rates, especially for gold and palladium.

Umicore uses the recovered metals in its own catalysis and performance materials businesses, as well as selling them on to external users. By doing this, the company is able to reduce the risk of supply disruptions and price fluctuations to its own business operations, as well as gaining a source of external revenue.

Savings are not confined to the simple and obvious benefits of reducing resource costs and avoiding waste disposal fees. A 'zero-waste' approach can also play a valuable role in stimulating innovation. As Robin Murray puts it, 'landfills and incinerators ask no questions'.[125] The traditional disposal approach tries to make waste invisible – but this misses an opportunity. Waste is a sign of failure and inefficiency, and efforts to reduce waste can improve the quality of the whole production system, tracing inefficiencies back to their source and resulting in cleaner, more innovative production methods.

New businesses, happy customers

We have seen how material efficiency measures can cut costs for businesses, making them more competitive. If resource prices continue to rise, as expected, this will be increasingly important in the future. But material efficiency also offers new business opportunities in areas such as waste reduction, materials recovery, repair and maintenance, lean manufacturing, design for durability and the 'product-service' model.[126]

The product-service model, which involves selling services such as 'mobility' rather than products such as cars, is of particular interest. In 2000, the market share of product-services was 10 per cent in the EU and 15 per cent in the United States, covering sectors such as the hire of office equipment, commercial vehicles and power tools.[127] Product-service helps to avoid waste because the service provider has an incentive to provide goods which are durable and easy to repair, upgrade and re-manufacture. Environmental benefits are greatest where the service is sold on a per-use basis, such as vehicle hire which is charged per kilometre driven so that suppliers have an incentive to supply fuel-efficient vehicles and users have an incentive to drive less. Examples include Xerox with its pay-per-page print services, Michelin with its tyre maintenance programme for fleet vehicles (charged per kilometre driven), and Safechem which leases industrial solvents, collecting used solvents for recycling. Safechem now offers a de-greasing service, charging per square metre of surface de-greased, giving the company an extra incentive to use as little solvent as possible and helping users to minimise the risks of using hazardous chemicals.[128]

Product-service offers a number of benefits to both the supplier and the customer. Customers no longer have to worry about the maintenance and disposal of the product, which is all taken care of by the supplier. Rather than being stuck with the same product for many years, customers can easily upgrade to a newer model, or change to one with different features, or even cancel the contract. The customer can also gain the opportunity to use a higher-quality product which they might not be able to afford as a one-off purchase. The product is likely to be more durable and reliable, avoiding the inconvenience caused by unexpected breakdowns.

From the business point of view, the product-service model promotes a closer relationship between the supplier and the customer, which is more likely to generate customer loyalty and lead to repeat business. Manufacturers also retain ownership of the valuable materials contained in their products, which can be recovered when the products are returned for reuse or re-manufacture – a particularly important benefit in view of likely future supply risks and price volatility for materials such as rare metals. Finally, the product-service model also gives a more predictable income stream to business. On the down side, start-up costs can be high and it will take some time to recover the initial costs of setting up the business. There is also a considerable culture change involved in switching from one-off purchases to a product-service model.[129]

In the UK, it is estimated that 20 per cent of current household expenditure on goods could be transferred to a product-service model, and there could be even greater opportunities in the business sector. Ericsson and Novatium are about to launch a global 'PC as a Service' business which they are targeting at 'the next billion users, especially in developing markets'.[130] By offering services such as surfing the internet, virus protection and software upgrades as a 'cloud computing' service run at centralised data centres, they hope to reduce battery use and increase computing speed for users – offering a computer start-up time of just 5 seconds, as well as the convenience of centralised file storage.

Similar benefits can occur when goods are sold with an accompanying service provision. The ongoing relationship between the customer and the provider means that goods are more likely to be taken back for recycling or re-manufacture at the end of their life. The supplier also has an incentive to make goods more durable and reliable, in order to minimise the costs of honouring the service contract as well as keeping the customer happy and encouraging brand loyalty. For example, IBM now make computers with reusable casings, and Kyocera make a laser printer with a silicon-coated drum which lasts for 300,000 pages, or about 5 years, compared with 5,000–10,000 pages for the conventional printer.

Even businesses that stick with the traditional sales model have an opportunity to engage more with customers to encourage them to minimise the life-cycle impacts of the goods they buy. Marks & Spencer (M&S) in the UK encourages customers to wash clothes at lower temperatures, and also offers a £5 voucher for customers who return used M&S clothes to Oxfam for resale in their charity shops.

Outdoor clothing company Patagonia, which already tries to minimise its environmental impact by using low-impact materials such as recycled plastic bottles, hemp and organic cotton, has recently taken the radical step of encouraging its customers to buy fewer clothes – only buying what they really need. The company also provides recycling bins in its stores, and offers free repair and resale services. Patagonia finds that this message has not decreased sales. The company reasons that customers are choosing to buy more durable and recyclable Patagonia products rather than cheaper 'throwaway' items from its rivals.[131]

The product-service model does face cultural and economic barriers – many consumers prefer to buy rather than rent, and the rental sector for goods such as televisions has declined as prices have fallen. However, price signals and awareness campaigns can be used to encourage greater uptake in sectors where product-service shows potential.

Conflicts

The link between material use and climate change is strong, and there are few direct conflicts. On the whole, saving materials saves considerable amounts of energy, cuts emissions from waste disposal and avoids emissions from food, wood and fibre production. However, some low-carbon technologies can lead to increased pressure on scarce resources. The impacts of biofuel production on demand for water, fertilisers, energy and land were discussed in Chapters 4 and 5. This section will discuss the dependence of various low-carbon technologies on a supply of rare metals.

There is also a potential conflict with the strategy of lifetime optimisation, in the case of goods that use a lot of energy, such as cars or refrigerators. As energy efficiency is continually improving, consumers need to balance the resource cost of buying a new appliance with the energy savings from switching to a newer model. One solution would be to design modular products, with each part being easily upgradeable, so that customers could simply replace the motor, leaving the body of the appliance intact, or install an existing engine into a new, lighter car body. DaimlerChrysler's Smart Car, for example, has been designed with interchangeable body panels and other parts that are easily replaceable.

Finally, if we succeed in restraining our total material consumption to a sustainable level, there could be major implications for employment, economic growth, trade, international development and poverty alleviation. Issues concerning trade and development are introduced below, and the discussion will continue in Chapters 7 and 8.

Rare metals for clean technology

Rare metals are increasingly used in a range of advanced low-carbon technologies.[132] These include:

- indium, gallium, tellurium, germanium and selenium for solar panels;
- gallium for energy-efficient LED lighting;
- platinum, palladium and rhodium for fuel cells, hydrogen and hybrid cars and emission control;
- rare earth elements for fuel cells, efficient lighting and cooling technologies and emission control;
- lithium, cobalt and tantalum for electric vehicle batteries;
- rhenium and ruthenium super-alloys for fuel-efficient aircraft and advanced wind turbines;
- neodymium and samarium for permanent magnets used in hybrid car motors and wind-turbine generators;
- cobalt, rhenium and platinum for use in the Fischer–Tropsch process for making liquid biofuels from biomass.

The move to low-carbon technologies will place increasing pressure on the availability of these resources, many of which are already in short supply. It is estimated that clean-energy technologies already account for 20 per cent of the demand for rare earth and other critical metals.[133] By 2040, the demand for gallium for solar panels in Europe alone could be seven times higher than current global production, and demand for indium and tellurium could also exceed supply by factors of 3 and 30 respectively.[134] On the other hand, this could be viewed as another reason why resource efficiency must be improved, in order to safeguard supplies of these key materials and enable our transition to a sustainable infrastructure. Many of the measures described in this chapter can be applied to enhance stocks of these valuable commodities – especially lifetime optimisation, reuse, re-manufacturing and recycling of electronic waste (including renewable energy equipment).

Trade and international development

The earth is already in ecological overshoot, and the situation is worsening as population and affluence grow. But there are considerable inequalities in resource use, with 20 per cent of the world's population consuming 80 per cent of global resources. On average, North Americans consume 88 kilograms (kg) of resources per person each day and Europeans 43kg, while Asians consume 14kg and Africans just 10kg. Despite this, there is a net transfer of material goods from developing countries to the West. Citizens of the EU each import almost 3 tonnes of material per year (net), while those in developing countries export a net average of 0.7 tonnes per year.[135]

If consumption is to stay within sustainable limits, then as developing countries consume more, Western consumers will need to consume less. One commonly quoted suggestion is that Western economies must dematerialise by a factor of 10 (see 'The way forward' below). But this presents us with a paradox: reduced consumption in the West could have a major impact on the economies of the less developed countries that rely heavily on income from exporting raw materials or agricultural commodities.

This is a complex area and there are no simple answers, but various options have been suggested that may help to ameliorate the situation – both by helping developing countries to 'leapfrog' to cleaner production methods which reduce the environmental and social impacts of producing commodities for export; and by reforming trade laws and tackling the causes of poverty so that developing countries become less reliant on exports of raw materials. This will be discussed in 'The way forward'.

The way forward

Cutting our use of raw materials offers a very wide range of co-benefits. It can save money for households and businesses; reduce the environmental and social impacts of resource extraction; avoid the impacts of waste disposal; reduce exposure to material supply disruptions and price fluctuations; and help to minimise worldwide conflict and unrest arising from rising commodity prices and shortages of food, land and water.

A consensus is emerging on a number of priority areas:[136]

- *Electronic waste.* This is a growing problem because of escalating consumption of electronic goods, the large 'ecological rucksacks'[137] attached to their manufacture, the use of increasingly scarce rare metals, the hazardous nature of the waste (containing toxic metals or refrigerants), and the increasing export of e-waste to developing countries. Recycling is technically challenging because of the need to separate metal, plastic, glass and other materials, but eco-design techniques can make reuse, re-manufacture and recycling much easier. There is also a need to set up better collection mechanisms for used goods.
- *Construction materials.* These form the bulk of waste in most countries – accounting for 33 per cent of total waste in Europe, for example, with mining and quarrying waste forming another 24 per cent, far outweighing municipal waste at just 8 per cent.[138] Although the waste is fairly harmless, it can cause problems where there is a shortage of landfill capacity. Reuse and recycling of construction materials can reduce the landscape damage associated with raw material extraction, and the high energy use and carbon emissions from making concrete, bricks, glass, metal girders and plastic window frames, although there are limits on how far materials can travel before any greenhouse gas savings are cancelled out by transport emissions. Only 7 per cent of the construction aggregates used in Europe are currently derived from recycled material.[139] There are promising opportunities to explore ways of reusing entire buildings; for example, using modular construction that allows internal walls to be moved and windows, doors and cladding to be easily replaced. This can allow buildings to be reconfigured or refurbished to meet changing needs and fashions instead of being demolished and rebuilt.
- *Food waste.* Large amounts of food are wasted, due mainly to storage problems in developing countries, and wasteful consumer habits and retail practices in developed countries. Yet one study estimated that food and drink production

accounts for 20–30 per cent of all environmental impacts in the EU.[140] Cutting this waste can alleviate a wide range of problems including high use of land, water, fertilisers and energy; greenhouse gas emissions; biodiversity loss; water pollution; and rising food prices.

- *Waste prevention and eco-design.* Although recycling is now well established in many countries, efforts at waste prevention have been limited. There is great potential for waste to be limited at source through eco-design of products and processes, more reuse and re-manufacturing, and reduction of consumer waste. In recognition of this, EU member states are obliged to establish waste prevention programmes by the end of 2013.
- *Development of new business models* such as product-service, and boosting the market for recycled materials.
- *Addressing consumption patterns.* The European Environment Agency finds that there is a 'huge, largely unused potential for encouraging environmentally less intensive consumption patterns'.[141] While attitude and lifestyle changes can be hard to achieve, the level of waste in affluent societies is currently so high that there is plenty of potential for 'quick wins' through very simple actions such as delaying the purchase of new products until the old ones are worn out.

The last point emphasises that we cannot reduce our total use of materials to sustainable levels through resource efficiency alone. We also need resource *sufficiency* – consuming as much as we need but no more – to keep total resource use within sustainable limits. Affluent societies will need to reduce their consumption levels and restructure their economies away from the current model of consumption-based growth. Developing countries need to be able to increase their living standards while coping with a potential decline in the income they currently receive from the export of raw commodities.

There is no point in pretending that an absolute reduction in our levels of material consumption will be easy. It will require a total change of attitude and culture, from our current throwaway society to one which values thrift and places higher value on non-material goods. This challenge will be addressed in Chapters 7 and 8. This section will summarise policies that can help to reduce our total use of raw materials to sustainable levels, and then look at ways of helping to mitigate the impacts on commodity-exporting developing countries.

Resource sufficiency: living within our means

Existing policy frameworks tend to undervalue resources and encourage over-consumption and waste. If we are to live within our environmental means, we need an integrated strategy to measure total resource use, set appropriate targets, and use a combination of price signals, regulation and other policies to keep resource use within safe limits.

Measure resource use

Many countries already measure their resource use, using the technique of Material Flow Analysis. This uses economic input–output tables and trade data to assess the flow of materials and resources within an economy and between countries. A suite of indicators has been developed (Table 6.7), ranging from simple indicators based on the total weight of materials extracted within a country to more advanced indicators that take account of exports and imports, or consider the environmental impact rather than just the amount extracted.

Table 6.7 Indicators of resource use

Indicator	Definition	Notes on use
Used extraction	Resources which are extracted and then used or processed.	Includes waste from processing, such as mining tailings.
Unused extraction	Resources which are moved during the extraction process, but not used or processed further.	Includes overburden shifted during mining and quarrying; crop waste left in fields after harvest; soil shifted during construction.
DE Domestic extraction	Sum of used resources extracted within a country.	Includes crops harvested, metal ores mined, fossil fuels extracted, etc. Excludes water, as that would dwarf other material flows.
DMI Domestic material input	DE plus imports.	All resources input to the economy.
DMC Domestic material consumption	DMI minus exports.	All resources used within a country.
RME Raw material equivalents	Upstream indirect flows of materials used to make an imported product, less the weight of the product.	The 'invisible' materials used and then discarded when a product is made in another country.
DMC_{RME} Domestic material consumption expressed in raw material equivalents	DMC plus RME: all materials used by a country, including those used to make imported products but which remain in the country of origin.	All resources used within a country plus the invisible materials associated with imports.
TMR Total material requirement	Used and unused domestic extraction, plus imports, plus used and unused indirect flows related to imports.	A measure of all the material required by an economy. Difficult to calculate – only done for a few countries and a few years.

Indicator	Definition	Notes on use
RMC Raw material consumption	DMC_{RME} less RME of exports.	Represents the total amount of global used extraction, including that from imports, used to provide goods for domestic consumers.
Ecological rucksack Also known as MIPS (material input per service unit)	Total weight of natural used and unused resources moved from their original place in the ecosphere during the life-cycle of a product or activity, including manufacture, use and disposal, but not included in the product itself. Split into biotic, abiotic, water, air and soil erosion, which should be expressed separately.	For imported products, this is equal to the indirect flows of used and unused materials. This indicator is mainly suitable for raw materials and products with a low level of processing – it is difficult to calculate for highly processed goods, and has only been calculated for a limited number of materials and products so far.
Ecological footprint	Productive land and sea area needed to provide biotic material resources (crops, timber, water, fish, livestock) and absorb waste and pollution (forest area needed to absorb carbon dioxide). Measured in global average hectares (gha).	The ratio between the ecological footprint and the actual area available gives an estimate of overshoot, but it does not cover non-renewable resources, or impacts for which there is no regenerative capacity (toxicity, eutrophication, etc.).
EMC Environmentally weighted material consumption	Material flows are multiplied by environmental impact factors for climate change, human health, land use, ozone depletion, eco-toxicity, ozone formation, resource depletion, acidification, eutrophication and radioactivity.	Goes beyond the weight of resource extracted to look at the level of environmental damage. Cannot be aggregated into a single indicator unless the different impacts are subjectively weighted.
HANPP Human appropriation of net primary production	Percentage of earth's annual production of biomass being taken by humans. Includes direct use such as crops harvested, and indirect such as biomass which can no longer grow because land use has changed.	Measures how intensely ecosystems are being used by humans, and so how much is available for other organisms. Uses remote sensing data to assess land use, together with agriculture and forestry statistics.
LEAC Land and ecosystem accounts	Measures how much land area is being used by different economic activities.	Covers impact on land use and biodiversity. Uses remote sensing data to assess changes in land cover. Can be linked to monetary valuations of ecosystem services.

Source: Based on Reisinger *et al.*, 2009

Ideally, indicators should include the impact of materials imported from other countries – not only the volume of imports, but also the indirect 'cradle-to-border' impacts that remain in the country of origin, such as mining waste. The European Environment Agency recommends the use of TMR (total material requirement) or DMC_{RME} (direct material consumption in raw material equivalents). However, it is hard to obtain data on indirect resource use in other countries, and so the most widely used indicator in practice is DMC (direct material consumption).

The next step is to move beyond the total weight of materials extracted and look at environmental impacts in more detail. The EU recommends that a basket of four indicators should be used to cover the full range of impacts: the ecological footprint, EMC (environmentally weighted material consumption), HANPP (human appropriation of net primary production) and LEAC (land and ecosystem accounts).[142] Unfortunately, it is very challenging to compile these indicators and few datasets currently exist. The OECD and Eurostat have produced guidance documents, but further work is needed to develop and standardise methodologies, improve understanding of the life-cycle environmental impacts of resource use and improve data quality for indirect impacts.[143]

Set targets

Most existing targets tend to focus on resource productivity; i.e. the amount of resources used to produce one unit of economic output (expressed as GDP). The *Factor Four* book, for example, used case studies to demonstrate that we could use existing technologies to double production whilst halving resource use, thus increasing resource productivity by a factor of 4.[144] This was succeeded by *Factor Five*, using the same approach.[145] Most of the examples in these two books dealt with energy use rather than material use.

In a separate initiative, the Factor 10 Institute draws a distinction between resource use by developed and developing countries. Based on the observation that 20 per cent of the world's population consume 80 per cent of total resources, it is proposed that rich countries should improve their resource productivity by a factor of 10 in order to allow poorer countries the ecological space to increase their consumption by a factor of 2–5.

A number of countries have already adopted resource productivity targets. Austria and the Netherlands are aiming for a Factor Four increase; Germany wants to double material productivity by 2020; and Belgium and Ireland are aiming to gradually decouple natural resource use from economic growth.[146] There is considerable scope for more countries to adopt targets: resource productivity in the EU averaged €1.3 per tonne of DMC in 2007, but it ranged from €0.3 to €2.5 per tonne in individual countries.[147]

The problem with resource productivity targets is that they do not necessarily limit consumption to sustainable levels, because efficiency improvements tend to stimulate economic growth which drives extra consumption (this is the 'rebound effect', which will be discussed in Chapter 7). An alternative approach is to set targets for resource use *per capita*, to try to keep consumption levels down. For example, the Factor 10 Institute and the Lindau Group propose cutting total global used and unused resource extraction (measured as TMR) from 20 tonnes to 6 tonnes *per capita* by 2050, in parallel with the Stern Report target to cut carbon emissions from 5 tonnes to 2 tonnes *per capita*.[148] The World Resources Forum proposes a less ambitious target: to stabilise resource use at the present level of 6–10 tonnes *per capita* (this excludes unused resources, which are roughly equal to used resources).[149]

Careful readers will have spotted that *per capita* targets avoid the thorny issue of population growth, and thus fail to guarantee a reduction of total impact to sustainable levels. The population issue is fraught with moral and ethical difficulty, and a detailed discussion is outside the scope of this book. However, while specific targets to limit population growth are problematic, it is worth noting that research shows that voluntary reductions in family size are linked to a number of desirable factors such as better education, more economic opportunities for women and reduction of child mortality rates.[150]

In order to keep consumption within sustainable limits, we need to set targets that address our total impact directly. Perhaps the most obvious goal is that our ecological footprint should stay within the Earth's carrying capacity. However, the ecological footprint deals only with renewable resources – separate targets would be needed to cover non-renewable resources such as metals and minerals. One study concludes that we need to set four targets simultaneously: resource extraction rates; carbon emissions; resource use *per capita* and resource use per unit of GDP.[151]

The benefits of resource efficiency for the economy mean that resource productivity targets are far less controversial than targets to limit absolute consumption levels. Despite this, some countries have adopted targets for an absolute reduction in resource use. Italy aims to reduce the TMR of the economy by 75 per cent by 2030 and 90 per cent by 2050. Denmark has a long-term goal to limit resource consumption to a quarter of the 2002 level.[152] Germany has a separate target for land use, aiming to reduce the loss of undeveloped land to housing and transport from 120 hectares per day in the 1990s to 30 hectares per day by 2020. These countries have low rates of population growth, which may help to make these targets more achievable.

Economic signals: get the prices right

Our current economic policies encourage an inefficient 'throwaway society', where resources are over-exploited, valuable raw materials are discarded as waste and ecosystems are undervalued. This is due to a combination of factors:

- The external environmental and social costs of resource extraction are not reflected in the market price of the materials.
- Governments often subsidise the costs of extracting resources (particularly for fossil fuels and agricultural production).

- Until the early 2000s, there has been a gradual long-term decline in most commodity prices as technological progress has made production easier.
- Labour is heavily taxed, but materials are not taxed at all, so labour costs are relatively high.

This policy framework gives little economic incentive to reduce waste. For example, high labour costs and low material costs mean that repair and maintenance are often more expensive than simply buying new goods and discarding the old ones. There is also little incentive for companies to collect their used products for re-manufacture.

Price signals can be used to reinforce a change in attitudes. The use of taxes, caps and subsidies to cut carbon emissions is discussed in more detail in Chapter 7. This section focuses on specific policies that can be used to reduce resource use and discourage waste.[153]

Waste disposal charges, in the form of landfill or incineration taxes, have been successfully applied in many European countries. In the UK, for example, the landfill tax has directly reduced the amount of waste sent to landfill.[154] Countries that are moving most quickly towards the goal of a 'recycling society' are those that have higher landfill taxes.[155] In addition, some countries charge householders through pay-as-you-throw schemes, with charges varying by weight, volume, number of bags, collection frequency or bin size. Over 2,000 communities in the United States have adopted pay-as-you-throw systems, with typical reductions of 25–35 per cent in the amount of waste generated.[156] In Europe, pay-as-you-throw has reduced waste generation and increased recycling rates in Austria, Belgium, Denmark, Italy and the Netherlands.[157]

Rewards have also been used, such as the deposit system for refillable bottles which is applied in Germany and the Netherlands. However, the deposit system faces barriers due to the increase in international trade in bottled drinks, which makes it difficult and costly to return bottles to the supplier.

Taxes or cap-and-trade schemes can be used to ensure that resource prices reflect their full environmental costs (see Chapter 7). Although these tools have mainly been applied to fuel and carbon emissions, they could be expanded to cover other resources such as minerals and metals.[158] In fact, taxes on the extraction of construction materials already exist in several countries including the Czech Republic, Italy, Sweden and the UK. However, although taxes would be expected to reduce resource consumption, only caps or quotas can guarantee that resource use is kept to a sustainable level. Paul Ekins and colleagues propose an international cap-and-trade scheme based on a target for all countries to converge on a TMR of 6 tonnes *per capita* by 2050. Imports from countries not agreeing to participate would have to be taxed at the border, but with exemptions for low consumers (countries using less than the current average resources *per capita* of the countries within the trading system).[159]

While economic instruments can be powerful tools for changing behaviour, there are some caveats in applying them to material extraction. First, raising the price of useful materials could be less politically acceptable than taxing waste and pollution. Second, if taxes or permits are based on weight, then resources with widely differing environmental impacts will be taxed at the same rate per tonne. This could result in a switch towards lighter resources which will not necessarily reduce environmental impacts (for example, from wood to plastic, which has higher toxicity and is more energy-intensive).

With all economic instruments, there may be regressive effects – taxes can mean that essential goods become unaffordable for those on low incomes, or that small companies struggle with material costs. To counter this, the tax revenue can be redistributed as income tax cuts or to fund efficiency improvements, so that the whole scheme remains revenue-neutral. This will be discussed further in Chapter 7.

Finally, economic instruments can also be applied to put a value on ecosystems (see Chapter 4). By making it clear that natural ecosystems have an intrinsic financial worth, and that they cannot be destroyed without compensation, the cost of extracting virgin resources from sensitive areas should increase, thus promoting resource efficiency and the use of less damaging materials.

Regulations and standards

Although price signals are essential to encourage resource efficiency and to keep consumption within sustainable limits, regulations and standards also have an important role to play.[160]

Regulation has been particularly successful in increasing recycling rates. European Directives set reuse and recycling targets for various materials: 70 per cent for construction and demolition waste; 100 per cent for batteries; 50 per cent for household waste paper, metal, plastic and glass (by 2020); 55 per cent for packaging waste (by 2006); and 85 per cent for end-of-life vehicles (by 2015). New proposals will increase the target for recycling waste electrical and electronic equipment to between 50 per cent and 75 per cent and, for the first time, impose a separate target of 5 per cent for reuse.[161] In addition, biodegradable waste (food, paper, wood) sent to landfill sites must be reduced to 35 per cent of 1995 levels by 2016.

Despite progress towards these targets, large quantities of recyclable material are still landfilled or incinerated. One way of tackling this is to ban the disposal of recyclable materials or unsorted waste. Landfill bans have been enforced in various countries, including Austria, Germany, Sweden, the Netherlands, the Flanders region of Belgium (see Box 6.3 above) and Massachusetts in the United States. In all cases, rates of material recovery increased, even though recycling rates were high to start with. In Germany, for example, the proportion of waste sent to landfill fell from 27 per cent to just 1 per cent, while the proportion recycled or composted increased from 51 per cent to 62 per cent.[162]

Landfill bans are most successful at increasing the recycling rate when there is a simultaneous ban on incinerating recyclable waste. Incinerators often act as a powerful disincentive to recycle, as they tend to require a constant guaranteed supply of high calorie feedstock (especially paper and plastic) in order to remain economically viable, and this can result in diversion of recyclable waste to incinerators.[163] Regulations could also be used to ensure that recyclable material is separated out at source, which gives a vast improvement in the quality of recovered material. And standards could be used to specify a minimum quantity of recycled content in certain goods, such as glass or cardboard packaging.

Regulations can also help to encourage the repair and reuse of goods. Extended Producer Responsibility (EPR), sometimes known as product stewardship or takeback legislation, forces suppliers to take back products at the end of their useful life. Coupled with charges for waste disposal, this can encourage producers to design products that can be more easily reused, re-manufactured or recycled. The WEEE Directive in Europe, for example, requires manufacturers and retailers to fund the collection and recycling of end-of-life electronic and electrical goods. Together with a sister directive on restriction of the use of hazardous substances, this drives manufacturers to phase out the use of toxic substances in their products in order to reduce end-of-life treatment costs.[164] Extended Producer Responsibility directives also apply to batteries, packaging and end-of-life vehicles. Other examples exist in Germany, Austria, Belgium, Japan, China, South Korea and some US states.[165] These EPR schemes tend to transfer the cost of waste management from the taxpayer to the consumer, via an increase in the product price, and should therefore act as an incentive for consumers to optimise the lifetime of their products.

Technical standards can also help to make goods more durable and easier to repair, upgrade or re-manufacture. One example already in progress is the development of a standard for a universal mobile phone charger, which will work with all makes and models of phones, so that consumers do not have to buy a new charger each time they buy a new phone. Most of the 51,000 tonnes of chargers produced each year currently end up in landfill.[166] Another example is the EU End of Life Vehicles (ELV) Directive, under which vehicles must be designed so that 85 per cent of the materials can be reused or recycled. Regulations could be used to ensure that equipment such as computers is designed to be easily upgradeable. Repairability standards could ensure that parts which tend to wear out are easily replaceable at moderate cost. And quality standards for recycled products and materials can boost consumer confidence and strengthen the market. All these initiatives can be supported with labelling schemes; for example,'easy to repair/ upgrade' labels.

The EU Ecodesign Directive of 2005 was originally intended to promote these kinds of measures, but by the time it was implemented in 2009, it dealt only with the efficiency of energy-using goods, not with material efficiency. A report for the European Parliament recommends that the Ecodesign Directive should be extended to include all products and all types of eco-efficiency.[167] The European Commission is also developing a waste prevention policy that includes product ecodesign; reduction of hazardous substances; promotion of more durable, reusable and recyclable products; and new consumption patterns. Examples of legislation to reduce consumer waste include bans on the use of plastic bags or on the sending of junk mail.[168]

Regulation can be used to stimulate companies to reduce their use of resources. Companies could be required to include a basic environmental audit in their annual reports, with estimates of resource use, waste generation and pollution together with an assessment of the potential for improved resource efficiency or cleaner production methods.

Finally, legislation is vital in protecting vulnerable ecosystems from the threats posed by resource extraction – for example, by designating conservation zones where development is restricted. Improving property and access rights, in order to give local people an incentive to manage their land sustainably, can also help to avoid the 'tragedy of the commons' where common resources are over-exploited because no one has long-term ownership of them. One potential barrier is the inclusion of 'stabilisation clauses' in international trade agreements, which require compensation if new legislation affects company profits. These agreements can inhibit national governments from setting up new social or environmental legislation.

Investment and research

Although there is tremendous scope to improve resource efficiency with existing technology, there is also a need for technical advances in areas such as eco-design, lean production and material recovery: recovering rare metals from electronic waste, for example. Recent OECD and EU reports recommend that government support for investment and research could be better targeted towards promoting eco-innovation; for example, by co-ordinating research programmes and offering grants, loans, tax breaks and technical assistance to companies or research groups.[169]

Education and awareness

Cutting material use will require a major culture change, and education and awareness programmes will be crucial. Just a few generations ago, thrift was a way of life for most people, but the arrival of cheap fossil energy led to the establishment of a wasteful consumer culture and a throwaway society. The repair sector, for example, has all but disappeared in many Western countries, where generating and throwing away waste has become a natural part of our everyday lives.

To counteract this, we need to start thinking of waste as an unnatural and avoidable concept. Rather than dealing with waste after it has been generated, we need to design out waste from the start of the product lifecycle. This will require:

- incorporating training on sustainability, resource efficiency and cleaner production into educational curricula, especially engineering, economics and business courses;[170]
- industry awareness, such as demonstration projects, high-profile awards, conferences, training schemes, advice centres, auditing and industrial symbiosis programmes;

- employee awareness, such as training, waste-reduction suggestion schemes and competitions, or linking bonuses to waste reduction;
- business awareness – information about new opportunities in material recovery, re-manufacture and eco-design and product-service business models;
- consumer awareness campaigns, promoting efficient use of products, waste avoidance, reuse, repair and recycling;
- better product labelling to help inform consumers about the environmental impacts of the products they buy, and to encourage businesses to look at the embedded impacts of their supply chains;[171]
- mandatory sustainability reporting for companies, to encourage them to identify opportunities to improve resource efficiency, as well as to help consumers choose products from better-performing companies;
- governments setting an example and boosting the market for resource-efficient goods and services – for example, by public procurement of recycled goods and by waste prevention in the public sector.

Developing countries: escaping the resource curse

We have identified a potential conflict in that cutting material consumption in the West could reduce the export earnings of developing countries that produce raw materials. Yet in many low-income countries, it has been shown that once depletion of resources and pollution are taken into account, export of commodities actually causes national wealth to diminish.[172] This is linked to a phenomenon called the 'resource curse', where over-reliance on export of raw resources inhibits the development of higher-skilled areas of the economy, especially in countries with weak 'grabber-friendly' governance.[173]

Trade laws can contribute to this problem. Western countries often impose high tariffs on imports of high-value-added goods, but low tariffs on imports of raw commodities. This has the effect of encouraging countries to export lower-value raw commodities such as cocoa beans and iron ore rather than high-value processed goods such as chocolate and steel. For example, 70 per cent of the value of Africa's exports is from minerals and fuels, and only 20 per cent from manufactured products. In contrast, around 80 per cent of exports from Europe and North America are manufactured products. This means that less income accrues to Africa for each tonne of material exported, meaning that more raw material has to be exported to earn the same income. In Africa, it takes 7 kilograms of domestic resources to produce \$1 of GDP, compared to less than 1 kilogram in Europe and North America.[174]

Another problem arises from export of subsidised products from the West. The OECD subsidised its agricultural products by \$265 billion in 2008. Europe, for example, imports a large amount of raw commodities for animal feed, such as soya from Latin America, but is a net exporter of milk and meat. With subsidies under the Common Agricultural Policy, these animal products are sold abroad at prices which undercut local producers in developing countries. Dairy farmers in Burkina Faso, Senegal and Cameroon have been put out of business by the import of subsidised milk powder from Europe. Although these imports do provide cheap food for the local population, they also destroy livelihoods. In recent years, the ability of African governments to control subsidised imports by imposing tariffs has been progressively reduced through enforcement of trade laws imposed by Western countries.

Reform of trade laws and an end to unfair subsidies would help developing countries to strengthen their economies and reduce their reliance on income from raw commodity exports.[175] It has been estimated that removing subsidies and tariffs on cotton alone would increase real incomes in Sub-Saharan Africa by $150 million per year.[176]

Eco-labelling, standards and certification schemes, together with awareness campaigns, can shift the market towards more sustainable goods, thus ensuring that developing countries extract maximum value from their exported resources at minimum social and environmental cost. Examples include the Marine Stewardship Council certification for sustainable fisheries; the Forest Stewardship Council scheme for timber; the Initiative for Responsible Mining Assurance; standards on food safety, pesticide use and the use of hazardous substances; and fair trade and organic certification systems.[177] Small companies and developing countries might need help with the costs of meeting standards and qualifying for labelling schemes – including 'capacity building' to help set up the national institutions needed to administer the schemes.

Finally, tackling the causes of poverty directly, through targeted aid programmes and debt cancellation, can also help developing countries to escape a dependence on resource exports. In order to keep up with debt repayments, many countries are forced to export large quantities of raw materials or cash crops, making them vulnerable to fluctuations in international commodity prices and encouraging over-production.

7 A stronger economy
Long-term stability and prosperity

Key messages

- *The old view of climate policy as an economic burden* is giving way to a new vision of a dynamic, prosperous green economy.
- *Economic co-benefits* include a probable net increase in jobs compared to business-as-usual; resource cost savings for businesses and households; increased productivity and competitiveness; new business opportunities; more innovation; and a secure and stable economy, protected from resource shortages and price shocks.
- *A low-carbon economy needs strong government support* including a clear regulatory and cost framework, ambitious long-term targets, and investment in education, research and infrastructure.
- *Resource use and carbon emissions need to be kept within sustainable limits* through caps and/or taxes. This will prevent increased consumption from wiping out resource efficiency savings.
- *Governments need to take action to protect vulnerable sectors of society* from the impacts of a low-carbon transition: retraining workers who lose their jobs; helping businesses to introduce low-carbon technologies; funding home efficiency improvements for low-income households; and helping developing countries to 'leapfrog' to sustainability.
- *In the long term, we need to restructure our economies away from a dependence on continuous growth in material consumption* so that we can achieve well-being for 9 billion people in a world of finite resources.

Can we afford to cut carbon emissions to sustainable levels? Fear of the economic cost of climate action remains a key obstacle to progress. Western economies have become dependent on lavish supplies of relatively cheap fossil energy to heat our inefficient buildings, transport goods around the globe, sustain our intensively farmed food chain and support the throwaway culture on which our model of consumption-based growth depends.

In the wake of the financial crash of 2008 and the ensuing global recession, many argue that we cannot afford to limit carbon emissions or to invest in green technologies – that there is a trade-off between economic progress and environmental protection, and that 'being green' is a luxury which we can only afford when the economy is booming. Yet there is a growing body of evidence to suggest that in fact the opposite is true: investment in clean, efficient, low-carbon technologies can create jobs, stimulate innovation and lead to a more prosperous economy. Continuing with a fossil-fuel-based economy, on the other hand, is likely to lead to ecological and economic disaster, as vital ecosystem services collapse with enormous costs to society – especially to the most vulnerable communities in low-income countries.

There is no denying that the transition to a low-carbon society will involve significant financial investment. Some of the changes we need – improved energy and material efficiency, for example – will quickly pay for themselves, but others will not. Our infrastructure was designed for a different era – one in which, at one point, it was thought that energy would become 'too cheap to meter'. We need radical changes – new energy supply and distribution infrastructure, better public transport systems and a massive overhaul of millions of poorly insulated and inefficiently heated buildings. We need a new industrial revolution to 'design out waste' from the start of the production process. These changes will cost money, but they can also generate long-term economic benefits.

In fact, the problem starting to emerge now is that resource efficiency can almost be *too* successful in strengthening the economy. The cost savings and productivity increases arising from efficiency improvements can result in additional economic growth that eventually wipes out the carbon savings originally achieved (this is called the rebound effect). Energy and material efficiency, traditionally thought of as environmental policies, can in fact be more effective as economic and social policies.

To counter this effect, an effective low-carbon strategy needs to limit total emissions, through caps or taxes. This introduces additional social and economic impacts – taxes can be regressive, for example, (hitting the poor hardest) and can also affect competitiveness. Further social impacts can arise due to the shift in employment towards low-carbon sectors and technologies. In the long term, we need to look at ways of restructuring our economies so that jobs and living standards do not rely on maintaining unsustainable levels of consumption-based growth. This chapter will look at the economic co-benefits of climate change policies, assess the complex trade-offs between environmental, social and economic factors and attempt to suggest a way forward that maximises the benefits and minimises adverse impacts.

Relevant climate policies

Which climate policies can yield economic co-benefits? Typically we might think of sectors that can create new jobs, such as renewable energy; or strategies that can improve industrial competitiveness, such as resource efficiency. But in fact we cannot isolate particular policies, technologies or sectors, because the economic impacts depend strongly on the exact combination of policies adopted – not just climate policies, but also the accompanying social and economic policies. This chapter will therefore focus on the overall policy framework rather than on particular technical solutions.

It can be useful to consider the policy framework in the context of the 'I=PAT' equation:

$$\text{Impact (I)} = \text{Population (P)} \times \text{Affluence (A)} \times \text{Technology (T)}$$

The equation simply says that the total environmental impact (carbon emissions in this case) is the product of the population (P), the average consumption of each person in dollars *per capita* (A) and the technology factor (T), which is the carbon intensity of the economy (tonnes of carbon per dollar of GDP).

To limit total carbon emissions, we can address either the left side or the right side of the equation. Policies to address the left side include caps or taxes on carbon emissions, which can be effective, but which tend to be politically difficult. Addressing the right side involves trying to reduce P, A or T, but without allowing the other two factors to increase too much. This is difficult because the links between P, A and T are complex and can be unpredictable: as one decreases, the others often increase.

With this framework in mind, we can identify four types of climate policy that have broadly different economic impacts.

Resource efficiency

Improving energy or material efficiency (thus reducing T) tends to deliver direct economic benefits by saving money for businesses and households, thus improving productivity and competitiveness, and stimulating growth. It also plays an important role in avoiding future resource shortages, thus contributing to a more stable economy and improved well-being. For this reason, improving resource efficiency has been a main strand of climate policy. However, as we will see in the section on 'Conflicts' below, the rebound effect means that cost savings from higher efficiency result in increased affluence (A), which can lead to extra consumption that wipes out some or all of the carbon savings. Other types of rebound are also possible: efficiency improvements such as increased agricultural productivity might cause the population to increase, for example. Increasing affluence, however, does at least mean that we can afford cleaner and more efficient technologies, thus helping to reduce T still further.

Low-carbon energy

Replacing high-carbon fossil fuels with low-carbon alternatives such as renewable or nuclear energy will reduce the carbon intensity, T, but not necessarily with direct economic benefits. New jobs will be created, but these will be offset to some extent by jobs lost in the high-carbon energy industry. Investment costs tend to be high: low-carbon energy can be more expensive than coal, oil and gas in the short term, though this may well be reversed in the long term, and the cost can be offset by the benefits of reduced pollution and improved energy security (see Chapters 3 and 5). If investment in low-carbon energy leads to short-term price rises, then additional policies will be needed to protect vulnerable energy users.

Lifestyle changes

Lifestyle changes such as eating less meat, driving less and buying fewer material goods (see Chapter 8) all have the effect of reducing carbon emissions per person – the product of A and T. These changes focus on consuming only as much as we need, and are therefore sometimes referred to as 'sufficiency' as opposed to 'efficiency'. The economic and environmental impacts depend to some extent on what we do with the money freed up by our reduced consumption – do we invest it, spend it on non-material services, give it away or choose to work less? Spending on services may seem an environmentally friendly option, for example, but will your yoga teacher spend her income on solar panels or on a flight to Tenerife? There may also be a 'sufficiency rebound' – if demand for resources falls, the price may fall, which means that consumers who could not previously afford the resources will start to consume more. This can improve well-being and social equity but may not reduce carbon emissions.

Caps and taxes

Caps (including emission trading schemes) and taxes can be used to limit carbon emissions, thus directly controlling the impact, I, on the left-hand side of the I=PAT equation. This is the most effective way of guaranteeing emission cuts, but there could be significant impacts on growth and employment, depending on the degree to which growth can be decoupled from carbon emissions (by reducing the carbon intensity, T). This makes caps and taxes politically difficult – they are often resisted because of the possible adverse impacts on competitiveness and profitability. However, taxes or trading schemes do generate a source of revenue that can be used to help businesses adapt to a low-carbon economy, and protect vulnerable energy users.

Co-benefits

For most of the twentieth century, policy makers viewed the economy and the environment as being in conflict. Regulations introduced to protect the environment, along with those to protect workers, were viewed by industry as a burden, or at best as a necessary evil. Yet this viewpoint is changing, as evidence mounts that the scale of economic activity now poses serious risks to the ecosystems on which our long-term prosperity depends. At the same time, green entrepreneurs have shown that sustainable businesses can flourish. Policy makers across the world are now falling over each other to produce new strategies for 'green growth' that show how sustainability can go hand in hand with job creation, innovation, competitiveness and economic development.

The European Union's *Roadmap for Moving to a Competitive Low-Carbon Economy by 2050*, for example, envisages that an annual investment of just 1.5 per cent of GDP will fuel the transition to a low-carbon society where:

> we will live and work in low-energy and low-emission buildings, with intelligent heating and cooling systems. We will drive electric and hybrid cars and live in cleaner cities with less air pollution and better public transport . . . By 2050, the energy sector, households and business could reduce their energy consumption by around 30% compared to 2005, while enjoying more and better energy services at the same time. More locally produced energy would be used, mostly from renewable sources. As a result, the EU would be less dependent on expensive imports of oil and gas from outside the EU and our economies would be less vulnerable to increasing oil prices.
>
> (European Commission, 2011c, web page)

As well as cutting greenhouse gas emissions to between 80 per cent and 95 per cent of 1990 levels by 2050, this strategy is expected to lead to increased economic growth compared to business-as-usual, an extra 1.5 million jobs by 2020, and annual savings of up to €320 billion on fuel costs and €88 billion in health benefits.

Similarly, the OECD has produced a strategy for green growth which expects environmental policies to result in increased employment, net savings of over $60 trillion in fuel costs, and enhanced productivity, innovation and economic stability (by reducing resource price volatility and environmental degradation). The United Nations *Towards a Green Economy* report extends the analysis worldwide, concluding that an annual investment of 2 per cent of GDP in greening the economy would result in higher growth and more jobs in the long term than a business-as-usual case, and showing how a green economy can be compatible with achieving the Millennium Development Goals for poverty reduction.[1]

Governments around the world are now using public investment in green technologies as a tool to stimulate their economies after the global recession. 'Green new deals' have been announced in various countries and regions including the EU, United States, China and South Korea. Out of a total of $2.8 trillion in stimulus funding, around 15 per cent is aimed at green sectors including energy efficiency in buildings, renewable energy and rail transport.[2]

This section will look at the potential economic co-benefits of climate policies, including jobs, cost savings, competitiveness, innovation, new business opportunities and economic stability.

Jobs

In 2009, over 200 million people worldwide were unemployed – around 6 per cent of the global labour force. Even having a job does not guarantee well-being. Around

630 million workers – a fifth of the world total – are the 'working poor', living with their families in extreme poverty on less than \$1.25 per person per day. Half of the world's workers – 1.5 billion people – are in 'vulnerable' employment, which includes informal work with low pay, no health insurance, no sick pay and often dangerous working conditions.[3]

What impact will climate policy have on employment? Fear of job losses has been a major barrier to climate action, and this argument is still being used to block climate policy in many countries. Certainly some jobs will be lost in high-carbon industries and in firms that do not adapt through greater efficiency or new business models. Yet many new jobs could be created in low-carbon sectors such as renewable energy, building efficiency, recycling, repair, sustainable agriculture and low-carbon transport.

This section begins by assessing whether a lower-carbon economy would be more labour-intensive, and then looks at current employment in low-carbon industries, before reviewing estimates of the overall effect of climate policy on jobs in the future. Finally, it considers whether low-carbon jobs differ in quality to the jobs they will replace.

Labour intensity

Labour intensity – the amount of human labour required per unit of output – has declined steeply over the last two centuries, as fossil fuels have been substituted for human energy. Mechanisation has greatly reduced the number of farm workers needed to produce a tonne of wheat, for example, and the number of construction workers needed to build a new road. In fact, our use of fossil energy has sometimes been equated to having 'energy slaves' – the average person today uses the equivalent of around 25 energy slaves to sustain their lifestyle, with inhabitants of developed countries using closer to 100 energy slaves.[4]

It might therefore be expected that a shift away from fossil fuels could reverse this trend, and in fact it does seem that low-carbon businesses are often more labour-intensive than their high-carbon equivalents (see Box 7.1). But this is not universally true – nuclear energy has a similar labour intensity to conventional fossil fuel energy production, for example.[5]

Box 7.1 Is low-carbon industry more labour-intensive?

Renewable energy

Compared to fossil fuels, renewable energy creates more jobs per dollar invested, per unit of installed capacity and per unit of power generated.[6] In the United States, investing in energy efficiency and renewable energy is estimated to produce from two to four times as many jobs per dollar invested compared to producing energy from oil.[7] Cultivation of biofuels, in particular, is very labour-intensive, requiring 100 times more labour per joule than fossil fuels. However, total employment can fall if smallholders are displaced to make way for biofuel plantations.[8]

Energy efficiency

Energy efficiency measures such as insulating houses are estimated to provide eight times more jobs per dollar invested than coal-based energy.[9]

Transport

Research suggests that public transport tends to create extra jobs compared to private transport: the new jobs created in the manufacture and operation of buses, rail vehicles and the associated infrastructure outweigh job losses in the manufacture of cars and trucks. One study estimated that an increase of 70 per cent in the use of buses and railways in the UK could create 130,000 new direct jobs, outweighing job losses of 43,000 in the automobile sector.[10] A German study found that using fuel taxes to stimulate a 70 per cent increase in public transport could create 338,000 new jobs, outweighing the loss of 130,000 jobs in the car manufacturing sector.[11] Thousands of new jobs could be created by retrofitting direct injection technology to the millions of motorbikes and mopeds in developing countries, which could halve CO_2 emissions and cut other pollutants by 90 per cent.[12]

Material efficiency

A material-efficient society would create new jobs in repair and maintenance, reuse, re-manufacturing, recycling and eco-design. On the other hand, there would be fewer jobs in manufacturing, mining, resource extraction and waste disposal. It could be argued that because raw materials and energy are being replaced by human labour, net job gains would be expected. This is borne out by a recent study that estimates that five different material efficiency strategies all result in a net gain in jobs.[13] Another study reported that for every 100 jobs created in recycling, only 13 jobs were lost in other sectors.[14] Recycling has also been shown to generate many more jobs than landfill and incineration of waste (Table 7.1).[15] The improved productivity resulting from resource efficiency will also create jobs elsewhere in the economy. Indirect jobs created like this tend to exceed the direct jobs lost.[16]

Agriculture

The labour intensity of agriculture has declined considerably as human labour has been replaced with machinery and chemical inputs such as fertilisers, herbicides and pesticides. The proportion of the global population employed in agriculture fell from 44 per cent in 1995 to 36 per cent in 2006. In China, it fell from 81 per cent in 1950 to 50 per cent in 2000, while in the United States, just 2 per cent were employed in agriculture in 2000. This has contributed to the global trend of migration out of rural areas to towns and cities, which can lead to unemployment. More sustainable farming methods can be more labour-intensive. For example, a study of over 1,000 organic farms in the United Kingdom showed that organic farms provided almost twice as many jobs per hectare as conventional farms, when adjusted for farm size.[17]

Table 7.1 Job creation from reuse, recycling and waste disposal in the United States

Type of operation	Jobs per 10,000 tonnes processed per year
Product reuse	
Computer reuse	296
Textile reclamation	85
Miscellaneous durable product reuse	62
Wooden pallet repair	28
Recycling based manufacturers – average	**25**
Paper mills	18
Glass product manufacturers	26
Plastic product manufacturers	93
Materials recovery facilities	10
Compost	4
Landfill and incineration	1

Source: Platt and Seldman, 2000, p. 27

Although energy-intensive industries often cite the threat of job losses to block the introduction of climate legislation, these industries are not labour-intensive. Amongst OECD countries for which data are available, the seven most carbon-intensive sectors (including transport, fossil fuel production, power generation, metals and minerals) produced 82 per cent of direct CO_2 emissions, but accounted for only 8 per cent of direct employment in the non-agricultural sectors.[18] In the EU, energy-intensive sectors account for 23 per cent of industrial value added, but only 18 per cent of industrial employment.[19] In addition, employment is declining in many heavy industries. In the steel industry, global employment fell from

2.3 million in 1974 to 0.9 million in 2000. Excluding China, the average number of employees needed per million tons of cement produced declined from 555 in 1980 to 272 in 2000.[20]

Figures on the labour intensity of different sectors today, however, do not tell the whole story. For most renewable energy and energy efficiency technologies (with the exception of biofuels), much of the labour requirement is for manufacture and installation of the technologies, and less is involved in ongoing operation and maintenance, so labour intensity could decline in the longer term – although a phase of intensive installation of new capacity will need to be maintained for several decades at least, in order to meet climate targets. In addition, many renewable energy technologies such as solar, wind, wave and tidal power are at an early stage of commercialisation and development – again implying that labour intensity could decline as the technologies mature and as economies of scale are achieved.

Because renewable energy can be more expensive than fossil energy at present and may require additional financial support from governments, it can result in higher energy prices in the short term. This can have a knock-on effect on employment elsewhere in the economy, by reducing consumer spending on other goods. In the long term though, the costs of renewable energy will come down, fossil fuel prices will go up and renewables will acquire a cost advantage. A modelling study for the European Commission found that even when the knock-on effects were taken into account, investment in renewable energy still generated net employment benefits.[21] Similarly, the net effect on employment in Europe of a 1 per cent annual improvement in energy efficiency has been shown to be positive, even after including factors such as reduced energy tax revenues.[22]

Because low-carbon sectors are more labour-intensive, investing in a greener economy is a good strategy for boosting employment in the short term; for example, to stimulate the economy and alleviate unemployment during a recession. Whether a shift to more labour-intensive industries would increase total employment in the long run, however, is more debatable. At the beginning of the Industrial Revolution, the Luddites feared that mechanisation would result in mass unemployment, but they were failing to take account of the impact of increasing productivity. The increase in labour productivity as a result of mechanisation fuelled economic growth which ultimately created more jobs overall.

Today, however, things have changed. The Luddites lived in an era when land, materials and fossil energy appeared to be inexhaustible, and by comparison, labour was scarce and expensive. As a result, governments focused on trying to improve labour productivity, which has increased by a factor of 3 to 4 in developed countries over the last 50 years, while material productivity doubled and energy productivity increased only slightly.[23] Yet today we live in a resource-limited world with unemployment problems, where it makes more sense to focus efforts towards increasing resource productivity. A 2009 study demonstrates that for an economy to be both socially and environmentally sustainable, resource productivity must grow faster than the size of the economy, but labour productivity must grow more slowly.[24]

Low-carbon jobs today

Low-carbon activities already form a dynamic and growing part of the global economy. Globally, the low-carbon and environmental goods and services sector was thought to employ around 28 million people in 2009.[25] In Europe, the eco-industry employed 3.4 million people in 2008, accounting for 1.7 per cent of paid employment (which is more than major sectors such as car manufacturing or pharmaceuticals) and producing 2.5 per cent of GDP.[26] Almost 900,000 people are employed in the eco-industry in the UK alone.[27] In Germany, eco-industry already accounts for 8 per cent of GDP, and this is predicted to rise to 14 per cent by 2030.[28] Globally, it is estimated that there are over 3 million jobs in renewable energy (Table 7.2), and there could be up to 20 million by 2030.[29]

Energy efficiency is another big source of employment. One estimate puts the number of jobs in the United States alone at 8 million – half of which are indirect jobs created in the supply chain or through the 'double dividend' effect whereby fuel cost savings enable additional economic growth. Retrofitting of buildings is a

Table 7.2 Estimates of jobs in renewable energy

	Global	Selected national estimates				
		China	US	Germany	Spain	Other
Biofuels	>1,500,000					Brazil 730,000
Wind power	630,000	150,000	85,000	100,000	40,000	Italy 28,000; Denmark 24,000; Brazil 14,000; India 10,000
Solar hot water	300,000	250,000			7,000	
Solar PV	350,000	120,000	17,000	120,000	14,000	
Biomass power			66,000	120,000	5,000	
Hydropower			8,000			Europe 20,000
Geothermal			9,000	13,000		
Bio-gas				20,000		
Solar thermal power	15,000		1,000		1,000	
Total estimated	>3,500,000			373,000	67,000	

Source: REN21, 2011

Notes: Excludes indirect jobs.

The 'total estimated' in column 2 applies to all rows (including biomass, hydropower, geothermal, etc., for which individual estimates are not available).

popular type of green stimulus funding, especially as it can be used to boost employment in the construction sector, which tends to be badly hit in recessions. In Germany, a large-scale building retrofit programme created over 140,000 jobs between 2001 and 2006.[30]

In the transport sector, electric and hybrid vehicles are finally taking off, with a flurry of new factories opening in the United States thanks to green stimulus funding. However, rail transport continues to decline in many countries. Railway jobs in China and India fell from 5.1 million to 3.3 million between 1992 and 2002.

The recycling sector is also a major employer. In 2004, recycling was estimated to generate 800,000 direct and indirect jobs in the EU-27, over a million jobs in the United States, 500,000 in Brazil and 10 million in China (1.3 million in formal recycling, with the rest in reuse or re-manufacture and 2.5 million in informal scrap collection). Re-manufacturing employed 500,000 people in the United States in 2003, and was estimated to account for 4 per cent of EU GDP in 2000.[31]

What effect will climate policy have on jobs?

Climate policy will have a number of impacts on employment:

- New jobs will be created in sectors such as renewable energy and energy efficiency.
- Jobs will be lost in carbon-intensive sectors such as coal mining or oil and gas production.
- Many jobs may change; for example, construction workers could build low-energy houses, electricians could install solar panels or factory workers could make electric instead of gasoline vehicles.
- In the absence of climate action, the adverse impacts of climate change (including floods, droughts and biodiversity loss) and resource scarcity are likely to destroy many livelihoods, especially in the agriculture, forestry, fishing and tourism sectors. On the other hand, adaptation to climate change could also be a source of employment (e.g. building flood defences).

Studies of employment impacts tend to neglect the last of these points, which is extremely difficult to quantify because future climate impacts are highly uncertain. A notable exception is the UN study *Towards a Green Economy*, discussed below, which attempted to include some of the impacts of environmental degradation in its economic model.

Although it is very difficult to model the effects of future climate policy, most recent studies seem to agree that on balance, the overall impact on employment will be fairly small, but that there will be a net gain in jobs compared to business-as-usual, of the order of 1–2 per cent. For example, consultancy GHK reviewed 36 different studies, of which eight predicted a decrease in overall employment (four of these were by the same organisation) and 19 reported positive effects on employment, with the remaining nine reporting little change or mixed results. They concluded that climate policy will probably have a much greater impact on the distribution of jobs between sectors and technologies than on the overall employment level.[32] A UNEP review also finds that most studies agree there will be a small positive change in total employment, including a study which suggested a net increase of 1.5 per cent for the EU-25.[33]

Examples of studies for individual sectors and countries include:

- The OECD *Towards Green Growth* report estimates that climate policy could boost the growth in employment in OECD countries over the period 2013 to 2030 from 6.5 per cent to 7.5 per cent if revenues from carbon pricing are used to cut labour taxes (this is called ecological tax reform – see Box 7.5, p. 280). This does not take account of the potential additional growth stemming from green innovation.[34]

- The Potsdam Institute estimates that increasing the EU emission reduction target from 20 per cent to 30 per cent below 1990 levels by 2020 would increase European investment levels from 18 per cent to 22 per cent of GDP, generating up to 6 million extra jobs and increasing GDP by up to 6 per cent ($840 billion).[35]

- The IEA estimates that 30,000 jobs can be created for every billion dollars invested in clean energy. A US study estimates that $100 billion spent on clean energy over 10 years could create 2 million new jobs, compared to just 500,000 jobs if the money were invested in oil- and gas-related industries.[36]

- The European Commission estimates that meeting the EU target for a 20 per cent share of renewable energy by 2020 would increase direct and indirect employment in the sector from 1.4 million jobs in 2005 to 2.8 million in 2020 and 3.3 million in 2030. Taking account of job losses in conventional energy, as well as the effect of higher energy prices, there would be a net gain of 410,000 jobs and an increase of GDP of 0.25 per cent in 2020, compared to a scenario with no government support for renewable energy.[37]

- UNEP predicts that there is potential to generate up to 2 million jobs worldwide in wind energy, over 6 million in solar PV and around 12 million in biofuels by 2030. Replacing traditional stoves with efficient biomass stoves in India would create 150,000 jobs. Over 800,000 jobs could be created in the United States and millions or tens of millions worldwide by promoting greener buildings. Retrofitting all the EU's residential buildings to reduce CO_2 emissions by 75 per cent could create 2.6 million jobs by 2030, at a cost of $4,300 billion.[38]

- A UK study estimated that 400,000 new jobs could be created by 2015 if plans to reduce greenhouse gas emissions were realised.[39]

- A study by the Climate Group estimates that India could create over 10 million green jobs and earn $130 billion from energy efficiency and renewable energy by 2020, helping to reduce reliance on coal imports.[40]

- A study by Friends of the Earth estimated that 500,000 extra direct and indirect jobs could be created in the EU if all countries increased their recycling rates to 70 per cent.[41]

- The European Trade Union Council predicted that 50,000 jobs could be lost in the EU iron and steel sector, out of a total of 350,000, as a result of climate legislation. But another study demonstrates that if 10 per cent more steel was made from recycled metal, there could be a net gain of almost 2,000 jobs (4,000 jobs lost, but 6,000 gained), plus an indirect gain of 1,800 jobs due to the supply chain effect.[42]

• A study in the UK estimated that increasing the amount of organic farmland from 4 per cent to 20 per cent would create over 70,000 new jobs.[43] Another study estimated that if the UK switched completely to organic farming, agricultural employment would increase by 70 per cent.[44]

The UN report *Towards a Green Economy* is the most comprehensive attempt to date to assess the economic impacts of greening the economy on a world scale. The report looks at three scenarios:

1   *business-as-usual* (BAU);
2   *green investment* (G2): extra annual investment of 2 per cent of world GDP directed at green technologies such as energy efficiency and renewable energy;
3   *conventional investment* (BAU2): extra annual investment of 2 per cent of world GDP following existing investment patterns.

By comparing BAU with G2, we can see the effect of investing an extra 2 per cent of GDP in greening the economy compared to continuing with business-as-usual. But since any scenario with extra investment might be expected to stimulate extra growth and employment, the study also includes the BAU2 scenario in which there is also an extra 2 per cent of investment, but it is allocated according to existing investment patterns.

The model predicts that in the short term, both of the investment scenarios (G2 and BAU2) deliver more jobs than the BAU scenario, but the green scenario produces slightly fewer than the BAU2 scenario because it restricts over-extraction in the fishing and forestry sectors. In the longer term, however, the business-as-usual scenarios suffer from resource scarcity, rising energy prices, declining fish stocks and poor soil quality, and the green scenario overtakes them (Table 7.3).

The estimates of the employment benefits of the green scenario are thought to be conservative because the model does not take account of the potential of ecological tax reform (see Box 7.5) to create additional jobs under the green scenario. More importantly, it does not include potential climate impacts such as floods, storms, droughts and rising sea levels, which are likely further to reduce employment under the business-as-usual scenarios. The benefits of climate policy in maintaining a stable climate could dwarf the co-benefits of job creation in new industries (see 'A secure and stable economy' below).

Job quality

We have seen how most studies agree that climate policy has the potential to stimulate net job creation – but will the new jobs be better or worse than those they displace in other industries?

Table 7.3 UN Green Economy model predictions for employment in 2050 (direct and indirect jobs)

Sector	Green investment compared to business-as-usual (BAU)	Green investment compared to conventional investment (BAU2)
Agriculture	+214 million	+47 million
Fishing	+9 million	+8 million
Forestry	+5 million	+5 million
Transport	+8 million	+10 million
Energy	+4 million	+4 million
Waste	+3 million	+2 million
Industry	−17 million	−25 million
Services	+30 million	−8 million
Total	+257 million	+43 million

Source: UNEP, 2011a

Many low-carbon jobs in sectors such as renewable energy or clean manufacturing will be highly skilled and well-paid. Some will provide the opportunity to learn new skills: construction workers could be trained to undertake building energy retrofits, for example. The jobs that will be lost, on the other hand, often have high industrial accident rates or poor working conditions, such as coal mining and oil extraction (see Chapter 5).

The trend towards clean manufacturing and waste prevention is likely to result in jobs that are cleaner, safer and more enjoyable, due to the phasing out of toxic substances and reduction of pollution, dust, noise and vibration from heavy machinery. A European study found that regular upgrades of processes and machinery are linked to extra training and an increase in employee skills and job quality.[45]

In the retail sector, a low-carbon economy could lead away from a predominance of low-paid jobs focused on selling large quantities of cheap goods towards a culture of 'quality retail' based more on selling a service. Staff selling or leasing more durable, higher-quality goods will need to be more knowledgeable, and will need to be able to advise on product maintenance, use (for example, how to minimise energy or water consumption), repair, upgrade and end-of-life take-back schemes for re-manufacture or recycling. There will be more focus on interacting with customers and building a long-term relationship, which should lead to higher job satisfaction. There could be more jobs in repair and maintenance, which should provide higher satisfaction levels, greater skill levels and better pay than the jobs they will displace in low-value manufacturing. In short, jobs in retail and manufacture could become a lot more interesting.

Another feature of low-carbon jobs is that they often rely less on the location of a specific resource (such as metal ores or fossil fuels) and more on the provision of government support, so they can be directed to areas suffering from high unemployment. Retrofitting buildings, for example, provides opportunities for local construction workers. Sustainable agriculture and renewable energy technologies can provide extra jobs in rural areas that are suffering from lack of work opportunities. Wind turbines can provide a stable supplementary source of income to farmers. Sale and installation of solar technologies such as PV panels and cooking stoves provides income for many people in developing countries such as India and Kenya.

In the United States, clean technology jobs are arriving in many of the 'rust-belt' towns that have been devastated by the decline of the US automobile industry. A concentrating solar dish has been designed specifically for production by US auto manufacturers, using existing production lines with surplus capacity, and securing up to 500,000 jobs. In Toledo, Ohio, one factory has switched from making car windscreens to making solar panels. An auto transmissions factory in Indiana is reopening as a PV plant; an auto parts company in Milwaukee will now build high-speed rail cars; a closed-down auto manufacturer will make Tesla's new all-electric Model S sedan; and redundant steel plants in Pennsylvania will be making wind turbine components.[46]

On the other hand, there are some low-carbon sectors where there is considerable cause for concern over job quality. Low-skilled assembly jobs are vulnerable to the same problems of low pay and poor conditions that exist in factories around the world. Rapid development of solar panel manufacturing in China, for example, has led to concerns over working conditions and pollution from silicon tetrachloride.

There are also major problems in informal recycling, which employs an estimated 2.5 million people in China alone. Some 700,000 of these are involved in recycling electronic waste, 98 per cent of whom work in small family workshops with little attention to health and safety. On top of its own domestic waste, China receives up to 70 per cent of global electronic waste shipments, with another 20 per cent going to India, Pakistan, Bangladesh and Myanmar – and many of these shipments are illegal. Although China has safety and environmental regulations similar to those in Europe, they are difficult to enforce in the fast-growing e-waste sector. High levels of heavy metals and organic contaminants from e-waste recycling operations have been found in soil and water close to agricultural land, and there is a risk that pollutants can enter the food chain. Workers often suffer diseases of the skin, stomach, lungs and other organs. Pay is low and workers are generally not covered by health insurance, unemployment protection, or pension plans.[47]

Ship breaking is also frequently carried out in developing countries under poor working conditions. Between 200 and 600 large ships are dismantled each year to recover steel, but workers can be exposed to hazards from asbestos and other toxic compounds such as polychlorinated biphenyls (PCBs).

Informal recycling of electronic waste is also a growing practice in Eastern Europe, where members of impoverished communities such as the Roma are involved in activities such as burning the plastic coating off electric cables to recycle the metal, risking health damage from inhaling dioxins, particulate matter and other toxic pollutants. The garbage pickers of developing countries are also a cause for concern, including the zabaleen in Cairo and the slum dwellers of India. These workers carry out a valuable and highly effective materials recovery service, but at great personal cost in terms of health impacts.

One solution is to enable informal recycling workers to form co-operative groups and to set up proper businesses with better recovery facilities and training. By selling direct to customers, the groups could also improve their income. Brazil now has 500 such co-operatives representing 60,000 people, and Colombia has 100 co-operatives which recover over 300,000 tons of material each year. In Brazil, a co-operative recycling plant set up in 2005 is expected to increase the income of trash scavengers by about 30 per cent, ending exploitation by unscrupulous middlemen.

Jobs in the biofuels sector can also be low-paid and dangerous, and may involve displacing local people from their land in order to establish plantations. Examples include the cultivation of sugarcane in Brazil and palm oil in Indonesia. Workers in these industries face the same problems as agricultural workers all over the world – only 5 per cent of the world's 1.3 billion agricultural workers have any legal health and safety protection. There are 3–4 million cases of agricultural pesticide poisoning every year, and 40,000 deaths, and the death rate at work is twice as high as in any other sector. Similarly, planting seedlings for reforestation initiatives gives jobs that are often seasonal, temporary and paid by piece rate, which results in long hours for low pay and under harsh conditions. The United Nations Environment Programme recommends that safeguards such as certification schemes should be implemented to ensure that biofuel and reforestation schemes comply with international labour laws, meet health and safety standards and protect the right to join a union.[48]

Cost savings

One of the most obvious co-benefits of climate action is the opportunity for households and businesses to save money by cutting energy use and reducing waste. These savings are likely to become even more important in future, as prices for energy, materials, food and water continue to rise. As we saw in Chapter 6, resource efficiency can deliver major cost savings, often at little or no cost, or with very short payback periods. Renewable energy systems can also deliver savings in the longer term as fossil fuel prices increase. Estimates of potential savings include:

- The OECD and IEA estimate that low-carbon energy systems could save $112 trillion in fuel costs between 2010 and 2050 for an initial investment of $45 trillion.[49]
- The UNEP report *Towards a Green Economy* estimated that an average annual investment of $38 billion in industrial energy efficiency over the period from 2010 to 2040 could halve energy consumption whilst saving $193 billion per year on average in the chemical sector, around $120 billion in the steel sector and $37 billion in the paper sector.[50]
- The IPCC estimated that by 2030, greenhouse gas emissions in the building sector could be reduced by almost a third by implementing measures that pay for themselves.[51]

- The Europe 2020 strategy envisages that meeting the EU goals for increasing energy efficiency by 20 per cent and increasing the share of renewable energy to 20 per cent could save €60 billion per year on oil and gas imports by 2020 as well as increasing energy security.[52] In the longer term, the EU 2050 road-map estimates that savings would average between €175 billion and €320 billion per year up to 2050.[53]
- Simple resource efficiency actions with a payback of less than a year could save UK businesses £55 billion per year whilst reducing carbon emissions by 13 per cent.[54]

Resource efficiency improvements often lead to wider benefits such as savings in work time and reduced wear and tear on equipment. A survey of 41 projects that improved the efficiency of industrial motors found that 22 reduced maintenance requirements, 14 improved productivity or product quality, and 8 reported lower pollution or savings in treatment chemicals. Another review of 54 energy efficiency technologies found that 20 also reduced material waste or air pollution, and 35 improved productivity or product quality.[55] A study of 52 energy efficiency projects found that including this type of co-benefit in assessments reduced the average payback period from 4 years to 2 years.[56]

A more surprising hidden benefit could come from improved working conditions, which can lead to increased productivity. One report used case studies to show that better lighting, more comfortable temperatures and lower noise levels in energy-efficient buildings typically increases worker productivity by between 6 per cent and 16 per cent. Labour costs in the commercial sector are typically at least 25 times energy costs, so these productivity improvements could far outweigh the energy cost savings.[57]

Savings can also come from cutting travel to meetings – using videoconferencing, for example; or encouraging rail travel instead of driving, which allows staff to continue working on the train. The UK government managed to save almost £14 million in fuel costs through reduced road travel in 2008/2009, and this also saved 1.7 million hours of staff time (equivalent to over 1,000 full-time staff working for a full year) on top of savings from reduced car purchases, repairs and administration.[58]

Many governments and businesses are investing in energy efficiency and renewable energy to protect themselves against future supply shocks and price rises. The EU 2050 roadmap, for example, stresses the importance of protecting the EU against future oil price rises. Oil prices doubled between 2005 and 2010, and the EU's import bill for oil and gas rose by $70 billion from 2009 to 2010. The 2050 strategy aims to decrease the import bill by a third, but without action it could double – a difference of €400 billion per year by 2050, which is the equivalent of 3 per cent of Europe's GDP.[59]

But fuel costs are not the only consideration. The other co-benefits of climate policy can also deliver financial savings. These include reduced air pollution, which can deliver health and ecosystem benefits that could even exceed the cost of climate mitigation in the long term (see Chapter 3); reduced deforestation, with natural capital worth an estimated $1–3 trillion each year being lost at present (see Chapter 4); and reduced impacts of oil spills and coal mine accidents – the BP Deepwater Horizon spill alone was estimated to cost at least $6 billion in immediate payments, and a fund of $20 billion has been set aside for further compensation claims (see Chapter 5).[60] In Chapter 8 we will see that many low-carbon lifestyle changes such as more cycling and walking and lower consumption of meat and dairy products can be beneficial to health, and that policies to cut vehicle use also lead to reduced noise, less congestion and fewer traffic accidents. Further health benefits arise from improvements to home energy efficiency – a warm, well-insulated home with less condensation will reduce illness and improve general health, especially for those on low incomes who cannot afford to heat or cool their homes adequately.[61] All these factors taken together will reduce national expenditure on healthcare, cut premature deaths and reduce working time lost to sickness.

Not all climate policies offer a short-term cost saving, of course. Many renewable energy technologies have high initial investment costs, although they may well offer cost benefits in the longer term. There is also the problem of the rebound effect – cost savings can be spent on other goods and services that wipe out the emission savings. These issues are discussed in the section on 'Conflicts' below.

Competitiveness, innovation and new business opportunities

Opponents of climate action like to portray it as a backwards step, leading to a future where miserable cave-dwellers huddle around flickering candles, wondering why they ever abandoned the great fossil-fuelled march of progress. In reality, however, a low-carbon future will require advanced technologies and innovative business models. High levels of ingenuity, driven by advances in science and engineering, will be needed to ensure a comfortable and sustainable life for 9 billion people in 2050.

A list of some of the technologies that will play a part should illustrate this point:

- wind, solar PV, solar thermal, wave, tidal and micro-hydroelectric power;
- second-generation biofuels from waste or algae;
- fuel cells; hydrogen, electric and hybrid vehicles;
- advanced building energy control technologies and micro-CHP;
- intelligent logistics and transport systems;
- smart grids, smart meters, smart appliances, advanced energy storage;
- eco-design, advanced material recovery, diagnostics and repair, bio-materials;
- industrial symbiosis; closed-loop manufacturing;
- new product-service models;
- climate-smart agriculture; forest restoration.

A low-carbon future will involve the rapid introduction of smart, clean, efficient and innovative technologies. There will be a major role for information and communication technology (ICT); for example, in industrial energy and emission control systems, smart grids and appliances, intelligent transport systems, building management systems and videoconferencing. One study estimates that ICT could deliver a 15 per cent reduction of business-as-usual emissions in 2020 – more than five times the carbon footprint of the entire sector.[62]

The new low-carbon sectors score highly on innovation. Between 1999 and 2008, the number of patents registered grew by 24 per cent annually for renewable energy, 20 per cent for electric and hybrid vehicles, and 11 per cent for energy efficiency in building and lighting, compared to a rate of 6 per cent overall for all technologies. Green technologies accounted for a quarter of all venture capital investments in the United States in the first half of 2010,[63] and nearly a fifth of all venture capital funding in China in 2006.[64] There is some evidence that innovation is higher where climate policy is stricter.[65]

There will be opportunities for innovative companies and entrepreneurs to open up new markets and explore new business models. The global market for environmental products and services is large and is growing fast. One estimate values the market at $1.3 trillion in 2005 and expects it to grow to over $3 trillion by 2020.[66] Another estimate puts the value in 2007 at over £3 trillion ($4.7 trillion), growing to an expected £4 trillion ($6.3 trillion) by 2015.[67]

Opportunities to invest in low-carbon energy are particularly promising. Investment in renewable energy (excluding large hydropower) rocketed from $30 billion in 2004 to $210 billion in 2010, exceeding investment in new fossil fuel capacity in the EU and the United States in both 2008 and 2009. Globally, almost half of the new generating capacity added in 2009 was renewable. From 2004 to 2009, average annual growth rates were 60 per cent for grid-connected solar PV capacity, 102 per cent for utility-scale PV and 32 per cent for wind power.[68]

The recycling and re-manufacturing sector is also growing fast. In the United States, this sector generates $236 billion per year, more than twice the revenue of the waste disposal industry, despite the fact that lower volumes of waste are handled – reflecting the economic value of the materials that are recovered. Three-quarters of this value is in product re-manufacturing, and 6 per cent in refurbishment of products.

New business models such as the product-service concept also provide an opportunity for entrepreneurs (see Chapter 6). Another new business model is 'performance contracting', where suppliers offer a service to help businesses cut waste, and then take a share of the savings. In the United States, energy service companies that earn most of their money by delivering efficiency services earned an estimated $3.6 billion in 2006.[69]

Often the business sector argues that climate policy will affect competitiveness. Nevertheless, leading companies are increasingly finding that sustainability is essential to maintain a competitive edge. Companies need to address efficiency in order to insulate themselves from rising resource prices and supply shortages. There is also an increasing risk of damage to reputation and image, as customers become more aware of environmental issues. Shareholders are becoming more concerned about the risks and opportunities arising from climate change, and proposals related to corporate social responsibility are being tabled more often at shareholder meetings.[70] Investors are pressing companies to disclose their climate-related risks;

and as a result, over 3,000 firms, including most of the world's largest companies, now report to the Carbon Disclosure Project (CDP). The CDP supplies information to over 500 investment firms which are in charge of over $70 trillion. Carbon Disclosure Project data even appear on Wall Street traders' terminals and on Google Finance alongside financial information. There is evidence that companies which take sustainability seriously are more successful. A study by consultants A.T. Kearney found that share prices of the top 99 companies in the Dow Jones Sustainability Index outperformed the industry average by 15 per cent in 16 out of 18 sectors.[71]

Leading businesses are starting to see that a strong regulatory framework can help them to gain a competitive advantage over rivals who are slower to adapt. In the *Vision 2050* report from the World Business Council for Sustainable Development (WBCSD), 29 major companies including Alcoa Inc., Rio Tinto, Toyota, Volkswagen, PwC, Sony, E.ON and Boeing warn that sustainability will not happen on its own, but if governments set the right framework, then companies will innovate.[72] In the UK, businesses comprising the Aldersgate Group have asked the government to make corporate carbon reporting mandatory. And a group of more than 70 large European companies have signed a declaration asking the EU to increase its target for carbon reduction from 20 per cent to 30 per cent by 2020 (Box 7.2).

Box 7.2 Joint Business Declaration: increasing Europe's climate ambition will be good for the EU economy and jobs

Declaration by the Climate Group, the Cambridge Programme for Sustainability Leadership and WWF Climate Savers

Businesses call for the EU to increase its greenhouse gas reduction target to 30 per cent to drive low-carbon investment:

- Climate action will boost economic growth and create new jobs.
- The EU needs the right policies to maintain its leadership and competitiveness in the global low-carbon economy.
- The EU must ensure energy security through greater low-carbon energy investments.
- The EU needs to invest now for tomorrow's technology and infrastructure to avoid high-carbon 'lock-in' and the financial risk of needing to engineer a rapid shift away from such stranded assets.

> - The recession has made emissions cuts easier and cheaper, but market incentives are required to spur action.
> - 'Carbon leakage' should be evaluated and concerns addressed based on real facts and data about competitiveness.
>
> The signatories together employ over 3 million people and have an annual turnover of €1 trillion. They include BT, Carrefour, Coca-Cola, Google, IKEA, M&S, Philips, Sony Europe, Unilever and Vodafone.

The companies call for a 'European policy framework that will spur innovation and investment, notably in renewables and energy efficiency, to ensure European energy security'. They cite research showing that increasing the climate target could avoid €45 billion in oil and gas imports in 2020, and €600 billion or more per year by 2050, as well as creating 6 million net jobs – a 'win-win-win for Europe'.[73]

Various studies have found that more sustainable countries are more competitive and have higher credit ratings and higher government bond yields.[74] Countries with long-established carbon taxes, such as Norway, Sweden, Denmark, Finland, Germany and the Netherlands, also have high GDP *per capita* and high well-being as measured by the Human Development Index (Table 7.4), and there is evidence that stricter environmental standards help companies and countries to become more competitive.[75]

A study by the GLOBE organisation of national legislators found that often the main motivation for national climate legislation is economic. Governments are realising that strong national action will yield co-benefits including energy security, resource efficiency, improved competitiveness and leadership in the new markets for environmental goods and services. South Korea's Green Growth Law, for example, was driven by the need to reduce energy costs to stay competitive, the opportunity to create jobs and the potential to gain a competitive advantage for new businesses in low-carbon industries.[76]

A secure and stable economy

A major benefit of moving to a low-carbon economy will be protection from resource price shocks and shortages, which lead to economic and social instability. In Chapters 5 and 6, we saw how fossil fuel and material prices are likely to continue to rise as scarcity increases. By reducing our reliance on these resources, through improving efficiency and adapting our economies to depend less on material consumption, we can achieve a more secure and stable economy.

A move away from a consumption-based economy could also help to provide jobs that are more resilient to economic fluctuations. An economy dependent on ever-increasing consumer spending on material goods has a high number of jobs in manufacturing and retail that are vulnerable to job losses during a recession.

Table 7.4 Countries with high environmental taxes tend to have high GDP and a high position on the Human Development Index (HDI)

	Cumulative environmental taxes 1994–2006 $/capita	GDP per capita, $, 2010	GDP per capita, rank in world*	HDI 2010	HDI, rank in world	IHDI (inequality adjusted HDI) 2010	IHDI, rank in world
Norway	16,300	85,389	5	0.938	1	0.876	1
Sweden	10,900	48,897	11	0.885	9	0.824	3
Netherlands	14,000	46,904	13	0.890	7	0.818	4
Germany	8,800	40,116	23	0.885	10	0.814	5
Switzerland	9,500	67,457	6	0.874	13	0.813	6
Denmark	22,000	56,245	8	0.866	19	0.810	10
Finland	13,500	44,378	18	0.871	16	0.806	11
Austria	12,400	45,181	17	0.851	25	0.787	16
UK	9,400	36,343	27	0.849	26	0.766	21

Source: Based on data from OECD/EEA, 2011, UNDP, 2011b and World Bank, 2012b

Notes: The table shows selected countries with relatively high environmental taxes, and shows that these are typically in the top 25 countries for GDP *per capita* and HDI. Environmental taxes include taxes, charges and fees for energy (transport fuel, gas, electricity, coal), whether or not based on carbon content, as well as water, pollution, resource use and waste disposal.

HDI is derived from a combination of lifespan, GDP *per capita* and education.

IHDI is HDI adjusted for income inequality.

*Top four countries for GDP *per capita* are tax havens.

Perhaps even more importantly, low-carbon policies will protect our economies against the catastrophic impacts of climate change, which could dwarf any other economic considerations. The most severe impacts are likely to be felt in Africa, Asia, Latin America and southern Europe, which are expected to suffer more heat waves, droughts and floods, leading to reduced agricultural yields, more forest fires and reduced tourism. Some regions may profit in the short term – a moderate rise in temperature could benefit parts of northern Europe, Russia and the Arctic regions, by opening up new land for farming, development and tourism, raising yields and allowing new crops to be grown. But rising sea levels, more storm surges and saltwater intrusion into surface water and groundwater will affect coastal cities around the world. Extreme weather events such as storms and floods will cause economic and social damage including loss of life, displaced communities and costly replacement of infrastructure. Species will be lost and rising ocean temperatures are likely to reduce fish stocks. In the long term, economies around the world are likely to suffer due to global food shortages, social unrest, conflict and mass migration away from the most badly affected areas.[77]

It is very difficult to quantify these impacts, but the UN report *Towards a Green Economy* has attempted to illustrate some of the implications of continuing with business-as-usual, compared to taking strong action to protect the environment and hence the economy. Table 7.5 shows the key predictions in the short term (to

2020) and the long term (2050) for the three scenarios described earlier in this chapter: business-as-usual (BAU), green investment (G2) and conventional investment (BAU2).

Both of the extra investment scenarios provide higher growth than the business-as-usual scenario. At first, the green scenario provides slightly less growth than the conventional investment scenario, but after 2020, it overtakes both the other scenarios. By 2050, GDP in the green scenario is 32 per cent higher than the business-as-usual scenario and 16 per cent higher than the BAU2 scenario, and is growing faster. Between 2030 and 2050, annual growth averages 2.5 per cent in the green scenario, but just 1.6 per cent under business-as-usual and 1.9 per cent in the BAU2 scenario.

The key point is that vital ecosystem services are preserved under the green scenario, whereas they are severely depleted under the business-as-usual scenarios – especially under BAU2, where the higher growth rates are fuelled by faster depletion of environmental resources. The business-as-usual scenarios suffer from lower soil quality, higher water stress and higher energy costs, which all act to slow growth. Under the green scenario, however, resources are conserved, providing a firm foundation for long-term growth. Water demand is 22 per cent lower and fossil fuel demand is almost halved compared to BAU2.

Energy efficiency plays a key role in the green scenario, reducing primary energy demand by 40 per cent compared to BAU2. Private cars account for only 33 per cent of passenger travel, compared to 62 per cent under business-as-usual; this shift to public transport, together with measures to cut transport demand, reduces the energy used in the transport sector by almost two-thirds. The share of renewable energy increases from 13 per cent to 27 per cent in 2050, and the share of fossil fuels in primary energy falls to 61 per cent compared to 83 per cent under BAU2; the share of fossil fuels used in electricity generation is just 34 per cent, compared to 64 per cent under business-as-usual. Nuclear power production is 38 per cent higher than under BAU2, with the share of nuclear in the energy mix doubling to 11 per cent. Biofuels provide 21 per cent of the world's liquid fuels, but these are mainly second-generation biofuels derived from agricultural and forestry residues.

The deforestation rate is halved under the green scenario, saving over 280 million hectares of natural forest. In addition, an extra 500 million hectares of new forest are planted, resulting in a total forest area 21 per cent higher than business-as-usual. Forests cover 35 per cent of the land surface, compared to 28 per cent under business-as-usual, and store over 500 billion tonnes of carbon, as well as preventing soil erosion and improving water availability.

Some of the most striking benefits are seen in agriculture. Crop production is 9 per cent higher in the green scenario, thanks to reduced water stress and the use of organic fertilisers, which improve soil quality by 27 per cent compared to business-as-usual. The yield per hectare is 10 per cent higher, meaning that 11 per cent less land is needed for agriculture – agricultural land occupies 37 per cent of the land surface compared to 42 per cent under business-as-usual. This will bring benefits for biodiversity.

Table 7.5 Key results of the UN Green Economy model

	Unit	2011	2020 BAU	2020 BAU2	2020 G2	2050 BAU	2050 BAU2	2050 G2
Additional investment	$ billion/year	0	0	1,798	1,788	0	3,377	3,889
World GDP	$ trillion/year	69	89	92.5	92.2	151	172	199
GDP *per capita*	$	9,992	11,698	12,205	12,156	17,068	19,476	22,193
Per capita GDP growth rate	%/year	1.8	1.7	2.1	2.2	1.4	1.7	2.2
Population below $2 per day	%	19.5	16.9	16.2	16	11.4	9.8	8.4
Total employment	millions	3,187	3,641	3,722	3,701	4,613	4,836	4,864
Agricultural production	$ million/year	629	690	718	744	849	913	996
Calories per person		2,787	2,802	2,946	2,955	2,981	3,273	3,382
Forest land	billion ha	3.9	3.9	3.9	4.0	3.7	3.7	4.5
Arable land	billion ha	1.6	1.6	1.6	1.6	1.6	1.6	1.5
Change in fish stocks	$ billion/year	-160	-134	-141	1	-88	-91	142
Change in fossil fuel stocks	$ billion/year	-1,212	-1,645	-1,788	-1,163	-4,312	-4,972	-979
Water demand	km³/year	4,864	5,737	5,792	5,375	8,141	8,434	6,611
Waste generation	million tonnes/year	11,238	11,775	11,864	12,084	13,201	13,505	14,783
Landfill	billion tonnes	8	9	9	7.7	12	12	2
Primary energy demand	mtoe/year	12,549	14,651	15,086	13,709	19,733	21,687	13,051
Energy intensity	mtoe/$ billion	0.18	0.17	0.16	0.21	0.13	0.13	0.07
Renewable energy share of primary demand	%	13	13	13	17	13	12	27
Fossil fuel share of primary demand	%	81	81	82	73	81	83	61
Nuclear share of primary demand	%	6	6	5	8	6	5	11
Fossil fuel CO_2	Gt/year	30.6	35.6	37.1	30.3	49.7	55.7	20.0
Footprint to biocapacity ratio		1.5	1.6	1.7	1.4	2.1	2.2	1.2

Source: UNEP, 2011a, Chapter 13

Note: mtoe = million tonnes of oil equivalent; Gt = gigatonnes (1 Gt = 1 billion tonnes).

Both the improved agricultural yield and the higher GDP *per capita* contribute to improved nutrition, which is 13 per cent higher in the green scenario than under business-as-usual, and 3 per cent higher than BAU2. The number of people living in poverty with incomes of less than $2 per day is 26 per cent lower than business-as-usual, equivalent to almost 270 million people, and 14 per cent lower than BAU2, equivalent to over 120 million people.

Perhaps most importantly, the green scenario reduces the ratio of the ecological footprint to the earth's bio-capacity to 1.2. Although this means that we would still be consuming 20 per cent more resources than can be replaced each year, this is far better than the projected ratios of 2.2 for business-as-usual and 2.1 for BAU2.

The green scenario gives a far greater chance of avoiding catastrophic and irreversible climate change. Under business-as-usual, emissions of carbon dioxide from fossil fuels increase to 138 per cent above the 1990 level by 2050, even though the carbon intensity of the economy has declined by 26 per cent. This means that greenhouse gas concentrations are heading for 1,000ppm by the year 2100 (from a pre-industrial level of 280ppm). There will be less than a 5 per cent chance of limiting the global temperature rise to 2°C, and a rise of at least 4°C is likely, which will have severe impacts on water supplies, sea levels, crop yields and biodiversity.[78] Even for a 2°C rise, between 15 per cent and 40 per cent of species face extinction. These impacts are not included in the model, but the potentially disastrous effects on the world economy are clear.

Under the green scenario, carbon dioxide emissions from fossil fuels in 2050 are 64 per cent lower than under the BAU2 scenario, and other greenhouse gases from land use are also reduced thanks to lower use of fertilisers, less deforestation and a smaller land area used for agriculture. This should limit greenhouse gas concentrations to 500–600ppm. If the benefits of carbon sequestration from organic agriculture are taken into account, concentrations could be as low as 450ppm, giving a chance of limiting the temperature rise to 2°C.

The UN report *Towards a Green Economy* has its limits – as with all models, it is impossible to take into account all the complexities of real life. However, it does at least illustrate the principle that the strength of our economies depends on preserving healthy ecosystems and other natural services, and maintaining climate stability. A low-carbon economy, combined with other green actions such as conserving fish stocks, can help to protect our economy from resource shortages, energy price rises, water stress and loss of agricultural production. Far from acting as a burden on the economy, low-carbon investment can safeguard jobs, reduce poverty, minimise conflict and underpin stable and secure economic development in the long term.

Conflicts

This book has argued that a strong climate policy is essential to secure the long-term sustainability of the economy, and has shown that it can result in higher employment, cost savings and improved resource security. However, there will be trade-offs to be made, and there are some possible negative effects, especially in the short term.

Climate action will involve high initial investment costs – could this have a negative effect on the economy in the short term? What about international competitiveness – will companies simply move to other countries to avoid measures such as carbon taxes? Will higher energy costs cause problems for people on low incomes? Will climate policy slow down economic growth and poverty reduction in developing countries? Climate policy must be designed with great care to minimise all these potential adverse effects and maximise economic and social, as well as environmental, benefits.

Resource efficiency involves a trade-off between cost savings and energy savings, because cost savings can be re-spent into the economy, creating a rebound effect. This leads to a wider question – can continuous economic growth ever be sustainable in the long term? These issues are discussed in more detail under 'Is endless growth possible?' and 'New economic thinking' below.

One of the most obvious impacts in the short term will be the inevitable redistribution of jobs from high-carbon to low-carbon activities, which could cause considerable social hardship. We will consider how to alleviate this and the other conflicts in 'The way forward' below.

The cost barrier: can we afford the future?

Many energy-saving measures pay for themselves within a few years, but this is not true of all climate investments. Low-carbon energy technologies such as wind, solar, wave, geothermal and tidal power have high upfront investment costs, although once the plant is built, the ongoing operation and maintenance costs are very low. By comparison, typical fossil fuel technologies have costs that are more evenly spread over the lifetime of the plant, with ongoing fuel and labour costs usually outweighing the initial investment. This means that even where renewable energy can be produced at a similar cost per kilowatt hour to fossil energy, the investor must be prepared to pay out almost the whole cost before the plant has even started operating.

In the longer term, however, renewable energy is predicted to become cheaper than fossil energy, as fossil fuel prices rise and the costs of renewable technologies fall due to technical improvements and wider deployment. For wind and solar energy, investment costs per unit have been falling by 10–20 per cent for each doubling of investment.[79] The price of solar PV modules fell from $3.50 per watt in 2008 to under $1.50 per watt in 2010, and is predicted to fall to $1 per watt by 2013.[80] Also, market prices do not take into account the external costs of fossil fuels such as air pollution – if these are included, many renewable energy sources are already cheaper than fossil fuels in terms of the net cost to society.[81]

Nuclear power poses slightly different investment barriers in that initial costs are high, construction times are long and there may be large and uncertain costs at the end of the life of the plant, during decommissioning and waste disposal. There is also a very small risk of very large costs in the event of a serious nuclear accident along the lines of Chernobyl or Fukushima.

Carbon capture and storage (CCS) involves an additional cost, but without the compensation of resource savings, health benefits or future price differentials. In fact, due to the additional fuel required to power the systems, rising fuel prices in the future will make this option even less competitive.

Low-carbon buildings and vehicles can also face high up-front capital costs, and sometimes the investor may not be able to reap the benefits of the low operating costs. For example, the developer of a low-carbon building would not benefit from low heating and lighting costs. Those benefits would pass to the building occupier, who may or may not be willing to pay more for the building because of this.[82] This effect can be magnified in the case of rented buildings or vehicles, when the owner does not benefit from lower fuel bills and so may be reluctant to invest in options such as insulation, solar panels or electric cars.

Other costs of climate policy will include paying countries to protect forests, or providing technical assistance to enable developing countries to switch to low-carbon energy. Energy or carbon taxes, however, do not count as a net cost because the revenue can be respent elsewhere in the economy.

What impact will these costs have on the economy as a whole? Most estimates conclude that the scale of the investments needed to prevent dangerous climate change is relatively small compared to the size of the world economy – of the order of 1–2 per cent of global GDP.

- The UN Green Economy model predicts that greenhouse gas concentrations can be kept to 450ppm by investing 2 per cent of GDP, which is around $1.3 trillion at present and represents less than 10 per cent of current annual investment. As we saw earlier, this leads to slightly lower growth and employment in the short term, but both are higher in the long term. Green investments also overtake conventional investments in the long term, outperforming them by over 25 per cent to 2050 and yielding an average of $3 for every dollar invested.[83]

- The Stern Review estimated that to stabilise greenhouse gas concentrations at 500–550ppm by 2050 would cost around 1 per cent of global GDP annually, which is equivalent to reducing the average annual growth rate between now and 2050 by just 0.12 per cent.[84] Stern later stated that in view of increased estimates of the damage of climate change, he now believed emissions should be kept to below 500ppm, which would cost 2 per cent of global GDP.[85]

- An IPCC review shows that most models estimate that it would take less than 1 per cent of GDP to stabilise greenhouse gases at 550ppm CO_2e, and 1–2 per cent of GDP to reach 400ppm.[86]

- The *OECD Environmental Outlook to 2050* estimates annual costs of 0.2 per cent of GDP to meet a 450ppm target, resulting in a global GDP 5.5 per cent below business-as-usual by 2050. To put this in perspective, world GDP would still be expected to quadruple over the same period.[87]

- The IEA estimates that 1 per cent of annual GDP would be required up to 2050 to halve energy-related CO_2 emissions between 2005 and 2050 and limit global warming to 2–3°C, but that this would be outweighed by savings in fuel costs and other benefits.[88]

- A review by consultants GHK estimates the cost as 1 per cent of annual GDP, but rising with delays in implementing policies, or where inefficient choices of policy are made.[89]

- The McKinsey cost curve report estimates annual costs of €250–350 billion, which is less than 1 per cent of forecast world GDP in 2030 and only 5–6 per cent of expected investments under a business-as-usual scenario.[90]

- The European Commission estimates the costs of achieving an emissions target of 80–95 per cent below 1990 levels by 2050 as being around €270 billion per year, equivalent to 1.5 per cent of GDP on top of the 19 per cent of GDP that is invested in infrastructure and new technologies annually. This would simply be a return to the investment levels before the economic crisis of 2008, and would still be far lower than the investment levels in China (48 per cent), India (35 per cent) and South Korea (26 per cent).[91]

Costs will rise steeply if action does not begin immediately. The IEA estimates that spending $1 on low-carbon energy today will avoid more than $4 of costs in the future,[92] and every year of delayed investment adds at least $500 billion to the price tag.[93] In fact, the IEA estimates that the costs of limiting the global temperature rise to 2°C rose by $1 trillion between 2009 and 2010 because of the weakness of the agreement reached at the 2009 Copenhagen climate change negotiations.[94] Delaying action until after 2020 – the timescale proposed for the new post-Kyoto global agreement – could increase the costs of meeting a 450ppm target by 50 per cent.[95] Delay carries the risk that we will become locked into high-carbon infrastructure – the inefficient buildings, high-carbon power stations and car-oriented transport networks being built today will last for decades. To meet the 450ppm limit, the IEA estimates that a third of coal- and gas-fired power stations must be retired before the end of their technical lifetime, and a third of these retired plants will fail to recover their investment.[96]

Set against the costs of investing in a low-carbon future are, of course, the costs of inaction. The Stern Review estimates the annual costs of continuing with business-as-usual as being equivalent to a permanent reduction of 20 per cent in average global consumption *per capita*, when conservative impacts on health and the environment are included as well as direct economic impacts, and when the effects of climate feedback loops (such as methane release from melting permafrost) are taken into account. The OECD puts the costs at 14 per cent of GDP.[97]

This estimate far outweighs the estimates of the costs of climate action – but perhaps this type of cost–benefit analysis is also missing the whole point of climate policy. In his book *Can We Afford the Future?*, economist Frank Ackerman argues persuasively that rather than adopting a narrow cost–benefit analysis approach, in which we only act if the predicted benefits outweigh the costs, we should instead treat the costs of climate mitigation as an insurance policy against the worst that can happen.[98] And in the case of climate change, the worst-case scenario involves social and economic disruption on a huge scale: mass extinctions, the inundation of coastal cities, severe global food shortages, civil unrest and mass migration of climate refugees.

Competition: a mass exodus to pollution havens?

There is a real fear that tough climate policy will result in many high-carbon businesses relocating their operations to 'pollution havens': countries with lower carbon or energy taxes, or weaker standards and regulations. Alternatively, domestic companies might lose out to foreign competitors who can flood the market with cheaper imported goods. This causes 'carbon leakage' as some emissions are displaced to other countries, as well as resulting in job losses in the countries with a more ambitious climate policy.

Despite the widespread concern over carbon leakage, evidence is now accumulating that the adverse impacts of climate policy on competitiveness are less significant than once thought.[99] Perhaps unsurprisingly, companies tend to overestimate the costs of compliance with environmental regulations. A study based on a decade of OECD and IEA policy analysis finds that: 'Adverse competitiveness impacts from climate change policy are generally limited to a small number of sectors representing a small share of economic activity in any national context.'[100]

It seems that the impacts of climate policy on individual firms depend on a whole variety of factors, such as how market power is distributed and how fast the industry is growing. Firms that respond by reducing their emissions can gain a market advantage if their products can be distinguished from those of competitors, perhaps by labelling or certification schemes, providing that customers are receptive. Some early studies concluded that environmental policy could reduce productivity, but later studies have shown that productivity benefits do start to appear after a certain time lag.

The threat of relocation to pollution havens has been found to be exaggerated. Although a pollution haven effect does exist, other factors such as the stability of the political regime, the presence of a skilled workforce, and transport and labour costs are found to be far more important to firms. In areas where jobs in heavy industry are declining, this is typically due to reasons such as increased automation, restructuring and outsourcing, rather than environmental regulations. One survey found that only 12 out of 224 permanent plant closures between 1980 and 1986 were partially motivated by environmental regulations.[101]

But although overall adverse impacts are small, they cannot be ignored. An OECD study found that if Annex 1 countries acted to halve emissions from 2005 to 2050, production in Annex 1 countries in 2030 would fall by 3 per cent and production in other countries would rise by 1 per cent, resulting in a carbon leakage rate of almost 6 per cent (in other words, 6 per cent of the emission savings in OECD countries would be cancelled out by increased emissions in other countries).[102] The European Union estimated that increasing its 2020 emission reduction target from 20 per cent to 30 per cent would result in a production loss of around 1 per cent for the most energy-intensive sectors (metals and chemicals), provided that special exemptions for energy-intensive industry stay in place.[103] This effect would be reduced, however, if major trading partners also adopt higher targets.

A range of measures can be taken to counteract the impacts of climate policy on competitiveness. These will be discussed in 'The way forward' below.

Fuel poverty: hitting the poor hardest

In the long term, well-designed climate policy should lead to lower energy prices as the costs of renewables decrease and the costs of fossil fuels increase. Nevertheless, some climate policies can contribute to higher energy costs in the short term. This could have a severe impact on people on low incomes, who spend a high proportion of their income on energy and may already be struggling to pay energy bills. Those affected may be unable to afford fuel and electricity for heating, lighting, cooking and transport, as well as being affected by the knock-on effect on prices for food and other goods. This section will focus mainly on the problem of 'fuel poverty', as it is known, in developed countries. The impacts of climate policy on developing countries are discussed in the next section.

Fuel poverty is the result of three factors – low incomes, energy inefficiency and high energy prices. The poor are often forced to pay higher prices for energy – for example, because energy suppliers insist that they use pre-payment meters for gas and electricity (to minimise the risk of non-payment). In the UK in 2009, customers using pre-payment meters were charged up to £255 more than those paying online by Direct Debit.[104] They may also live in low-quality housing, which is often badly insulated, draughty or damp, and they are less able to afford efficient appliances. In the UK, 27 per cent of those in fuel poverty live in homes with poor energy efficiency. Also, their heating needs may be higher; for example, in the case of the elderly, sick, disabled or those with very young children, who are more likely to be at home all day and need warmer homes.

Fuel poverty is strongly linked to health problems. Asthma can be triggered by exposure to mould that thrives due to condensation in cold, damp homes. Long-term exposure to very cold or hot conditions can cause heart and lung problems including strokes, heart attacks, chest infections, bronchitis and pneumonia. Living in cold conditions can also cause joint problems, eczema and psychiatric problems such as anxiety and depression. The elderly, sick and disabled are particularly vulnerable – and these groups are more likely to be living in fuel poverty.

The UK has a particular problem because many houses are poorly insulated. In England, the death rate increases by 18 per cent over the winter months, compared with increases of 10 per cent in Finland and 11 per cent in Germany and the Netherlands – countries with similar climates. Over 36,000 excess winter deaths occurred in the cold winter of 2008/2009. Fuel poverty is defined to exist when 10 per cent or more of household income is needed to heat the home to a comfortable temperature.[105] Between 1996 and 2003, the number of UK households living in fuel poverty fell from over 6 million to just 2 million, largely due to rising incomes and falling energy prices; but since then, the number has increased to over 5 million in 2009 – almost 20 per cent of all UK households – as energy prices rose steeply.[106]

Although the problem has not been widely studied in other countries, fuel poverty is also estimated to affect 15 per cent of households in Belgium, 6 per cent in France, 9 per cent in Spain and 11 per cent in Italy.[107] The problem will worsen as energy prices continue to increase. In Europe, domestic gas prices rose by an average of 18 per cent and electricity prices by 14 per cent from 2005 to 2007.

Climate policy can certainly result in higher energy prices; for example, through the introduction of carbon or fuel taxes, carbon trading schemes, removal of energy subsidies and funding of low-carbon energy such as nuclear power, renewables or carbon capture and storage (CCS). In the UK in 2011, climate policies added 5 per cent to domestic gas prices and 15 per cent to domestic electricity prices, reflecting the cost of the European Union Emissions Trading Scheme, the Carbon Emissions Reduction Target (under which suppliers have to fund energy efficiency improvements for customers) and the Renewables Obligation, which forces electricity suppliers to generate a certain percentage of their power from renewable sources. By 2020, this is expected to rise to 7 per cent for gas and 27 per cent for electricity prices, as climate policy is extended to include support for CCS and renewable heat sources.[108]

However, climate policy is not solely to blame for energy price rises. Other factors include the increasing cost of fossil fuels (see Chapter 5). A UK report in 2011 attributed just £75 of the £455 rise in average fuel bills over the last 6 years to support for energy efficiency and renewable energy, with the rest being due to rising gas prices and other costs.[109]

Fortunately there is an obvious solution which can cut both fuel poverty and climate emissions: schemes to improve home energy efficiency in low-income households. This will be discussed in 'The way forward', below.

Home energy costs are not the only aspect of fuel poverty. Transport fuel prices may increase as a result of climate policy, affecting both the costs of running a private vehicle and of using public transport. This could affect low-income households in rural areas, potentially limiting their opportunities to travel to work, school, shops and leisure activities. This could be countered by policies to make public transport cheaper, faster and more reliable, especially if low-income users are allowed free or subsidised transport.

Developing countries: catch-up or leapfrog?

Could climate policy damage economic growth and poverty reduction in developing countries? There are a number of areas for concern:

- Carbon reduction targets or carbon pricing schemes (such as caps or taxes) for developing countries could restrict their development and increase fuel poverty (because fossil fuel prices would increase).
- Pressure to reduce fossil fuel subsidies could increase poverty and hinder industrial development.
- Exports from developing countries could fall if developed countries insist on strict efficiency standards, or impose border carbon taxes, which are viewed by some developing countries as 'green protectionism'.[110]

- Reduced global demand for raw materials as material productivity improves could affect exports from resource-rich countries (see Chapter 6).

These fears are certainly not unfounded: modelling by the OECD suggests that halving emissions by 2050 by implementing a global carbon pricing scheme would result in a shift of production to more energy-efficient regions, with the output of European heavy industry rising by 2 per cent, while the output from the rest of the world falls by 16 per cent.[111] However, the first point to note is that many developing countries face devastating economic and social impacts if climate change is not restrained: falling crop yields; desertification; rising sea levels and increased storm surges; and dwindling water supplies for those dependent on glacial melt-water. Low-income countries cannot afford costly adaptation measures such as relocating vulnerable communities or building flood defences, and they will struggle with the costs of repairing damage from more frequent storms, floods or fires. With millions already living in poverty, rising food prices and water shortages will be devastating, probably leading to widespread civil unrest and mass migrations of 'climate refugees'.

Despite these threats, many developing countries remain concerned that climate action could affect jobs, growth and poverty reduction, and argue that the industrialised countries should shoulder most of the responsibility because of their far higher historic emissions and *per capita* emissions, and their greater ability to pay.

The unavoidable reality is that if developing countries follow industrial nations down the same fossil-fuel-based development path, it will lead to ecological catastrophe – a temperature rise of 4°C or more, coupled with rapid resource depletion and loss of biodiversity. [112] If everyone on the planet is to achieve a reasonable standard of living, developing nations must be able to bypass fossil fuel dependency and 'leapfrog' to a sustainable low-carbon economy based on clean, efficient technology.

Although this is a challenging task, many low-carbon strategies are well suited to the needs of developing countries. Renewable energy technologies can play a key role in providing affordable energy to remote rural communities (see Chapter 5), cutting indoor air pollution (see Chapter 3) and providing local jobs in manufacture, sales, installation and maintenance.[113] Developing countries also have a strong tradition of recycling and repair, though health and safety issues in the recycling industry need to be addressed. And as we saw in Chapter 4, climate policy aimed at preventing deforestation, promoting reforestation and encouraging sustainable agriculture also has substantial benefits for developing countries, helping to protect water supplies, improve soils, avoid landslides and reduce reliance on expensive imported fertilisers. The report *Towards a Green Economy* estimated that investing in sustainable agriculture could boost rural incomes in Sub-Saharan Africa from an average of $500 in 2005 to over $1,100.[114] On the other hand, over-use of first-generation biofuels as a climate change strategy can have adverse impacts on land use and food prices.

Developing countries that are at an early stage in constructing their infrastructure could move straight to a more efficient system, avoiding the problems of Western nations that are to a large extent 'locked in' to a carbon-intensive infrastructure of fossil fuel power plants, energy-guzzling buildings and sprawling car-dependent towns and cities. This would reduce fuel import costs. One study found that cities that have invested in good public transport systems spend only around 5 per cent of their wealth on transport, compared to around 15 per cent for those with car-based infrastructure.[115] Adopting a more technologically advanced development path, based on low-carbon technologies, can also help developing countries to escape the 'resource curse' (see Chapter 6), enabling them to improve their competitiveness, exploit new markets and export higher-value-added products. China, despite a heavy reliance on coal, has already installed more wind capacity than Germany and the United States, and could soon become the leading global supplier of wind and solar technology.

A strong international policy framework will be needed to enable developing countries to maximise their potential to leapfrog to a low-carbon economy. This will be discussed in 'The way forward' below.

Cost savings versus carbon savings: the rebound effect

There is a trade-off between cost savings and carbon savings due to the rebound effect (Box 7.3). The greater the cost savings from an efficiency measure, the more money there is available to spend on other goods and services, which can create further carbon emissions.

Box 7.3 The rebound effect

The rebound effect is a phenomenon where increased efficiency ultimately leads to greater energy or material consumption, via a number of direct and indirect mechanisms.

Direct rebound

As efficiency improves, the price of the energy service decreases, so consumers may choose to buy more of that service. Examples include:

- *Energy-efficient lighting* makes light cheaper, so people might fit more light bulbs in each room, or be less careful about switching off unused lights.
- *Fuel-efficient cars* mean that more people can afford to drive cars, and those that already own cars can afford to drive further, or buy larger cars or add extra services such as air conditioning.
- *Better insulation* makes heating a home cheaper, so householders may choose to keep their homes warmer rather than turning down the heating, thus taking comfort benefits instead of cost and energy savings.

This can also apply to material efficiency. For example, product light-weighting could make televisions cheaper because less material is used in manufacture. As a result, some families might buy extra televisions, or families that could not previously afford a television might buy one.

Indirect rebound

Consumers can spend the money they have saved from energy or material efficiency improvements on other goods or services such as a new item of furniture or a holiday. This could result in either higher or lower carbon emissions than before, depending on how the money is spent.

Economy-wide rebound

An overall increase in resource efficiency can stimulate the whole economy to use more resources. This happens in two ways:

- *Resource prices fall.* If the efficiency improvements reduce the demand for resources, the price may fall, which can encourage more consumption. For example, if demand for electricity falls as lighting efficiency improves, the price of electricity may also fall and that could stimulate greater use of other electrical appliances.
- *Shift towards energy-intensive sectors.* Energy efficiency will benefit energy-intensive sectors of industry more than other sectors. For example, if steel making becomes more efficient, the cost of steel will fall relative to less energy-intensive materials such as wood, and steel could displace wood for some applications (e.g. furniture).

Rebound is expressed as the percentage of the expected energy savings that are cancelled out by increased energy use. Rebound of over 100 per cent means that the initial energy savings are completely wiped out by increased consumption – this is the Jevons Paradox, sometimes called 'backfire'. It was first noted by William Jevons in 1865, when he observed that the consumption of coal in England shot up when the early Newcomen steam engine was superseded by James Watts's more efficient design. The Watts engine reduced coal consumption per unit of output by two-thirds, yet total coal consumption in England increased by a factor of 10 between 1830 and 1863.[116]

Ironically, rebound is greatest for the most successful 'win-win' efficiency strategies, such as those that simultaneously cut energy and material use, avoid pollution, improve product quality and save working time, because the cost savings are greater. It is also greater for improvements to technologies that are widely used throughout the economy – such as steel making, pumps or motors – or those that can lead to new applications of the technology in different sectors or markets. In the case of the Watts steam engine, for example, the new engines were not only more efficient, but also smaller, so that they could be used for ships and rail engines as well as in industry. A modern equivalent might be the massive uptake of mobile phones and laptop computers, as they became not just more efficient, but also smaller, faster and cheaper, allowing more users to afford them and more applications to be introduced (such as storing and transmitting photographs, music and videos).[117] The 'PC as a service' model being launched by Ericsson, which was mentioned in Chapter 6, may well offer opportunities to cut the material and energy intensity of computer services, but it also aims to expand the market to a billion new users in developing countries – a move that will have obvious social and economic benefits, but which is highly unlikely to result in a net decrease in material or energy use.

It used to be thought that the rebound effect would be limited because eventually 'saturation' would occur. For example, back in the 1950s we might have thought that household electricity use would eventually stabilise when every home had a TV, refrigerator and vacuum cleaner. Yet we now live in a world where many homes have TVs in every bedroom, large fridge-freezers, and a plethora of new appliances including microwave ovens, dishwashers, computers and food processors. Improved energy efficiency stands no chance of offsetting the additional energy used by this army of appliances.

One remarkable study showed that the efficiency of lighting in the UK improved by a factor of 1,000 from 1800 to 2000, yet the use of lighting increased by a factor of 6,500 over the same period, so that overall energy use increased by a factor of six.[118] In the EU-27, household electricity consumption for lighting and appliances increased by 20 per cent between 1990 and 2006, despite efficiency improvements, because more appliances were purchased (Figure 7.1, see plate section).

Over the same period, better insulation and more efficient boilers reduced the energy needed to heat each square metre of floor space by 10 per cent, but this was offset by a 23 per cent increase in the floor area of the housing stock, so that total energy used for heating actually grew by 10 per cent (Figure 7.2, see plate section). Similarly, fuel consumption in cars increased by 20 per cent between 1992 and 2007 despite improvements in fuel efficiency, because car ownership increased and more distance was travelled (Figure 7.3, see plate section).[119] Similar trends have been observed in the United States – the energy input per unit of GDP halved from 1974 to 2007, but total energy use grew by 37 per cent.

Opinion is divided on the overall significance of the rebound effect. It is difficult to determine how much growth in consumption is due to rebound and how much is due to other factors driving economic growth, such as non-efficiency-related technical advances. Some authors argue that efficiency improvements will always lead to economic growth that wipes out the initial emission savings.[120] Others conclude that rebound will reduce some, but not all, of the possible carbon savings.[121] For example, a UK study estimated that if consumers spent the money they saved from three simple energy-saving actions (turning down the heating by 1°C, cutting food waste by a third, and walking instead of driving for journeys under 2 miles) according to their typical spending patterns, the rebound effect would be 34 per cent. In other words, a third of the energy they had saved would be cancelled out by the energy used to provide the goods and services they bought with their savings.[122] Another study at a global scale predicted that current policies aimed at cutting energy use by 10 per cent by 2030 would actually only achieve half of the energy-saving target, but would succeed in increasing global GDP and employment by 0.2 per cent.[123] In developing countries, and among low-income consumers, the rebound effect is likely to be larger because energy costs are a greater proportion of household income, and there is a large unmet demand for energy services and products.

In fact, the evidence is clear: even though resource efficiency has increased steadily in most countries, and the carbon intensity of the global economy has declined (until 2010, when this trend reversed), total global carbon emissions have continued to climb due to population growth and increasing consumption *per capita* (Figure 7.4, see plate section). At least some of this growth is due to the rebound effect.

Rather than trying to measure the rebound effect precisely, it is more important to acknowledge that it exists and to put in place additional policies, such as caps or taxes, to keep resource consumption and carbon emissions at sustainable levels. This will be discussed in 'The way forward' below.

Is endless growth possible? The 9-billion-tonne hamster says no

We have discussed how climate action can result in more competitive industries and new market opportunities, creating jobs and, according to studies such as the UN report *Towards a Green Economy*, producing stronger economic growth than under business-as-usual. Another school of thought, however, questions whether infinite economic growth is possible on a finite planet. Given that all economic activity involves converting natural resources into goods and services, endless growth will eventually outgrow the capacity of the planet to support our economy. As we have seen in previous chapters, natural limits are already being exceeded:

- Our ecological footprint is 35 per cent bigger than the carrying capacity of the planet.
- Sixty per cent of ecosystem goods and services are degraded.
- Peak Oil is imminent, and supply constraints are approaching for other key metals, minerals and fuels.
- Seventy-five per cent of world fish stocks are fully exploited or over-exploited.
- Water supply will satisfy only 60 per cent of demand in 20 years.
- Although chemical fertilisers have improved agricultural productivity, soil quality has fallen by 10 per cent since 1970.

- Deforestation is increasing by 15 million hectares per year.
- We have already breached three of the nine 'planetary boundaries' within which humanity can operate safely: climate change, biodiversity and nitrogen pollution (Figure 7.5, see plate section).

Consider this in the context of the current rate of growth of the world economy. A growth rate of 3 per cent a year, typical of developed economies, means that the economy doubles in size every 25 years. A growth rate of 10 per cent a year, as seen in some emerging economies, means that it doubles every 8 years. And during each doubling period, an economy consumes more resources than during all the previous years put together. Over the last 25 years (1985–2010), the world economy has doubled in size in real terms and the OECD predicts that it will double again by 2050.

The New Economics Foundation points out that in nature, growth gradually slows down and then stops as organisms reach maturity. A hamster doubles in size every week until it reaches puberty, but if it continued growing at this rate then after a year it would weigh 9 billion tonnes, and would consume the world's entire yearly output of maize in a single day.[124]

Conventional economists generally argue that improvements in technical efficiency, together with development of substitutes for depleted resources, will allow us to continue this path of endless growth. Indeed some take this view to extremes, claiming that we could mine asteroids, migrate to other planets or even that, as the economist Robert Solow (who developed neo-classical growth theory) once claimed, 'The world can, in effect, get along without natural resources'.[125]

In order to decouple economic growth from environmental damage, however, we would need a massive, unprecedented increase in resource efficiency. Global carbon intensity fell by an average of 0.7 per cent per year from 1990 to 2007, but this was outpaced by annual average population growth of 1.3 per cent and *per capita* consumption growth of 1.4 per cent. To limit greenhouse gas concentrations to 450ppm, we must cut global carbon dioxide emissions from 30 billion tonnes in 2008 to just 4 billion tonnes by 2050. At current rates of economic growth, that means cutting world carbon intensity from 768 gCO_2 per dollar of GDP in 2007 to just 36 gCO_2 per dollar by 2050 – a decrease of 7 per cent per year, which is ten times faster than the historic rate. An even faster rate of improvement, 11 per cent per year, would be necessary if developing countries were to reach the levels of income that developed countries enjoyed in 2007 by 2050. And if growth were to continue at 2 per cent per year up until 2100, global carbon intensity will ultimately have to drop to just 6 gCO_2 per dollar – less than a hundredth of the current value.[126]

The UN report *Towards a Green Economy* appears to show a sustainable future of continued growth, but the green scenario only cuts carbon dioxide emissions to 20 billion tonnes in 2050 – still five times the target of 4 billion tonnes (although the report states that carbon sequestration from organic agriculture could make up the difference). Carbon intensity (based only on fossil fuel combustion) falls to 100 gCO_2 per dollar by 2050, compared to the target of 36 gCO_2 per dollar (including land use), and the ecological footprint still exceeds bio-capacity by 20 per cent.

Similar conclusions hold for other resources. One paper notes that a factor 10 improvement in resource efficiency, long held up as an appropriate target for developed countries, would actually have to become a factor 45 improvement by 2050 if growth continues at 3 per cent per year.[127] Another study concludes that even if developed countries were to cut their resource use *per capita* by two-thirds, allowing developing countries to converge on the same consumption level, total resource use would still be unsustainable.[128]

What will happen if we fail to achieve the massive decoupling of the economy from material consumption that is needed for sustainability? The only option would be to abandon the pursuit of endless economic growth, and try to restructure the global economic system in such a way as to allow a reasonable standard of living for everyone within a stable economy. This idea has been debated for some time among ecological economists, and is finally starting to reach the mainstream. The UNEP *Green Jobs* report, for example, questions whether 'a system of unbridled consumption . . . can ultimately be sustainable even with "leaner" ways of producing. This calls into question basic precepts of the economic system'.[129] The report discusses the need to reduce high consumption levels in developed countries in order to allow poorer countries to consume more, but it stops short of questioning the idea of continued economic growth. The concept of restricting the size of the economy to a constant level (sometimes called a 'steady-state economy') or even *reducing* the size of the economy to a sustainable level (termed 'de-growth') is taboo. Most policy makers still prefer to rely on the prospect of technical progress and voter-friendly win-win efficiency solutions to solve global problems.

Nevertheless, low-carbon policies focused solely on technical progress are unlikely to reduce resource use and emissions to sustainable levels. Ironically, the 'Green Economy' is in danger of being too successful for its own good. Efficiency, innovation and secure energy supplies will stimulate further growth, raising living standards, but also increasing consumption levels. A strategy for sustainability will need to limit total consumption in some way, whilst harnessing technical progress to improve living standards within these limits – as we will discuss in the next section.

The way forward

We have seen that a low-carbon economy can stimulate innovation, improve productivity, and in the long term, can outperform a business-as-usual economy that is reliant on diminishing supplies of fossil fuels and inefficient resource use. But we have also seen that there is a certain degree of trade-off between cost savings, carbon savings and welfare. This section will look at the way in which a well-designed policy framework could balance these conflicts, and deliver a secure, prosperous low-carbon economy.

Nature does not negotiate, so the starting point must be to ask how we can keep our economy within sustainable physical limits. Within those limits, we then need to maximise well-being by mobilising investment towards sustainable low-carbon technologies, and minimising adverse impacts on businesses, workers, low-income households and developing countries. Finally this chapter discusses the need to restructure our economies so that they no longer depend on continuous growth in material consumption.

Safe carbon limits

There are three main policy options for restricting carbon emissions to safe levels:

1 caps;
2 taxes;
3 voluntary restraint.

Caps are the only option that can guarantee that resource use or emissions will be kept to a sustainable level. Various options exist:

- *Regional cap-and-trade schemes* such as the European Union Emissions Trading Scheme (ETS): governments set a cap and companies then buy or are allocated carbon permits which they can trade amongst themselves.
- *Personal carbon quotas*: every citizen is allocated a certain amount of carbon which they trade in when buying fuel or electricity, but they can buy or sell allowances if they use more or less than the allotted amount.
- *'Cap-and-share' or 'cap-and-dividend' schemes*: these aim to simplify the system by auctioning permits to the relatively small number of companies involved in extracting or importing fossil fuels, and returning the revenue as an equal share to all adults in the country.[130]

The common feature of these methods is that they reward low consumers of carbon-intensive fuels and penalise those who consume more than average. For cap-and-trade schemes and personal carbon quotas, companies or individuals who consume less than average will be able to make a profit by selling their surplus permits. For cap-and-share schemes, the price of fuel would increase to cover the cost to the companies of buying permits, but this would be offset by the dividend returned to all citizens, so that low consumers make a profit and high consumers lose out. This would provide a strong incentive to reduce the use of carbon-intensive fuels.

The main problem with caps is that policy makers tend to set the cap too high by giving away too many initial permits, in response to political lobbying. In the EU carbon trading scheme, the cap for the period 2008–2012 was set at only 2 per cent below 2005 carbon emission levels, and emissions actually increased during the early phases of the scheme, with carbon prices crashing to less than €1 per tonne at one point. To be successful, the cap should be set low enough to keep carbon emissions to a sustainable level; permits should be auctioned instead of given away; and loopholes that allow emissions to be offset against poorly verified schemes in other countries should be tightened.[131] Some commentators have suggested that a World Environment Organisation should be created to define the level at which resource extraction and emissions should be capped in order to ensure sustainability.[132]

Another issue is that if the cap is implemented only in richer countries, then there is the possibility of global rebound, where falling fuel demand in the richer countries could lead to lower global fuel prices, which would encourage additional consumption in poorer countries. This could cancel out carbon savings, or even lead to a rise in emissions, if consumers in poor countries were driving less efficient vehicles, for example, although there could clearly be economic and social benefits from increased access to energy. This emphasises the importance of enabling all countries to progress to a low-carbon infrastructure (see 'Developing countries' below).

The second option, taxes, can be applied to fuel, energy, carbon or other resources such as water or construction minerals. Carbon, fuel and energy taxes have been used in various European countries for many years, though it has proved impossible to agree a common tax regime at international level. Although taxes send a strong economic signal to consumers, they do not necessarily guarantee that carbon emissions will be capped at a sustainable level. One strategy would be to increase taxes in line with efficiency improvements, so as to 'lock in' carbon savings.[133] For example, if the fuel efficiency of cars increased by 10 per cent, then gasoline prices could also be increased by 10 per cent. This should cancel out the rebound effect by removing the temptation for consumers to drive more, thus ensuring 10 per cent carbon savings. Although there would be no direct cost savings for consumers, the tax revenue would be available for redistribution, perhaps as help for firms, households and developing countries to switch to low-carbon technology. However, taxes would have to be constantly adjusted in line with efficiency improvements, and even then this cannot guarantee that emissions would be kept to a sustainable level. An alternative suggestion is to fix a gradually rising trajectory for resource prices many years in advance, along the lines of China's Five Year Plans. Although this may seem like a radical suggestion for a society accustomed to leaving prices to be determined by market forces, there are parallels with proposals to strengthen investment in low-carbon technologies by setting a floor price for carbon, in the UK, for example.

The final option, a voluntary cut in consumption, is challenging because it requires a dramatic culture change. However, there is evidence that awareness-raising techniques can encourage householders to cut their energy use: smart meters and more informative bills have been shown to reduce energy use by 5–15 per cent, for example. One US study found that telling people how their energy bills compared with those of their neighbours, and also giving them tailored energy-saving tips, reduced consumption by 2–5 per cent.[134] Interestingly, this can lead to a 'boomerang' effect where customers who are consuming less than average actually increase their consumption slightly. This effect can be counteracted by putting smiley or sad faces on the bills, depending on whether energy use was above or below average.[135]

Awareness could also help to limit indirect rebound effects due to consumers spending their energy cost savings on more energy-consuming goods and services, and with tackling perverse behaviour such as the 'feel-good' effect that leads some consumers to be less careful about saving energy if they install energy-saving equipment.

Companies and governments also need to be careful not to encourage perverse behaviour. One notorious example was the short-lived offer by Tesco to give their customers 60 free air miles if they bought an energy-efficient bulb – an offer that would have far outweighed the savings in lighting energy. Another example is the fact that building new roads to reduce congestion will increase traffic, as drivers travel further if journey times are reduced. A more effective approach to traffic reduction would be to take away road space through allocating one traffic lane to buses, bicycles or pedestrians, thus discouraging private car use.[136]

It is important to emphasise that awareness techniques are unlikely to work unless backed up by caps and/or taxes, because free-riders would increase their energy consumption as prices fell.

Could resource scarcity eventually act as a cap on consumption? Although this is possible, it is highly undesirable – economic chaos and social hardship would result as food and energy prices spiral out of control. Scarcity can also accelerate the development of low-quality resources with higher environmental impacts, such as tar sands and Arctic oil. Lastly, we are unlikely to run out of resources fast enough to save the planet! We can only burn less than half of our proved reserves of fossil fuel, for example, if we are to avoid dangerous levels of climate change (see Chapter 5).

Mobilising investment in low-carbon technologies

Having set safe limits for carbon emissions, we then need to deliver the maximum possible well-being within those limits. This means setting a clear policy framework to direct investment and consumer spending towards sustainable low-carbon technologies, so that we can optimise the energy services delivered per tonne of carbon emitted. For example, if consumers save money because cars become more fuel-efficient, an optimal policy framework would encourage them to spend their savings on home insulation rather than on a bigger car.

At present, however, a number of barriers inhibit investment in low-carbon technologies:

- Many have high capital costs, even though operating costs may be low.
- Many are new technologies, so prices are currently high although they will decline in future.
- Market prices ignore external costs, such as the health costs of fossil fuel pollution.
- Fossil fuels often receive larger subsidies than renewable energy.
- Existing infrastructure such as the electricity grid, transport fuel supply network and urban planning is geared towards the use of fossil fuels.
- New entrants to the market may struggle to compete with established companies, and there is a lack of awareness of the benefits of low-carbon technologies.

For a chance of avoiding severe climate change, we need to act fast to lower these barriers. Delay will result in higher costs in the future, and risk lock-in to high-carbon infrastructure. The level of investment required is relatively modest compared to the size of the global economy – most estimates, as we saw earlier, are less than 2 per cent of world GDP per year. Some nations are already exceeding this – South Korea is investing 3 per cent of GDP ($36 billion) in a Green New Deal, creating 960,000 jobs on low-carbon projects.

Investment in some sectors, such as renewable energy, is growing, but far more is needed, especially in low-carbon transport and buildings. The IEA warns that governments need to quadruple the amount of new spending on clean energy and energy efficiency if climate targets are to be met.[137] However, governments also need to mobilise private sector investment, which accounts for 86 per cent of global financial flows.

The first step is to recognise that environmental costs – estimated at a staggering $6.6 trillion in 2008, which is 11 per cent of global GDP – are not included in market prices, and so the benefits of low-carbon energy, such as reduced pollution, are generally ignored.[138] It is probably technically impossible to achieve 'true cost pricing' in the sense that all environmental and social costs and benefits are fully included in the market price, including air, land and water pollutants, greenhouse gases, landscape damage and accident risks. Valuing these impacts involves highly subjective decisions over, for example, the value of a human life, or the 'willingness to pay' of individuals to protect a view of the countryside. In the case of climate change, it is almost impossible to put a value on a tonne of carbon, given the uncertainty over future impacts. However, prices can be crudely adjusted to compensate for external costs using caps (because they put a price on carbon emissions), taxes, subsidies and regulations.

We have already discussed the use of caps and taxes as tools for keeping carbon emissions to safe levels. It is important to emphasise that cap-and-trade schemes will not succeed in driving investment in low-carbon technologies unless the cap is set low enough. Over-generous caps result in the traded price of carbon being well below its actual environmental cost. Cap-and-trade schemes can also result in large carbon price fluctuations, which can discourage investment in low-carbon technologies. One way of getting around this problem is to set a floor price for carbon, which effectively combines the efficiency of carbon trading with the stability of carbon taxation. The OECD estimates that setting a carbon price to match a target of 550ppm CO_2e in 2050 would quadruple investment in energy R&D and renewable power generation.[139]

Subsidy reform will be essential if we are to shift investment towards low-carbon technologies. Fossil fuels received $409 billion in consumption subsidies (reduced prices to consumers) in 2010 compared to $66 billion for renewable energy.[140] Consumption subsidies exist mainly in oil- and gas-producing countries (Iran, Russia, Saudi Arabia) and developing countries (India, China, Indonesia).[141] In addition, many governments provide production subsidies in the form of tax breaks, grants and other assistance to oil, gas and coal companies.[142] The G20 has already agreed to phase out 'inefficient fossil fuel subsidies that encourage wasteful consumption', and there is pressure on other countries to follow suit – it is estimated that this could reduce global carbon emissions by 7 per cent by 2020[143] or 10 per cent by 2050.[144]

Subsidies in other sectors can also have damaging effects on the environment. In 2002, there were thought to be over $1 trillion of subsidies applied globally, three-quarters of which were in OECD countries.[145] Energy, water, agriculture and fisheries subsidies add up to 1–2 per cent of world GDP, and often encourage over-production, environmental damage and waste.[146] However, removal of subsidies can be politically difficult, due to vested interests and to the problem of a disproportionate effect on the poor, who spend a higher proportion of their income on food and fuel.

Fossil fuel subsidies, however, are an inefficient way of helping the poor. The IEA estimates that only 8 per cent of subsidies reach the poorest 20 per cent of consumers (Figure 7.6). In Iran, for example, which sells gasoline for 10 cents per litre and diesel for just 2 cents per litre (at a cost to the country of $66 billion, or 20 per cent of GDP), 70 per cent of the subsidies go to the richest 30 per cent of consumers.[147] In fact, the IEA estimates that it would cost only $48 billion per year to end global energy poverty by 2030 – less than an eighth of the amount spent on consumption subsidies.[148]

Subsidies also encourage high levels of inefficiency: industrial energy use per unit of GDP in Iran has been increasing by 1.6 per cent per year since 1990. Discounts for large electricity users are particularly perverse: it would make more sense to charge *more* per unit above a certain threshold, to encourage efficiency. Subsidies

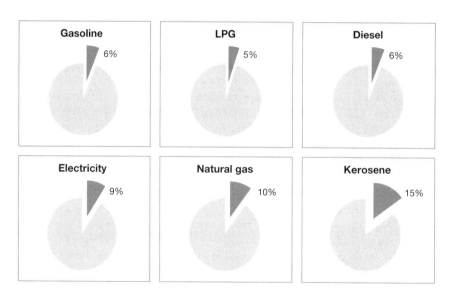

Figure 7.6 Percentage of fossil fuel consumption subsidies in 2010 reaching the poorest 20 per cent of consumers

Source: Adapted from IEA, 2011a

also act as a barrier to the wider uptake of renewable energy – which otherwise could be the most cost-effective way to provide energy for the poor in remote off-grid areas. Subsidies take up money that could be spent directly on services to help the poor – some countries spend more on subsidies than on public health. And they can escalate alarmingly as fossil fuel prices rise – in 2007, for example, the increase in oil prices led to an average increase of subsidy payments equivalent to 1.5 per cent of GDP for emerging and developing economies.[149]

Many countries are therefore choosing to gradually eliminate fossil fuel subsidies, and are redirecting the money to support the poor through cash payments, preferential tariffs, pro-poor health and education services, or subsidies for clean, affordable energy services such as home insulation, efficient stoves and small-scale renewable energy systems. For example, when energy subsidies were cut in Indonesia in 2005 and 2008, direct payments of $30 were made to 15 million poor households every quarter. In Gabon, subsidy reform was accompanied by a scheme to provide micro-credit for women in rural areas; and in Ghana, school fees were eliminated and extra healthcare was provided in poor areas.[150]

Subsidies can be redirected to promote low-carbon technologies. Feed-in-tariffs, currently used in more than 60 countries,[151] have been shown to be the most successful support measure for renewable energy, and are credited with the emergence of Germany's strong renewable energy industry. Low-interest loans have also been successful in encouraging householders to install solar panels via the 100,000 Solar Roofs Program in Germany and the New Sunshine Program in Japan.

The scale of support directed to different low-carbon options – such as wind, solar, biofuels, nuclear, resource efficiency, carbon capture and storage or geo-engineering – might vary depending on the potential carbon savings, but also on the co-benefits and drawbacks of each technology. A framework for comparing different technologies will be presented in Chapter 9.

Policy makers need to integrate caps, taxes and subsidy reform into a supportive policy framework to help mobilise business investment. Although businesses do not like regulations that are too prescriptive – specifying exactly what technologies to use, for example – there is evidence that they are happy to invest in low-carbon options when the right targets and incentives are in place, allowing them the freedom to innovate to find the most cost-effective solution. Government support can encourage businesses to undertake longer-term investments, spread the risks of investing in new technologies, and prevent promising technologies from falling into the 'valley of death' between initial development and sales. However, it is vital that governments should demonstrate a clear long-term commitment to a low-carbon strategy. Sudden policy changes such as the recent cuts to solar feed-in tariffs in the UK can undermine investor confidence. An integrated policy framework, combining many of the measures discussed in the preceding chapters, could include:

- clear long-term targets to limit emissions and resource use to sustainable levels, using caps or taxes;
- redirecting subsidies from fossil fuels to low-carbon technologies; e.g. through feed-in tariffs, grants or low-interest loans;

- ambitious and continually tightening energy efficiency standards for buildings, appliances and vehicles, to encourage innovation;
- certification and labelling to build a market for sustainable products and ensure that low-carbon goods meet high quality and ethical standards;
- regulation to prevent free-riders; e.g. by protecting natural capital such as forests, and preventing waste of valuable resources (e.g. through a ban on landfilling recyclable material);
- government funding for R&D in key technologies such as smart grids, energy storage and low-carbon vehicles – energy efficiency and renewable energy have received only 20 per cent of cumulative funding in IEA countries since 1974, although the situation is now improving (Figure 7.7, see plate section).

Cushioning the transition

In the long term, a low-carbon economy should provide more secure and affordable energy services; but in the short term, energy prices may increase as caps or taxes are applied to limit fossil fuel consumption and subsidies provided to stimulate a shift to sustainable energy infrastructure. During this transition, governments can help businesses to adapt, support workers who lose their jobs, protect low-income households and help developing countries to leapfrog to sustainability. With the right supporting policies, a low-carbon economy can improve welfare and reduce poverty; but without these policies, there could be severe social impacts, which will reduce public support for climate action.

Enabling businesses to adapt

Strong and well-designed climate policy can be a stimulus to innovation and greater efficiency, opening up new markets and driving process and product improvement. However, carbon-intensive firms that are unable or unwilling to adapt may lose out. This has led to divisions in the business community, with many leading businesses campaigning for tougher climate targets (see Box 7.2 above), while representatives of carbon-intensive industries are resisting climate targets and citing the threat of relocation.

Various measures to counteract adverse effects on businesses are possible, including:

- *Exemptions* from carbon or energy taxes, or reduced rates or partial rebates for high-carbon sectors. This is the most commonly adopted measure, with over 1,000 examples found by the OECD.[152]
- *'Grandfathering'* of permits for emission trading schemes, where free permits to produce carbon are allocated according to the existing emissions of each company.
- *Tariffs* on imported goods produced in countries with lax regulations, or *border taxes* based on the embodied carbon of imported goods.

However, these measures have drawbacks. Tax exemptions reduce the effectiveness of emission taxes, by shifting the incentive to reduce emissions from high-carbon sectors, which are responsible for most of the emissions, to low-carbon sectors, and reducing the price signal that can encourage customers to switch to lower-carbon alternatives. As a result, the costs of carbon reduction are likely to rise, as the most cost-effective opportunities usually lie within high-carbon sectors. The OECD has estimated that exemptions for energy-intensive industries could increase the costs of limiting greenhouse gases to 550ppm by over 50 per cent in 2050.[153] One solution is the output-based refund system, where firms have to compete for rebates by reducing their carbon intensity, with firms in the top half of the efficiency table receiving rebates based on their difference from the average carbon intensity.[154]

Grandfathering of permits can appear to reward heavy polluters with a windfall profit and penalise early movers who have already taken steps to reduce their emissions. It also provides little incentive to reduce emissions unless it is clear that it will be gradually phased out and replaced by auctioning of permits. The European Union Emissions Trading Scheme is moving towards auctioning, but there are plans to issue free permits to the most efficient companies in energy-intensive sectors, in order to provide extra incentives for carbon reduction whilst minimising carbon leakage to other countries.

Tariffs and border taxes could reduce carbon leakage, but carry a risk of retaliation by countries whose exports are affected, potentially triggering trade wars. Careful design would be needed to avoid conflict with World Trade Organization (WTO) rules. Also, estimating and verifying embodied carbon for a vast range of imported products would be an administrative nightmare, meaning that this system would probably only be practical for a limited number of standardised commodities such as steel or cement. Even then, costly monitoring and verification systems would be needed in both the importing and exporting countries.

An OECD study finds that rather than allowing exemptions, it is far more effective to focus efforts on trying to increase the number of countries taking part in climate action. For example, if the EU acted alone to halve emissions from 2005 to 2050, 11.5 per cent of emissions would leak to other countries; whereas if all Annex 1 countries acted together, then the leakage would fall to 7 per cent. Border taxes could reduce this leakage significantly – from 11.5 per cent to just 3 per cent for the EU.[155]

If a large international coalition cannot be built, there may be a need for transitional measures to protect companies in the short term – especially if countries are acting unilaterally.[156] Policy makers can:

- Have clear policy targets and set timetables for phasing in legislation well in advance, to reduce uncertainty and enable long-term planning.
- Allow a suitable transition period to allow companies to adapt to new regulations. German wastewater charges, for example, were announced a few years before actually being collected and had their strongest steering effect during the announcement period when the charge was still zero.[157]

- Ensure that transaction costs are as low as possible, by aiming for policies that are simple and cheap to administer.
- Provide support for adaptation, such as grants or loans for low-carbon technologies; R&D, demonstration, dissemination and best practice programmes; and environmental auditors and advice agencies to help firms cut carbon cost-effectively. This can be funded through revenues from carbon taxes or permits (auctioning of permits under a well-designed cap-and-trade scheme could generate $30–250 billion per year in the United States and almost $70 billion in the EU).[158]
- Improve the political acceptability of any border taxes that are applied by recycling tax revenues into providing technical assistance to developing countries to switch to low-carbon production methods.
- Build support for policies by emphasising the potential for 'win-win' solutions and co-benefits such as energy and resource cost savings, productivity improvements and improved company image.
- Encourage companies to set up carbon accounting systems to find the most cost-effective ways of cutting carbon, and to fully exploit the opportunity to improve their image with customers and shareholders via environmental reporting schemes such as the Carbon Disclosure Project or the Global Reporting Initiative.
- Use exemptions, free allowances or border taxes as transitional measures only, and phase them out as soon as possible.

These measures can help to smooth the transition to a low-carbon economy, but competitiveness impacts are inevitable to some extent. The whole point of a low-carbon policy is to ensure that carbon-efficient activities are more competitive than high-carbon activities. And while companies might be expected to pursue win-win opportunities voluntarily, government intervention will be needed to push carbon savings beyond this level – providing benefits for society as a whole, but creating winners and losers at the level of the individual firm. The ultimate aim of policy makers should be to enable firms to adapt as painlessly as possible to the new reality of a low-carbon economy; and to encourage a co-ordinated response among a wide coalition of countries, so that costly and ineffective special exemptions or tariffs to protect domestic industries can be phased out as soon as possible.

Empowering workers

The low-carbon economy will create new jobs and transform existing ones, but there will also be job losses in carbon-intensive sectors. It is essential to protect the workforce during this period of change, and to ensure that they have the skills to take advantage of new opportunities. In principle this can be done (for developed countries) by following the principle of 'flexicurity' pioneered by Denmark and now adopted by the European Commission in response to the employment challenges of globalisation and rapid technical change. Flexicurity is a combination of flexibility and security, including solid social protection for workers coupled with life-long learning and job mobility. It includes 'active labour market policies' which help people to adapt to a pattern of changing employment that will include spells of unemployment, retraining and reintegration into the workforce. Government support includes help with careers advice, retraining, relocation and childcare as well as welfare payments during periods of unemployment. However, it should be noted that flexicurity has been criticised as being expensive – in Denmark, it accounts for over 4 per cent of GDP – and it is not applicable to countries which have not yet developed a welfare state.[159]

Training and education are crucial. Most obviously, workers who lose jobs in carbon-intensive industries such as coal mining, oil and gas production or heavy industry will need retraining so that they can move to new jobs. Other workers will find that their jobs may change as their industry adapts to low-carbon production, and they need to learn new skills. Builders, plumbers and electricians will need to learn how to install heat pumps and solar water-heating systems, for example.

On top of this, there are already shortages of skilled workers in a number of areas vital to the low-carbon economy, such as renewable energy and resource efficiency.[160] To some extent, this is part of a wider shortage of graduates and apprentices in science, technology, engineering and mathematics that is beginning to emerge in Europe in particular. In 2008, for example, there were 64,000 job vacancies for engineers in Germany, causing the environmental sector major problems and costing the German economy over €6 billion.[161]

To allow workers to make the most of the opportunities offered by the low-carbon economy, governments will need to work with employers, trade unions, schools, colleges and universities to encourage:

- *New university and college courses* in areas such as renewable energy technology, eco-design, sustainable agriculture, carbon footprinting and energy management.
- *Training courses* to help workers with general skills such as welding or construction adapt to new areas such as wind turbine manufacture or low-carbon retrofitting of buildings. It is important to keep fees at an affordable level – there are reports that electricians would like to train as solar PV installers, but cannot afford the €2,000 fees, for example.[162] The American Recovery and Reinvestment Act of 2009 includes a $500 million programme of grants for clean energy workforce training.[163]
- *Mainstreaming* low-carbon principles into general education and training. It is pointless to consider the low-carbon economy as being separate from the conventional economy. In the future, every job will be a low-carbon job. This will mean updating the curriculum of many mainstream subjects, including engineering, design, architecture, agriculture, urban planning, accountancy, politics, economics and business management.[164]

Investing in retraining programmes is a highly cost-effective way of reaping the benefits of a low-carbon economy. For a start, training and education will improve the overall productivity and competitiveness of the workforce, improve employment flexibility and mobility, and provide extra interest and job satisfaction for workers. In many cases, workers will be able to switch to new jobs that use their core skills with relatively little retraining. For example, we saw in the section on 'Job quality' that many US factories are switching from car production to wind or solar manufacturing. Similarly, workers from the oil and gas sector already have skills in welding and surface treatment that are valuable to the wind turbine industry.

Action is also needed to make sure that low-carbon jobs are high-quality jobs. We have seen that some sectors such as biofuel cultivation and informal recycling have health, safety and environmental problems, or provide jobs that are poorly paid and temporary. To counter this, it is important to encourage the adoption of international standards for protection of workers, such as those enshrined in the International Labour Organization's five principles (workers' rights, decent work, social protection, social dialogue and sustainable business), or the UN Global Compact in which companies agree to support ten principles in the areas of human rights, labour standards, the environment and anti-corruption.[165] Other useful strategies include the formation of workers' co-operatives such as those formed by trash scavengers in Brazil and Colombia.

Looking after low-income households

Climate policies that increase energy prices can tip households into fuel poverty. The most effective strategy to address this problem is to target home renovation programmes on low-income households. Cold, damp, draughty houses can be upgraded through better insulation, draught proofing, double- or triple-glazed windows, more efficient boilers and renewable energy technologies such as solar panels. In hot climates, improvements can include shading from trees, blinds or canopies, reflective paint on walls and roofs, natural ventilation and more efficient air conditioning systems. In the UK, efficiency improvements kept 80,000 households out of fuel poverty from 2007 to 2008 alone.[166] The government put in place programmes such as Warm Front, which offers grants for insulation and more efficient heating systems. A home renovated under the Warm Front scheme could save £400 from the annual energy bill.[167]

Reduced fuel use can offset energy price rises. The UK government has estimated that the combined effect of all its climate policies – those that increase energy prices as well as those that save energy – will be to reduce the average annual household energy bill by 7 per cent (£94) in 2020.[168] As well as financial savings, there will be health benefits from living in warmer homes. It is estimated that every £1 spent on alleviating fuel poverty saves the UK National Health Service over £30 in the long term.[169]

Although renovation is the best long-term solution, additional measures may be necessary to protect vulnerable households in the short term. These can include:

* direct cash payments such as the UK Winter Fuel payments;
* 'social tariffs', which offer lower energy prices to low-income households (used in six EU countries);

- emergency measures to ensure that homes are not disconnected if they fail to pay their bills, and to provide crisis loans and help with rescheduling debt repayments;
- a switch away from tariffs that become cheaper as more energy is used, to those that become more expensive for higher energy users (normally the wealthy); however, this would need to ensure that low-income households with unavoidable high energy use (e.g. because of a sick or disabled occupant) were not penalised.

One study found that renovation will not be sufficient to eliminate fuel poverty in the UK: this could only be achieved by increasing incomes.[170] One strategy would be to recycle carbon taxes into income tax reductions for poor households, as part of a process known as 'Ecological Tax Reform' (see Box 7.5, p. 280). This could also help to address the knock-on effect of fuel prices on food and other costs.

Developing countries

A strong international support programme could enable developing countries to 'leapfrog' to a sustainable development path based on clean, efficient, low-carbon technologies, helping to reduce the problem of carbon leakage to other countries. This will require fair targets, funding for low-carbon infrastructure, targeted poverty reduction programmes and careful reform of subsidies and trade laws.

SETTING FAIR TARGETS

In order to protect climate stability, we need to cut greenhouse gas emissions to between 1 and 2 tonnes CO_2e *per capita*. How can that target be achieved in a way that allows developing countries to eliminate poverty and achieve living standards comparable with those of industrialised nations?

Current emissions *per capita* range from 25 tonnes in the United States and 10 tonnes in the EU to 5 tonnes in China, 2 tonnes in India and under a tonne in many African countries.[171] Developed countries are responsible for most of the historic emissions, with the United States emitting 30 per cent of cumulative emissions between 1900 and 2004, and Western EU countries responsible for another 21 per cent, compared to just 9 per cent for China and 2 per cent for Africa.[172] Because of this, developing countries argue that there should be a 'common but differentiated responsibility' (Principle 7 of the 1992 Rio Convention). In other words, all nations must work together to protect the environment, but developed nations should accept tougher targets because they are responsible for most of the damage and they are more able to pay for the necessary changes.

One way of setting targets is to set a global cap on emissions and then allocate a certain number of tradable carbon permits to each nation. There are different ways of sharing out the permits, shown here roughly in increasing order of benefit to developing countries:

1   grandfathering (in proportion to emissions in a base year);
2   auctioning (countries pay to buy permits; this is equivalent to a global carbon tax);
3   *per capita* (equal permits given to every person in the world);
4   ability to pay (permits given to each person in inverse proportion to GDP *per capita*);
5   historical responsibility (inversely proportional to historical emissions);
6   business-as-usual (developing countries are allowed to continue their current rate of carbon growth; developed countries share out the remaining permits among themselves).

The benefit to developing countries varies between countries. For example, Africa benefits most from the 'ability to pay' method, whereas China and Latin America lose out. South Asia (including India, but not China) does best under the '*per capita*' rule. Interestingly, an OECD model predicted that because developing countries have so much to lose from climate impacts, by 2100 they would actually be better off under *any* of these scenarios than under a business-as-usual scenario with no further climate action.[173] Perhaps in recognition of this, many developing countries are already adopting voluntary targets, though these tend to be directed at improving efficiency and reducing carbon intensity rather than limiting total emissions.

TECHNOLOGY TRANSFER

Many developing countries will be unable to switch to a low-carbon development path without considerable technical and financial assistance. In the past, technology transfer from developed to developing countries consisted mainly of direct aid funding from national governments or agencies such as the World Bank for large infrastructure projects such as hydroelectric schemes or coal-fired power plants. Today, however, there is far more emphasis on sustainable technology on a variety of scales, covering a wide range of applications such as solar panels, industrial energy efficiency and cleaner production methods, or sustainable agriculture and forestry schemes. A wider variety of organisations is involved, including private companies, charities and NGOs, and there is more emphasis on transferring skills and knowledge as well as equipment, and on embedding the capacity to manufacture, install and operate the technology locally, thus generating jobs and benefiting the local economy.[174]

Potential funding sources include the Clean Development Mechanism, the Green Climate Fund, which aims to provide $100 billion per year eventually, and the Global Environment Facility. Over $30 billion has been pledged to various funds to date (2012), of which just $13 billion has been provided so far and $2 billion actually spent on projects.[175]

There is considerable potential to redirect existing investment and development funding towards low-carbon technologies. For example, World Bank funding for fossil fuel projects from 1994 to 2004 was $26 billion, compared to $1.5 billion for renewable energy and energy efficiency. Similarly, the US Export-Import Bank directed 93 per cent of the $28 billion it spent on loans and guarantees for energy-related projects between 1990 and 2001 to fossil fuel projects and only 3 per cent to renewable energy projects. In terms of overseas development assistance, OECD countries spent an average of $420 million per year from 1999 to 2003 on large hydropower projects, compared to $130 million for other renewable energy technologies.[176]

Technology transfer can be delivered through co-operative international research and training centres, such as the UN network of National Cleaner Production Centres, as well as through scholarships and exchange programmes, and via direct investment in clean technologies.[177] Developing countries can also benefit from an effect called 'technical spill-over' where innovation by developed countries gradually spreads to other countries; for example, as R&D results are published and low-carbon technologies improve and become cheaper.

Technology transfer can benefit both the donor and recipient countries, by cutting global environmental degradation and resource depletion at the same time as reducing poverty, promoting stability and opening up new markets. Developing countries that learn how to manufacture clean technology can often undercut the price of OECD countries, as in the case of China, which is now a major exporter of wind and solar technology. But there is a stumbling block – companies may be reluctant to share technology in case the design is copied by manufacturers in the recipient country using 'reverse engineering'. This has discouraged Western suppliers from installing their latest designs of wind turbines in China, for example. Laws on trade-related intellectual property rights (TRIPS) need to be strict enough to ensure that developers can profit from their investments, but flexible enough to allow widespread access to clean technologies for developing countries. Certification schemes also need to be developed to ensure that products conform to basic quality standards – cheap, low-quality products (such as exploding light bulbs or inefficient solar panels) can give the whole industry a bad reputation and damage the market for green goods.

POVERTY REDUCTION

There is concern that climate targets could adversely affect efforts to reduce poverty in developing countries, by restricting economic growth in the countries themselves and perhaps affecting trade with developed countries. Yet although growth and trade have lifted millions out of poverty over the last few decades, the current system is not sustainable. For every $100 of global growth from 1990 to 2001, just $0.60 contributed to reducing poverty among the poorest billion people, who live on less than $1 per day. At that rate of trickle down of wealth to the poor, it would take 15 planets' worth of resources to get everyone onto an income of at least $3 per day.[178]

An alternative approach is to tackle poverty directly, through aid for specific critical areas such as education, health, sustainable energy and agriculture. It has been argued that relatively small amounts of aid targeted at key areas such as vaccination and malaria eradication can overcome the barriers that prevent countries from becoming self- reliant.[179] Cancellation of unpayable debt and reform of unfair trade rules are also crucial (see Chapter 6).

Climate action that protects natural assets, such as improved resource efficiency, sustainable agriculture and reforestation, can also play a strong role in poverty alleviation, as many people in developing countries depend directly on agriculture, fishing and forestry, and water is often scarce.

Provision of affordable low-carbon energy, such as solar panels, bio-gas schemes and efficient cook-stoves, can be highly effective in reducing poverty and increasing welfare at the same time as generating jobs and economic benefits for local people. However, it can be difficult for entrepreneurs in developing countries to access appropriate funding, and it would be useful to try to reduce the administrative barriers facing small companies trying to access finance from sources such as international climate or development funds.[180] Micro-finance initiatives such as the Grameen Bank and Kiva.org can also be very successful in enabling people to afford low-carbon technologies (see Chapter 5, Box 5.4). Around 10 per cent of the poorest families in Africa and the Middle East had been provided with micro-loans by 2007, but there is tremendous untapped potential.[181]

FRUGALITY IN THE RICH WORLD

Although developing countries face the prospect of losing export income if improved resource efficiency leads to a fall in demand for raw commodities (see Chapter 6), it is interesting to note that the G77 group has actively called for the rich world to use fewer resources. Writing on behalf of the G77, President Lula of Brazil said:

> Sustainable development not only requires a transfer of clean technology and substantial resources to developing countries so as to make it possible for them to grow and to preserve and avoid damage to the environment at the same time, but also a change in the unsustainable patterns of consumption, production and trade in developed countries, which harm the environment.
>
> (Luiz Inácio Lula da Silva, President of Brazil, 'Foreword' in Ahmia, 2011)

This acknowledges that natural ecosystems are being degraded and finite resources depleted at such a rate that rich countries will need to cut their consumption if other countries are to be able to raise their living standards. Of course this flies in the face of standard economic theory that states that growth is a 'rising tide that lifts all boats' – but conventional economics fails to take account of any physical limits to consumption.

Can voluntary frugality or 'sufficiency' by the rich really help the poor? The argument is that if the rich buy less, the global demand for resources will fall, and therefore the price will also fall. That means that consumers who could not quite afford to buy certain goods before (in economic jargon, the marginal consumers) will now be able to afford them. So the resources 'freed up' by voluntary frugality by the rich are now bought instead by slightly poorer consumers. This is the rebound effect, but on a global scale: Blake Alcott terms it the 'sufficiency rebound'.[182] It can be likened to the process of trying to squash a balloon – as consumption is reduced in one part of the global economy, it balloons out in another part. So although frugality may fail to achieve much reduction in environmental damage (or could even increase damage, if less efficient technologies are used in developing countries), it can act as a means of redistribution. It would, however, take a long time and a high degree of frugality by both rich and middle-class consumers before the benefits finally reached the poorest on the planet.

An interesting study by Steinberger and Roberts investigates whether we could actually achieve a reasonable living standard for everyone on the planet by redistributing resources from richer to poorer consumers. They use the Human Development Index (HDI) as a measure of living standards: this is derived from a combination of lifespan, GDP *per capita* and education level. A graph of HDI against *per capita* energy use for each country (Figure 7.8) shows that HDI increases with energy use up to a certain threshold, beyond which further increases in energy use produce little improvement in the HDI. This suggests that overall levels of well-being could be improved by redistributing energy use from high consumers to low consumers.

The interesting point is that back in 1975, the average global energy use *per capita* was below the level correlated with a 'reasonable' value of the HDI, which is considered to be 0.8 (on a scale of zero to 1). In other words, the world was not using enough energy to guarantee a reasonable standard of living for everyone on the planet, even if all the benefits were shared out equally. But since 2005, the average energy use has been above the level needed to achieve an HDI of over 0.8. That means that if it was possible to distribute economic benefits evenly to everyone on the planet, no one would have to fall below this reasonable standard of well-being.[183]

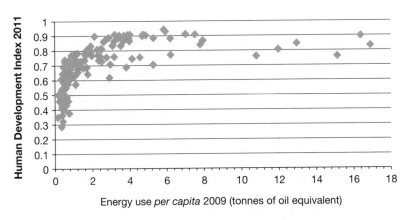

Figure 7.8 Energy use is correlated with the UN Human Development Index (HDI) only up to a certain threshold

Source: Based on data from IEA, 2011b and UNDP, 2011b

Steinberger and Roberts conclude that further global economic growth is not essential to achieve development and poverty eradication – the same result could be achieved in a more sustainable way by redistribution of resources between the poor and the rich – and that implies the need for the rich to consume less. But at our current level of energy use, our carbon emissions are not sustainable. We therefore need to cut total global emissions as well as redistributing resources. Efficiency improvements will be needed to ensure that cutting emissions does not reduce well-being below the 'reasonable' value.

New economic thinking

It should be clear by now that only an integrated approach can succeed in balancing the environmental, social and economic effects of climate policy. A well-designed cap-and-trade system coupled with a carbon floor price is a good start, with the revenue from auctioning permits being used to help firms and households adopt low-carbon technologies. Technology transfer is essential to stop global rebound leading to higher emissions in other countries. Finally, education and awareness are crucial to encourage consumers to limit their consumption and channel their spending towards low-carbon goods and services, but this will not succeed without caps or taxes to limit the potential for free-riders to undermine efforts by other citizens.

Now, however, we need to consider the implications of carbon and resource limits for the economy. Although most studies indicate that a resource-efficient green economy can provide more jobs and higher growth than business-as-usual, it is also apparent that growth cannot continue indefinitely into the future. Unless we can decarbonise our economy ten times faster than at present, we may have to accept limits on growth. If we do not accept these limits, resource shortages and environmental degradation will eventually start to limit growth whether we like it or not – and the result is likely to be catastrophic for the economy, society and the environment.

Over the last century, our society has become fixated on the idea of economic growth, and with good reason. Growth has brought tremendous improvements to living standards for millions of people all over the world. Technical advances have so far managed to bypass resource limits, and economists use this past achievement to perpetuate the claim that growth can continue forever. Politicians rely on continuing growth to allow them to deliver benefits to voters – better services, higher incomes, more jobs. Our only experience of negative growth is the chaos and hardship of recessions. As a result, the idea of implementing policies that could limit growth has become a taboo subject which is never discussed by the mainstream economic and political community.

Perhaps part of the problem is the way in which we define the success of our economies using just one single measure – growth in GDP. Gross Domestic Product measures only the quantity, not the quality, of economic activity. Earthquakes, fires, floods and car crashes all contribute to GDP, due to the cost of reconstruction activity. Similarly, the costs of cleaning up after a nuclear accident or oil spill, and the legal costs of prosecuting violent criminals are included.

But unpaid work such as growing your own food, caring for children and elderly relatives or voluntary work is excluded. Gross Domestic Product also fails to measure real household income, physical and mental health, education, poverty, social relationships, natural or man-made wealth and the well-being of future generations. A country that is rapidly depleting its natural resources by chopping down its forests or selling off its mineral wealth could have a high growth rate, but that growth is not sustainable and may not even benefit the majority of the country's inhabitants.

These problems were recognised by Simon Kuznets, the economist who set up the US national accounting system that was based on measurement of GDP. On presenting his work to Congress in 1934, he warned that 'The welfare of a nation can scarcely be inferred from a measurement of national income as defined above.' In 1962, he repeated his warnings, writing that 'Distinctions must be kept in mind between quantity and quality of growth, between its costs and return, and between the short and the long run. Goals for more growth should specify more growth of what and for what.' Robert Kennedy echoed these concerns, saying that 'GDP measures everything, in short, except that which makes life worthwhile.'[184]

Unfortunately these warnings went unheeded, and for the last 80 years, economic policy has been geared towards increased growth at any cost. Only recently have policy makers acknowledged the need to develop wider indicators of economic and social progress. Some alternative measures are listed in Box 7.4.

Box 7.4 Beyond GDP

Alternatives to conventional GDP can be broadly categorised as follows.[185]

Corrected GDP

These include the Genuine Progress Indicator (GPI) and its predecessor, the Index of Sustainable Economic Welfare (ISEW); various types of Green GDP developed by national governments; and the World Bank's Genuine Savings. These measures all start with GDP and 'correct' it by subtracting estimates of costs such as crime, pollution and resource depletion and adding on estimates of the value of benefits such as public infrastructure and unpaid household work. The drawback is that it is hard to place a value on non-marketed assets and activities, but these indicators do at least flag up the difference between 'real' and reported GDP.

Alternatives to GDP

Instead of measuring economic activity, these measure environmental, social or human capital. Examples include the ecological footprint and various measures of happiness or subjective well-being. Well-being and happiness could be seen as the most direct indicators of the success of economic progress, but they are hard to measure objectively and can be affected by cultural differences between countries.

Composite indicators

These combine various economic, environmental and social indexes into a single number. Examples include the UNDP's Human Development Index (HDI), which combines life expectancy, education level and GDP *per capita*; the WWF's Living Planet Index which combines the ecological footprint with biodiversity data; and the New Economics Foundation's Happy Planet Index which combines life expectancy, life satisfaction and ecological footprint.

Indicator suites

These report a number of separate indicators. Examples include the suite of 48 indicators used to measure progress towards the eight UN Millennium Development Goals; the Calvert-Henderson Quality of Life Indicators and the National Income Satellite Accounts developed by the UN, OECD, World Bank and International Monetary Fund.

There is public support for moving beyond GDP, with polls showing that more than two-thirds of EU citizens feel that social, environmental and economic indicators should be used equally to evaluate progress.[186] Following a series of reports and initiatives by the French government (the Stiglitz Report), the European Commission, OECD, the UN, the World Bank and the GLOBE organisation, there has been considerable progress recently.[187] The European Commission is developing a suite of indicators including:

- a comprehensive environmental index (including climate change, air and water pollution, biodiversity, waste, resource use and water);
- quality of life and well-being;
- inequality, poverty and social exclusion (including education, health, life expectancy);
- a sustainable development scoreboard;
- environmental limits, with threshold levels, tipping point and danger zones.

There are plans for all EU member states to use this system by 2013. In total, 24 countries now use some form of natural capital accounting, including Canada, Mexico, Costa Rica and South Africa. A major new development is the approval by the UN Statistical Commission of a 'System of Environmental and Economic Accounting' (SEEA), which enables countries to measure and record their natural capital such as mineral reserves, fish stocks, forests, soils and water supplies, as well as other relevant information such as energy use, resource use and pollution. The World Bank has set up the 'WAVES' partnership (Wealth Accounting and Valuation of Ecosystem Services) to promote the use of natural capital accounting alongside GDP.[188]

By expanding the definition of progress beyond GDP to consider other aspects of well-being such as health, happiness, education and a flourishing environment, we might be able to start imagining a future built around growing the quality rather than the quantity of economic activity. Rather than simply urging people to spend as much as possible in order to boost the economy, without really caring what they spend it on, we would design an economic framework that favours goods and services that will contribute to a sustainable future and maximise the social benefit provided by each unit of resources. We would need to abandon policies that deliberately encourage consumption-based growth. These include subsidies for larger families; long working hours and later retirement ages; excessive advertising; weak regulations for social and environmental protection; planned obsolescence; and a borrow-to-spend culture. We will look at the lifestyle implications of these changes in Chapter 8.

An economy that abandoned the endless pursuit of growth at all costs would not stop developing. But rather than growing in size, the emphasis would be on growing in quality – doing more with less. Human ingenuity and technical progress would be more in demand than ever, as we replaced an unthinking reliance on 'trickle down' with carefully designed policies directed at ending poverty, improving welfare and ensuring adequate employment in a world of finite resources. The economy would not be frozen and static – it would be constantly developing and improving in a state of 'dynamic equilibrium'.

Because of our fixation on growth as the means to solve all problems, we have not even started to discuss and research the best method of adapting our economy to fit within the ecological limits of the planet. Yet there are many areas that urgently need to be debated, including the implications for employment, public spending, pensions and the money supply. Without very carefully designed policies to cope with these impacts, the effect of lower growth or zero growth is likely to be disastrous.

Perhaps the most serious potential side-effect is the risk of unemployment. Economic growth generates new jobs to offset the effect of increasing labour productivity. If we are to avoid mass unemployment, we may need to adapt our working patterns to share work around more evenly – for example, by encouraging shorter working hours – and explore strategies such as job guarantee schemes and ecological tax reform (see Box 7.5).

Box 7.5 Jobs in a lower-growth economy

Conventional economists aim to keep unemployment at a certain 'normal' level called the NAIRU, or 'non-accelerating inflation rate of unemployment', which should allow them to keep inflation at a stable rate. If unemployment falls below this level, then there is a danger that wages will increase faster than productivity, which will increase inflation. The argument goes that government action to reduce inflation (such as cutting public spending and raising interest rates) can then lead to a higher unemployment rate than before. The NAIRU is generally at least 3–4 per cent (although it has tended to decrease over time in some countries), but it depends on economic growth to stop it from increasing.

Economist Philip Lawn has looked at three strategies for maintaining employment in a sustainable lower-growth economy:[189]

Basic income

All citizens receive a fixed basic income (in addition to any employment income), regardless of age or circumstances. Lawn recommends a level of around 40 per cent of the minimum wage, which is roughly equivalent to unemployment benefit in most countries. The idea is that this would reduce the need and desire for paid employment, enabling citizens voluntarily to reduce their working hours and also to contribute to society in other ways, such as voluntary work or caring for relatives. The drawback is that it risks rewarding free-riders.

Job guarantee

Anyone would have the option of taking up a state-provided job at a minimum wage. Advocates say this approach would be ideally suited to socially useful work such as environmental protection, as this would avoid competition with the private sector. Workers will be able to acquire new skills and maintain their employability, thus boosting labour productivity. There is a danger that a job guarantee could trigger inflation, because of the increased government spending on the scheme, but Lawn argues that this can be countered by an increase in interest rates, which will cause the number of private sector jobs to fall and the number of 'job guarantee' jobs to rise until inflation is stabilised. This is likely to result in more people on the job guarantee scheme than would have been unemployed, but it is argued that this is better than having a pool of unemployed people.

Ecological tax reform (ETR)

This involves recycling environmental taxes into reductions in labour or income taxes; for example, by reducing national insurance contributions by firms. The idea is to shift taxes from 'goods' such as employment and income onto 'bads' such as pollution and resource depletion, resulting in more jobs and more efficient resource use.[190] But Lawn considers it unlikely that conventional ETR can achieve full employment or sustain it over time, because aggregate demand will not be high enough and inflation will become a problem. He also favours resource caps over taxes, because taxes do not limit the size of the economy. A cap will automatically lead to efficiency, but efficiency will not lead to a cap. The problem with a cap, however, is that it will lead to higher resource prices in the short term, leading to inflation.

Lawn recommends combining these three approaches into the following strategy, which he calls 'ESNAIBER' (ecologically sustainable non-accelerating inflation buffer employment ratio).

1   Reduce labour taxes, to encourage employers to maximise employment.
2   As this does not guarantee full employment, set up a job guarantee scheme that includes a range of part-time jobs (working between 1 and 5 days a week, with half days allowed) that still include full benefits (holidays, sick pay and pensions). This will force the private sector to offer similar flexibility, encouraging work sharing. Pay a minimum wage of 60 per cent of the minimum liveable income (full-time equivalent).
3   Top up the job guarantee scheme with a basic income to reflect the value to society of unpaid work such as household work and caring. This should be 40 per cent of the minimum liveable income. Some people will voluntarily choose to work less, which will limit the GDP needed to achieve full employment and reduce pressure on the job guarantee scheme. This is effectively a private sector wage subsidy that should help to boost employment because the minimum wage can be 40 per cent less than without a basic income.
4   Set up a cap-and-trade scheme to limit resource use to a sustainable level. Permits should be auctioned and the revenue recycled to low-income households, environmental rehabilitation and to part-fund the job guarantee programme.

Lawn claims that this scheme will automatically adjust the ratio of private sector jobs to guaranteed jobs in a way that stabilises inflation. The proportion of people on the job guarantee scheme will be higher than the normal unemployment rate (NAIRU) in the short term, but lower in the long term as higher resource prices stimulate technical progress. The price paid for resource permits will automatically deflate the economy by the right amount to achieve stability, and governments will no longer need to control inflation rates through interest rates and tax rates.

Another critical issue is the role of the money supply in economic growth. It has been argued that the standard practice of allowing banks to create money out of thin air by issuing loans that are not backed pound-for-pound by the bank's assets (known as 'fractional reserve banking') promotes growth, by constantly expanding the money supply. It also makes growth indispensable because if the economy stops growing, people will be unable to pay off the interest on their loans. Therefore it is argued that a non-growing economy would not work without an alternative system, such as a return to the creation of money by governments spending directly into the economy.[191] Conventional economists have avoided debate over this important issue.

This lack of debate is in fact the main obstacle to the achievement of a prosperous low-carbon economy. Most policy makers ignore the rebound effect, placing their faith in technical progress to avoid dangerous climate change. They ignore the figures showing that economic growth is outpacing reductions in carbon intensity. They refuse to consider the prospect of abandoning the all-consuming drive for endless growth. They rely on macro-economic models that ignore resource constraints and environmental impacts.[192] And because of all this, there is virtually no mainstream debate or research on the complex economic issues that would have to be resolved if we were to successfully match the size of our economy to the resources available on our finite planet.

Yet there are signs that this may be changing. China recently became the first country to announce that it was reducing its growth targets (from 8 per cent to 7 per cent over the next 5 years) in order to protect the environment. He Jiankun, director of the low-carbon energy lab at Tsinghua University said: 'In the 12th five-year plan we need to consider quality and efficiency of economic growth. We need to change from a big economy to a strong economy.'[193]

And some members of the business community are ready for the challenge, as demonstrated by this comment from a *Vision 2050* discussion workshop in China: 'We need to change the value sets. For instance, currently a reduction in GDP is seen as a sign of government failure. In the future, reduction in GDP, while improving quality of life, could be seen as a success.'[194]

This is radical thinking, indeed, for a group of large international corporations. But it reflects the mounting evidence that growth is no guarantee of poverty reduction or of full employment – these problems are best addressed through specific policies. A model of the Canadian economy, for example, predicted that a non-growing economy could be successful provided that it was accompanied by

Table 7.6 Cuba uses active social policies to maintain high well-being with a smaller economy

	Cuba	*Spain*	*UK*	*US*
Life expectancy at birth (years)	78.5	80.7	79.3	79.1
Adult literacy rate (%)	99.8	97.9	99	99
Infant mortality rate (per 1,000 live births)	5	4	6	8
Electricity consumption (kWh *per capita*)	1,152	6,147	6,233	13,701
CO_2 emissions (tonnes *per capita*)	2.2	7.9	9.1	19.5
GDP *per capita* ($ PPP)	6,876	31,560	35,130	45,592

Source: Bell, 2010, using data for the year 2007, taken from the UN *Human Development Report 2009*

anti-poverty programmes, income support, shorter working hours, targeted technical investment and habitat protection.[195] A number of European economies have almost stopped growing in size, with low population growth and low GDP growth, but still with good social conditions and high well-being indicators.[196] And there is the example of Cuba, which was forced to adapt to a sudden halving of GDP when cheap oil imports from Russia ended in the early 1990s. Cuba reacted by introducing high energy efficiency, cheap public transport and sustainable local food production while prioritising health, education and social security (see Table 7.6). Although this was not an easy transition, in 2006, Cuba was the only country that achieved sustainability, reaching a high Human Development Index while keeping its ecological footprint within its share of the earth's bio-capacity.[197]

Finally, a move away from consumption-led growth can offer lifestyle benefits, as we will see in Chapter 8.

8 Health and well-being
Benefits of a low-carbon lifestyle

> **Key messages**
> - *Low-carbon lifestyles* can help to tackle growing public health problems related to physical inactivity, poor diet and stress, which cause half of all attributable deaths.
> - *Active travel*, such as walking or cycling instead of using a car, can improve health and fitness for the increasing number of people with sedentary lifestyles, reducing the risk of heart disease and strokes.
> - *A low-carbon diet* – reducing over-consumption of meat and dairy produce, and eating less processed food and more fresh fruit, vegetables, grains and pulses – is better for health.
> - *Less materialistic lifestyles* – buying less and working less – can be associated with reduced stress, more well-being and stronger communities.
> - *Current habits, infrastructure, culture and social norms often make it hard to shift to more sustainable lifestyles.* Stronger policies are needed to promote active travel, better diets and less over-consumption. The last of these will require a deep change to the prevailing economic orthodoxy which depends on consumerism to fuel growth.

In Chapters 5, 6 and 7, we saw that technology alone cannot solve the world's environmental problems. Population growth and increased consumption are outpacing technical improvements. It could even be argued that the sudden discovery of abundant supplies of cheap, clean energy would be disastrous, as we would then be able to extract the earth's remaining mineral and biological resources even faster! It is hard to escape the conclusion that the high-consumption lifestyles prevalent in affluent communities cannot be sustained in the long term. Affluent societies will need to consume less so that sufficient resources remain available for future generations, and for the many communities around the world where under-consumption, not over-consumption, is the problem. In parallel with technical progress, we need cultural and economic change towards a more sustainable system, and that means addressing the issue of human behaviour.

At one level, behaviour change is a very cheap, simple and effective means of tackling environmental problems. It costs nothing to turn off an unused light or appliance, for example, or to choose to eat less meat or cycle instead of driving. On the other hand, there are enormous political and cultural barriers to behaviour change. Governments can influence behaviour to some extent, through 'soft' measures such as awareness campaigns or through more coercive measures such as regulations and pricing schemes. Some options would have a minimal effect on lifestyle (e.g. encouraging people to eat more local and seasonal produce), but others would have a big impact on many people (e.g. large taxes on transport fuel). There can be considerable resistance to what many see as an attack on personal freedoms. The challenge for this type of policy lies in addressing deeply engrained habits and expectations of the right to the limitless use of energy and resources.

Despite the enormity of the challenges involved, changes in lifestyle can deliver a surprising array of co-benefits. The health benefits of walking, cycling and eating less meat and dairy produce are fairly well known, but there can also be benefits from adopting less materialistic lifestyles. The pressure of living in a society where we are expected to work long hours in order to be able to afford to buy ever more material goods often causes stress, and reduces the time available for family, friends and leisure. Moving to a new 'buy less, work less' paradigm can lead to healthier lifestyles and stronger communities, with more time available to spend caring for family members, contributing to the community and enjoying sport, art and other leisure activities.

Relevant climate policies

It is fair to say that there has been very little effort to date to persuade the public to voluntarily reduce consumption levels. At the time of the 1970s oil price shocks, there were campaigns to encourage simple energy-saving measures such as turning off lights, and since then this approach has extended to include waste prevention and recycling. These interventions are generally accepted by the public and by business, because they save money and improve competitiveness (see Chapters 5, 6 and 7). But politicians tend to shy away from measures that could be perceived as meddling in personal lifestyle choices, such as discouraging flying, driving or meat-eating. In fact, further expansion of air travel and the road network are still widely promoted as being good for the economy. Even more significantly, it is hard to imagine any government actively encouraging people to buy less. They generally do precisely the opposite – we are all urged to spend money in shops in order to keep the economy going.

Nevertheless, this chapter will examine three options that could achieve both carbon savings and well-being benefits: eating more healthily, driving less and adopting less materialistic lifestyles.

Eating better

Food production, processing and distribution account for 16 per cent of direct global greenhouse gas emissions, and emissions from deforestation to clear land for farming add another 18 per cent.[1] In Chapter 4, we saw how climate-friendly farming can provide co-benefits for the environment; and Chapter 6 discussed the benefits of cutting food waste. This chapter will look at the ways in which a low-carbon diet can be healthier. These include:

*   cutting over-eating;
*   eating less meat and dairy produce;
*   eating less highly processed fatty and sugary food (which is energy-intensive to produce);
*   eating more fresh, local and seasonal food.

Cutting over-eating will cut all the environmental impacts of food production at a stroke: greenhouse emissions, biodiversity loss and pollution. Shifting to a lower-meat diet can also have a significant impact, as meat and dairy production accounts for 80 per cent of agricultural emissions (see Chapter 4). Figure 8.1 shows that

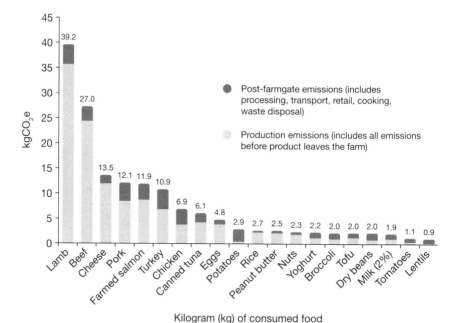

Figure 8.1 Life-cycle greenhouse gas emissions for different food types

Source: Hamerschlag, 2011. © Environmental Working Group (http://www.ewg.org); reprinted with permission

Note: For grain-fed animals raised in the United States. Includes food waste.

greenhouse gas emissions per kilogram from meat and cheese are far greater than those from cereals, vegetables and pulses. These impacts are rising – global meat production is projected to more than double from 230 million tonnes in 1999/2001 to 465 million tonnes in 2050, while milk output is set to almost double from 580 million tonnes to 1040 million tonnes.[2] A study in *The Lancet* finds that global average meat consumption was 100g per person per day, varying widely from 200–250g per person per day in rich countries to 20–25g per person per day in poor countries. To tackle climate change and provide public health benefits, they recommend cutting global average meat consumption to 90g per day, shared more evenly between rich and poor countries, and with not more than 50g per day coming from red meat from ruminants.[3]

As well as reducing meat and dairy consumption, carbon reductions can be achieved by choosing more fresh, local and seasonal produce, which will avoid 'food miles' emissions associated with transporting food around the globe, and energy emissions from producing food out of season in heated greenhouses or from cooling and freezing food for longer storage. There is a growing interest in local food initiatives such as farmers' markets, growing your own food in gardens or allotments, and community food schemes, often linked to the 'Transition Towns' and 'Slow Food' movements. As well as health and well-being benefits, this type of initiative can provide social benefits such as affordable fresh food, and can help to strengthen local communities.

Driving less

The transport sector produces 11 per cent of global greenhouse gas emissions, and these emissions are growing.[4] By cutting car use, we can tackle climate change and also gain a wide range of co-benefits including cleaner air, less congestion and noise and fewer accidents.

In Chapter 3 we saw that cutting air pollution from road transport has very significant benefits for health. Perhaps surprisingly, the potential health benefits from a switch to more physically active alternatives to car use, such as walking and cycling, are even greater. Other co-benefits include safer and quieter streets, and more opportunities for social interaction.

Policies to achieve this include those that discourage car use, such as higher fuel taxes, parking restrictions and congestion charging; and those that encourage alternatives, such as investment in public transport and safe walking and cycling routes. Measures to reduce transport demand are also important, such as encouragement of car-sharing, videoconferencing and tele-working. Urban planning has a vital role to play: journeys are shorter and walking and cycling easier if homes, schools, shops and workplaces are close together. However, it can be difficult to achieve this in areas where the infrastructure has evolved to depend on car use.

Less materialistic lifestyles

In Chapter 6 (Figure 6.1), we saw that 60 per cent of global greenhouse gas emissions arise from the manufacture and sale of material goods – food, clothing and other products – outweighing direct emissions from our homes and cars. Apart from the climate impact, the manufacture of these goods also creates pollution and waste, uses scarce resources and drives environmental degradation. Buying greener products, such as recycled paper or energy-efficient appliances, is not enough – we also need to cut the total amount of 'stuff' that we buy. This will require a considerable culture change, given that our society is conditioned to see shopping as both a fulfilling leisure activity and an economic necessity, but there is evidence that less materialistic lifestyles can be associated with greater happiness.

As discussed in Chapter 7, if we buy less, then we may also have to work less, by reducing our working hours, in order to avoid increased unemployment. Again, this runs counter to the prevailing economic orthodoxy that has become entrenched over the last few decades. However, there are signs that many people would welcome down-shifting to a way of life that leaves more time for social relationships and leisure activities.

Co-benefits

Health and fitness

While around 1 billion people around the world do not have enough to eat,[5] another billion suffer from the opposite problem – they eat too much (or too much of the wrong type of food) and are physically inactive, leading to health problems such as obesity, heart disease, strokes and cancer. There was a time when these problems were largely confined to the rich industrialised countries, but today they are increasingly prevalent in the developing world. Urbanisation and mechanisation are leading to more sedentary lifestyles, and people are eating less fruit and vegetables and more energy-dense foods that are high in fat, salt and sugar, but low in essential nutrients. This is partly because this 'junk food' is often cheaper, more heavily advertised and more easily available than healthy food; and partly due to a decrease in the time available for cooking and food preparation as more women find jobs outside the home.[6]

Figure 8.2 (see plate section) compares the estimated percentages of deaths that can be attributed to specific risk factors (which are about 73 per cent of total deaths) in high-income and lower-income countries. Unhealthy diet and lack of exercise cause two-thirds of attributable deaths in high-income countries, with an annual death rate of 4.0 per thousand members of population; but lower-income countries are not far behind, with almost half the attributable deaths at a rate of 3.2 per thousand.

It is striking that poor diet and lack of exercise were believed to be the main cause of 21 million deaths in 2004 – far more than environmental pollution. Of these, 7.5 million were from high blood pressure, 2.8 million were from being overweight or obese, 2.6 million were from high cholesterol levels and 3.2 million from physical inactivity.[7] In fact, heart disease and strokes are the leading causes of death in the over-45 age group.[8]

Figure 8.3 (see plate section) shows that while under-nutrition is still a major problem in low-income countries, a significant proportion of the population in these countries is now at risk from over-eating. In 2008, around 35 per cent of all adults aged 20 and over were overweight, including 12 per cent who were obese. In the upper-middle income group, 25 per cent were obese.[9] In addition, 31 per cent of adults aged 15 and over were insufficiently physically active.[10]

The burden of disease from unhealthy lifestyles is increasing – obesity more than doubled between 1980 and 2008,[11] and the WHO estimates that deaths from non-communicable diseases will increase by 17 per cent over the next 10 years.[12] In the UK, for example, obesity levels rose by 38 per cent from 2003 to 2007.[13]

The economic costs are considerable – both in terms of direct healthcare and in lost working days. One study estimates that physical inactivity alone cost the UK National Health Service almost £11 billion per year.[14] Another review suggested that obesity and people being overweight cost China almost $50 billion in 2000 (4 per cent of GDP) – most of which was from lost working days – and the costs would exceed $110 billion by 2025.[15]

The widespread prevalence of problems arising from poor diet and lack of exercise suggests that there is a tremendous potential for health co-benefits from climate policies that promote active travel and diets lower in animal produce, as we will see in the following sections.

The low-carbon diet

Low-carbon diets tend to be healthier. First, we can cut over-eating, which is now a widespread problem. In the EU-27, for example, the average calorie intake has been 36 per cent higher than the recommended level since the early 1990s.[16] More than half of the EU population is estimated to be overweight, a quarter is obese, and the numbers are rising. In a way, over-eating is even worse than simply throwing food away, because of the negative impact on health.

Second, affluent consumers can eat less meat and dairy produce, replacing it with cereals, pulses, nuts, fruit and vegetables. Although animal produce is an important source of protein and nutrients such as iron, calcium, vitamin B12 and zinc, it is also high in saturated fat. Many countries exceed the recommendation that saturated fat should form no more than 10 per cent of our calorie intake. In the Nordic countries and France, for example, saturated fat forms 13 per cent of calorie intake on average. Too much saturated fat is a major cause of heart disease, strokes, high blood pressure and some cancers, as well as contributing to obesity and therefore to related problems such as diabetes. High levels of consumption of red meat, especially processed meat, have been linked to cancer and heart disease.[17]

Even a minor change such as cutting saturated fat intake from 13 per cent to 12 per cent of total energy intake and replacing it with unsaturated fat would result in almost 10,000 fewer coronary heart disease deaths and 3,000 fewer stroke deaths in the EU each year.[18] The benefits of a more radical shift away from animal produce and towards fruit and vegetables could be considerable, given that around 2.6 million deaths worldwide each year are attributed to high cholesterol levels and a further 1.7 million to inadequate fruit and vegetable consumption.[19] A study for the UK estimated that 45,000 deaths could be prevented each year if meat consumption was reduced by 83 per cent, to around 210g per week – equivalent to two or three small servings.[20] Another study found that replacing one portion of red meat per day with other protein sources could reduce mortality risk by 7–19 per cent.[21] Obviously these benefits apply only to cases where there is over-consumption. Meat and dairy produce continues to form an important source of nutrition for many people, and can play a major role in improving health in communities where under-nutrition is a problem.

Similarly, cutting our consumption of highly processed 'junk food' such as sweets, crisps, biscuits and sugary drinks can cut carbon emissions (from sugar cane and beet cultivation, production, packaging and distribution) as well as improving health. High sugar and fat consumption is linked to increasing rates of many health problems including diabetes mellitus, which is expected to affect 300 million adults worldwide by 2025, up from 135 million in 1995. Diabetes gives an increased risk of cardiovascular and other circulatory diseases, and is aggravated by high-energy high-fat diets that encourage excess fat deposits, but it is less prevalent in people who follow a diet low in red meat and saturated fat and high in fibre, fruit and vegetables.

If everyone in the world followed recommendations for a healthy diet, up to 14 million premature deaths (a quarter of all deaths) could be avoided each year – that is the number of deaths attributed to diet-related risk factors: high blood pressure, high blood glucose levels, obesity and being overweight, high cholesterol levels and inadequate fruit and vegetable consumption.[22] This would also save billions of dollars in healthcare costs and lost working days. The World Health Organization identifies the rising levels of diet-related disease as a major burden on health services and economies in the coming decades.

There can also be cost savings for consumers, from cutting over-consumption of food and from replacing meat and dairy products with cheaper vegetarian equivalents such as beans and pulses. However, although low-income households in many developed countries eat more red meat, processed meat and junk food than the general population,[23] and thus stand to gain the most from changing their diets, they are also more likely to face difficulty in accessing affordable healthy food such as fresh fruit and vegetables. Some communities live in 'food deserts' where there are few local shops apart from those offering tinned, dried and highly processed food.

In Chapters 4 and 6 we saw that rising food prices are partly due to the increasing scarcity of fertile agricultural land and water, and in future there could also be shortages of phosphorous for fertilisers. Around 70 per cent of agricultural land is currently devoted to the meat and dairy food chain, with a third of arable land being devoted to animal feed crops.[24] In addition, meat production requires around

100 times more water input per kilogram than grain production.[25] Cutting meat and dairy consumption would be expected to ease these pressures, potentially leading to a fall in global food prices as well as cutting air and water pollution and reducing over-grazing and deforestation. This leads to the appealing prospect that under-nutrition, over-consumption, resource scarcity and environmental damage could all be tackled at the same time.

Further co-benefits for health could be obtained by a shift to a more organic diet, due to reduced consumption of pesticide and herbicide residues on food. There is also some limited evidence for increased nutritional benefits for some organic foods, such as milk. The scale of any benefits is likely to be small compared to the benefits of a shift from animal to vegetable produce. The main co-benefits of an organic diet are related to environmental impacts (see Chapter 4) and animal welfare. There could also be some minor health benefits from policies to encourage a move towards more local and seasonal produce, which could allow communities to access fresher and more nutritious food with fewer additives and preservatives.

Despite the strong evidence of the health benefits of a better diet, it is not easy to persuade people to change their eating habits. Salty, sweet and fatty food is appealing to the human taste and even those most at risk often find it very difficult to cut down. Possible approaches will be discussed in 'The way forward' below.

Active travel

In all countries of the world, there is a steady trend for people to become less physically active, for a number of reasons. Jobs in modern societies are generally more sedentary, as mechanisation replaces physical labour. Increasing wealth means that more people can afford to use cars. Increasing ownership and use of appliances in the home also decrease the daily activity involved in housework and cooking, as well as encouraging sedentary leisure activities such as watching TV and using home computers. Finally, urbanisation coupled with car-oriented development patterns have made it hard for many people, especially those on low incomes, to access safe and pleasant green space for outdoor exercise. Many inhabitants of towns and cities find themselves surrounded by hot, dirty, noisy, crowded and unsafe streets and are deterred from exercising outside by fear of crime, road accidents and pollution.

This increasing inactivity is contributing to an 'epidemic' of lifestyle diseases that has been identified by the WHO as a major challenge for modern health policy. The WHO recommends that adults aged between 18 and 64 should do at least 150 minutes of moderate physical activity or 75 minutes of vigorous activity per week, in bouts of at least 10 minutes duration. Policies that aim to encourage people to replace short car journeys with walking, cycling or public transport (which tends to be more active than driving, due to the need to walk or cycle to and from the bus stop or train station) can be a very effective way of addressing this problem. There is a large potential to shift to more active transport modes as many car journeys are very short – in Europe, for example, a third of all car journeys are shorter than 2 kilometres.[26] Active travel has a number of advantages as a means of introducing exercise into busy modern lifestyles:

- *Time-saving.* Many people find it difficult to make time for regular exercise. Walking and cycling, on the other hand, can often be fitted into the daily routine with little time penalty. For example, a 15-minute car journey through a congested town centre could be replaced by a 30-minute bike ride cutting through parks and back roads, thus achieving the recommended daily exercise target for just an extra 15 minutes. Some short journeys can even be quicker by bike or on foot, especially in congested areas and where finding a parking space is a problem.
- *Habit-forming.* People who struggle to find the will-power or motivation to drag themselves off to the gym at the end of a hard day's work can find it easier to build regular exercise into their lifestyles by commuting to work by bike or walking to the local shops.
- *Cheap.* Many sporting activities are expensive, involving club memberships, court fees, tuition fees and the need for specialised clothing and equipment. Walking, on the other hand, is completely free, and cycling requires relatively little expenditure. Both also save money compared to driving, although this may or may not be true for public transport, depending largely on government policy.
- *Suitable for all ages and abilities.* People who are very unfit or overweight might be deterred from signing up for a gym or sports club, or from more vigorous exercise such as running, but walking and cycling are perfect low-impact activities. Speeds and distances can be built up gradually until fitness improves sufficiently to try other sports and activities. Walking obviously requires no special training, and cycling requires only basic road safety training.
- *Outdoor.* Walking and cycling allow people to be in the fresh air and to be in contact with nature. There is some evidence that outdoor exercise creates greater feelings of revitalisation and decreases tension, confusion, anger and depression, compared with indoor exercise.[27]
- *Sociable.* Walking or cycling around your local neighbourhood, or using local public transport, gives more opportunities for chance meetings with friends and neighbours than driving.

The health and social benefits of regular activity are considerable. Exercise reduces the risk of obesity, heart disease, high blood pressure, high cholesterol, strokes, diabetes and some types of cancer. It can also strengthen bone and muscle structure and reduce the risk of falls, fractures, osteoarthritis and osteoporosis, especially in the elderly. Physical inactivity is estimated to be the main cause for around 21–25 per cent of breast and colon cancers, 27 per cent of diabetes and approximately 30 per cent of ischaemic heart disease, which is the biggest cause of death globally.[28] Being overweight or obese is estimated to cause 44 per cent of the diabetes burden, 23 per cent of the ischaemic heart disease burden and between 7 per cent and 41 per cent of certain cancer burdens including breast, colon, prostate, endometrium, kidney and gall bladder cancer.[29]

Regular physical activity is also associated with better mental health – it alleviates stress and anxiety, promotes sleep, improves cognitive function, reduces the risk of developing depression, and is as effective as drugs or psychotherapy in reducing the symptoms of depression.[30] Research indicates that it is not the type or intensity of exercise that matters, but the frequency – so regular walking or cycling as part of a daily travel pattern would be ideal. Mental health problems affect 10 per cent of adults worldwide, account for one-third of all years lived with a disability and impose a growing economic and social cost, both in direct healthcare costs, in loss of working time and through related problems such as substance abuse.[31]

There is also evidence that moderate exercise such as walking and cycling can help to reduce tobacco cravings. Smoking causes over 5 million deaths per year globally, which is 9 per cent of all deaths, but physical activity can increase the chances of giving up smoking by 24 per cent according to one small study, which could lead to tens of thousands of extra smokers stopping every year in the UK alone.[32]

Exercising for 30 minutes most days of the week cuts all-cause mortality rates by 30 per cent. A growing body of research shows the health benefits of active travel. A large-scale study in Copenhagen found that cycle-commuting cut the risk of premature death by a third,[33] and another review found that cycle-commuting is linked to higher fitness, lower death rates and lower cancer risk, with a positive relationship between the amount of cycling and the size of the health benefits.[34] A large literature review spanning 14 countries, all 50 US states and 47 of the 50 largest US cities found that active travel is linked to reduced rates of obesity and diabetes.[35]

A modelling study of London estimated that doubling walking levels and increasing cycling levels eight-fold (to match the highest levels achieved in other European cities such as Amsterdam and Copenhagen) could cut carbon dioxide emissions by 38 per cent whilst avoiding 528 premature deaths that would arise from insufficient physical activity and 21 deaths from air pollution annually. This would be offset by an extra 11 lives lost through road accidents. The scenario involves cutting car use by 37 per cent, but the authors point out that 55 per cent of all distance travelled by cars in London is on trips shorter than 8 kilometres, and 11 per cent is on trips less than 2 kilometres, suggesting that this target is feasible.[36]

Walking and cycling to school are ideal for helping children to meet the WHO guidelines of at least 60 minutes moderate to vigorous physical activity per day for those aged between 5 and 17. Many children live within walking or cycling distance of school. Childhood obesity is on the rise, affecting an estimated 42 million children under the age of five in 2010.[37] It leads to a higher chance of obesity, disability and premature death in adulthood, as well as to symptoms such as breathing difficulties, increased risk of fractures, hypertension, early markers of cardiovascular disease, insulin resistance and psychological effects.

Of course, there are also situations where the use of a private car can be far more convenient, such as for journeys involving longer distances, steep hills or destinations that cannot be easily reached by public transport. Other barriers include bad weather, awkward or heavy luggage, physical ability and time constraints.

Some people choosing to walk or cycle may face greater risks from road accidents, crime or air pollution – especially in some developing countries where street crime is a particular problem. Many potential problems can be reduced by carefully designed policies to make active travel a safer, quicker, cheaper and more pleasant experience. An integrated package of complementary measures is the most effective approach, typically including:

- physical infrastructure such as wider pavements (sidewalks), benches, cycle paths, footpaths; secure covered cycle parking, cycle repair stations; green space and tree shading to make journeys more pleasant; safer junctions (such as advance stop lines for bikes); more pedestrian road crossings; and signs, maps and information boards to help people find their way around on bike or on foot;
- urban planning measures to limit 'sprawl' and ensure that key facilities such as shops, schools, leisure facilities, workplaces and health centres are within walking distance of housing;
- a cheap, efficient and extensive network of public transport options with good interconnections;
- options to encourage links between public transport, walking and cycling, such as well-signed walking routes to and from bus and train stations; bike racks on trains and buses; and 'cycle stations' offering cycle hire, repair services, cycle parking, lockers and showers at bus and train stations;
- traffic-calming and traffic-limiting measures such as speed and parking restrictions, low emission zones, car-free zones, road and parking pricing (including congestion charging) and conversion of traffic lanes into cycle paths and walkways;
- information campaigns to highlight the health benefits of active travel, and inform people about local routes, facilities, public transport options and other initiatives such as car-sharing or bike-hire schemes;
- awareness-raising events such as bike-to-work weeks, car-free days and mass cycle events such as the 'Ciclovías' in Bogotá which attract 400,000 riders on closed roads every Sunday;
- support for local community groups that can help to promote active travel, such as cycling clubs, health professionals and schools, and encouragement for local businesses to provide cycle parking, showers and incentive schemes;
- cycle training and advice, especially for school children.

Research shows that the general 'connectivity' of a neighbourhood – the number of street intersections and safe road crossing points, for example, as well as the potential for taking short-cuts (such as through a park or along a footpath or cycle path) – is important in encouraging active travel, because it shortens walking and cycling distances and allows more choice of routes.

Urban density is also a key factor. Studies have shown that people living in areas of urban sprawl are more likely to be overweight or obese, suffer more from high blood pressure and other chronic diseases, and experience greater traffic fatalities, especially as pedestrians. People in more 'walkable' areas tend to have healthier body weights and better mental health – recent studies have found that they spend 35–49 minutes more per week on physical activity, and that they are 20–50 per cent more likely to meet physical activity guidelines. Doubling urban density can cut traffic by 5–12 per cent and potentially up to 25 per cent according to US studies. However, increasing urban density can also increase concentrations of pollutants, even if total emissions are reduced.[38]

Cyclists and walkers tend to face lower pollution concentrations than drivers, because they are slightly further away from the pollution source, but inhalation levels can be higher if the level of physical exertion causes deeper breathing. This can be minimised if cycle lanes and footpaths are sited as far as possible from the road, and there are opportunities to take alternative routes away from main roads. On the whole, the health benefits of cycling and walking have been found to far outweigh any additional risks from accidents and pollution.[39]

As more and more people switch to active travel modes, there is evidence that a cultural tipping point may be reached where cycling and walking are seen as being normal, safe and even fashionable. As the number of people opting for active travel increases, pressure builds for councils to provide more support. And the presence of more people on the streets can help to deter crime and allay the fear of crime, as well as providing more opportunity for social interactions.[40] Cities with strong supporting policies can achieve far higher rates of active travel than other cities. In the Netherlands and Denmark, for example, cycling accounts for up to a third of trips, whereas in the United States and southern Europe, it is only 1–2 per cent of trips (see Box 8.1).[41]

The costs of policies to promote active transport can be offset by savings in healthcare costs. In the UK, for example, the annual public health costs attributable to overweight and obesity are £5 billion, and this is projected to double by 2050. The wider costs to society and business, largely as a result of losses in working time and productivity, are estimated to reach £50 billion per year.[42] Businesses that take measures to encourage their staff to use active transport, such as providing showers or bike parking, are likely to reap the benefits – it has been shown that cyclists take fewer days off work due to sickness, for example.[43] One study estimated the net benefit of a single cycle route in Bruges in Belgium to be €5.5 million in healthcare savings, improved road safety and benefits for tourism and the local economy.[44]

Safer, quieter streets

Apart from the considerable health benefits of a shift to active transport, there will also be a reduction in noise, congestion and accidents. Road accidents kill almost 1.3 million people worldwide each year, and 20–50 million are injured or disabled. They are the leading cause of death for those aged between 15 and 29, and the second highest cause of death for those aged between 5 and 14. This typically costs countries 1–3 per cent of their gross national product, with an estimated annual global cost of $500 billion. If current trends continue, the number of deaths could almost double by 2030.[45]

More than half of those affected by road accidents are vulnerable road users – cyclists, walkers and motorcyclists. Fear of accidents can be a major barrier to encouraging more people to switch to active transport modes. As more people shift their journeys from car to active transport, however, pollution and accidents will fall. Studies have found that if more people cycle instead of driving, the risk of accidents and exposure to pollution falls for each individual traveller, but the total number of accidents will increase simply because of the larger number of walkers and cyclists. One study found that doubling walking rates would cut accident risk by 34 per cent for each walker, but lead to a 32 per cent increase in total accidents involving walkers.[46]

Yet road safety can be improved considerably with careful policy design – there are almost six times more fatal cycling accidents per kilometre in the United States than in the Netherlands, for example. Highly effective safety measures include speed limits, physical traffic-calming options such as speed bumps and chicanes, wide pavements, bike lanes (especially those separated from traffic by a raised kerb), safe pedestrian crossings, cycle training for children, and strict enforcement of drink-driving laws. For every 1 kilometre per hour cut in speed, there is a 2 per cent reduction in the number of crashes. For example, improving junctions in Copenhagen cut the number of seriously injured cyclists by 42 per cent, and introducing nine new 20-mile-per-hour zones in the London borough of Camden cut traffic injuries in those areas by 58 per cent.[47] Strong road safety measures can create a 'virtuous circle' of benefits – as the accident rate falls, more people may be encouraged to walk or cycle, and the reduction in traffic will cause the rate to fall still further. Traffic calming also increases the speed of cyclists relative to drivers, thus acting as a further incentive to cycle or walk,[48] and motorists tend to become more aware of cyclists and walkers when there are more of them on the roads.

Some examples of successful initiatives to increase urban cycling whilst improving safety are shown in Box 8.1.

Box 8.1 Cycling success stories

Groningen, the Netherlands

Groningen has perhaps the highest known cycling rate in the world: 40 per cent of all trips are made by bike, and this ratio has been stable since 1990. Safety levels have continued to improve – serious injuries halved between 1997 and 2005. This small town of 180,000 inhabitants has a network of over 200 kilometres of cycle paths, shortcuts and bridges that is completely separated from road traffic. There are also bike-friendly junctions featuring advance stop lines, priority for cyclists at traffic lights and even four-way green lights for cyclists at some junctions. There is extensive bike parking at train stations and bus stops, guarded parking at schools, traffic calming to 30 kilometres per hour in most residential streets, car-free zones and limited parking in the town centre, and mandatory cycle training for school children.

Amsterdam, the Netherlands

Amsterdam also boasts high cycling rates: the share of trips rose from 25 per cent to 37 per cent between 1970 and 2005, and serious injuries fell by 40 per cent between 1985 and 2005. This was achieved through a very similar mix of measures to those used in Groningen, including a separate network of 450 kilometres of cycle lanes, bridges and shortcuts together with safer junctions, traffic calming, bike parking and cycle training. Increasing parking fees cut car trips by 14 per cent and increased cycling by 36 per cent – bikes are now used more than cars in the city centre.[49]

Copenhagen, Denmark

Copenhagen also has a long tradition of encouraging bike use, with a 70 per cent increase in bike trips between 1970 and 2006, accompanied by a 70 per cent decline in serious injuries between 1995 and 2006. The share of trips by bike rose to 38 per cent for adults over 40 in 2003. This has been achieved through a massive expansion of cycle paths separated from the traffic by a kerb, reaching 345 kilometres in 2004, together with bike-friendly junctions, traffic calming, guarded bike parking, car-free zones in the city centre, mandatory cycle education, and the pioneering City Bike system which offers 2,000 free bikes at 110 locations in the city. There is also a survey of cyclists every 2 years to gather feedback.

Berlin, Germany

Cycling is more of a challenge in large, busy cities, but in Berlin, a city of over 3 million inhabitants, the number of bike trips almost quadrupled between 1970 and 2001. The share of bike trips doubled from 5 per cent in 1990 to 10 per cent in 2007, while serious injuries fell by 38 per cent between 1992 and 2006. This was achieved through tripling the network of separate cycle lanes from 271 kilometres in 1970 to 920 kilometres in 2008, whilst introducing traffic calming on 72 per cent of residential streets, providing over 22,000 bike parking spots at bus and train stations, providing bike rental at rail stations and making cycle training mandatory for school children.

London, UK

In London, the number of bike trips doubled between 2000 and 2008, following the introduction of congestion charging in 2003 and other supporting measures such as cycle lanes, bike-friendly junctions, cycle parking, traffic calming, training for school children, free route maps and shortcuts for cyclists and walkers. At the same time, cyclist injuries fell by 12 per cent. The share of bike trips increased from 1.2 per cent to 1.6 per cent between 2003 and 2006 – although that is still a low level compared to Berlin, for example, at 10 per cent. More recently (in 2010), a popular bike-hire scheme was introduced.

Paris, France

Cycling in Paris more than doubled from 1 per cent in 2001 to 2.5 per cent in 2007, and increased again by 46 per cent from June to October 2007 after the introduction of the Vélib bike-hiring scheme. Cycle paths and cycle parking were tripled, and the city introduced 38 '*Quartiers verts*' ('Green zones') with traffic calming to under 30 kilometres per hour, narrowed roadways, wider pavements, car-free zones and six 'civilised travel corridors' restricted to motor vehicles. Free car parking has been eliminated and there are cycle training programmes, free route maps, a route advice website and bike-engraving to discourage theft.

US towns: Portland, Oregon and Boulder, Colorado

Cycling in the United States faces the twin problems of greater urban sprawl and lower motoring prices. However, two towns have demonstrated successful programmes. With extensive bike lanes, bike parking, bike-to-work days and community outreach programmes, the town of Boulder has increased the share of trips by bike from 8 per cent in 1980 to 14 per cent in 2006. In Portland, the share of workers commuting by bike rose from 1.1 per cent to 6 per cent between 1990 and 2008, thanks to a four-fold increase in cycle lanes together with bike-friendly junctions, Bike Sundays (mass cycling events on closed roads), bike racks on buses and trains, and education and marketing programmes.

Others

Other towns and cities with high levels of bike use include Odense in Denmark (25 per cent share in 2002), Freiburg in Germany (27 per cent share in 2007) and Münster in Germany (35 per cent share in 2001) – all of which have deployed almost exactly the same combination of measures as those listed above for Groningen. Larger cities face bigger barriers to cycling, but Bogotá in Colombia managed to increase the bike share of trips four-fold, reaching 3.2 per cent in 2006, following the introduction of bike lanes and car-free zones; and Barcelona in Spain doubled cycling rates in just 2 years (reaching 1.8 per cent in 2007) with the introduction of the Bicing bike-hire system and other measures.[50]

Another benefit that is often overlooked is a reduction in road traffic noise. Noise affects health and well-being – it disrupts sleep, causes annoyance, reduces performance at work and affects the learning ability of children. Excessive noise exposure can trigger the release of stress hormones, and has been linked to high blood pressure and increased risk of heart disease.[51] Road traffic accounts for almost 90 per cent of excessive urban noise. Over half of people in large towns and cities in the EU are exposed to long-term daytime road traffic noise levels above the WHO guideline of 55 decibels (dB), and more than a third suffer night-time noise above 50 dB,[52] well above the guideline level of 40 dB and described as 'increasingly dangerous to health' by the WHO.[53] It is estimated that traffic noise may account for over 1 million years of healthy life lost annually in Western Europe, and the damage has been valued at €40 billion, projected to increase by €20 billion by 2050.[54]

There is also evidence that more people engaging in active travel can help to reduce fear of crime, and improve social life and community atmosphere. Good urban planning that provides short trip distances, safe walking and cycling facilities and pleasant space for pedestrians (such as parks, squares and wide pavements) can provide opportunities for social interaction that improves relationships and trust between neighbours, and improves community surveillance.[55] Safer, quieter streets with less traffic, traffic calming and speed-limiting measures can also help to encourage outdoor play amongst children, which can contribute to their health and fitness as well as enhancing social development and fostering community links between local families.

Active travel could also reduce traffic congestion, by decreasing car journeys. Congestion is an increasing problem as urbanisation continues and traffic demand grows. As well as creating more pollution, stress for drivers trapped in traffic and extra road accidents as frustrated drivers try to make up lost time, there are significant economic costs in terms of wasted time, lost business revenue and increased fuel consumption. There can even be deaths if emergency services are unable to get through traffic.[56] One estimate puts the cost of wasted time and fuel from congestion in the United States at $115 billion in 2009.[57] In the EU, congestion is said to cost around 1 per cent of GDP, and under current trends, this cost is estimated to increase 50 per cent to reach €200 billion by 2050. However, some measures to encourage active travel (such as on-road cycle lanes) can result in a decrease in road space, and others such as pedestrian crossings could slow down traffic, both of which could offset the impact of a reduction in traffic volume.

In summary, a shift away from car use to making more journeys on foot, bike or public transport can not only provide health benefits for individual travellers, but can also benefit the whole community by cutting road noise, accidents and congestion. But these benefits are unlikely to be achieved without carefully planned policies to make active travel safer, more convenient and more appealing. With good urban planning, a virtuous circle can be achieved where more people switch to active travel, the number of cars on the road falls, the streets become quieter, cleaner and safer and this encourages yet more people to start to walk and cycle.

Stronger communities

Many of the lifestyle changes that help to address climate change also bring benefits for community cohesion and social interaction. We have shown how active transport can lead to less traffic and safer, quieter, cleaner streets, encouraging people to get out more and increasing the opportunities for outdoor play. On top of this, designing urban areas to be more walkable and improving public transport can improve access to services for those who cannot drive or do not own a car. In the United States, 21 per cent of those aged over 65 do not drive, and these people take 15 per cent fewer trips to the doctor and 65 per cent fewer trips to family and friends because of their lack of access to alternative transport options.[58]

Social and economic benefits for local communities can also arise from policies that promote local and seasonal food. There is a growing interest in community food initiatives such as farmers' markets, farm shops, community-supported agriculture, local food co-operatives, organic box delivery schemes, allotment gardening and community-run shops and cafés. Those involved are typically motivated by the desire to produce fresh, affordable local food, often using sustainable methods such as organic farming. As well as reducing greenhouse gas emissions from food processing and transport, and providing a cheap source of nutritious produce for local people, these initiatives have obvious benefits in forging social links, encouraging community interaction and strengthening local economies. By selling food directly to local consumers, local producers can keep a greater share of the profit and less money is diverted to distant shareholders of multinational food corporations. A greater proportion of the sale price of the food therefore stays within the community, which is particularly beneficial in deprived areas. Face-to-face contact between producers and consumers enables consumers to find out more about how their food is produced, and helps producers to gather feedback about their products and build customer loyalty.[59]

Social interaction is also a key feature of 'collaborative consumption', which is an umbrella term for the rapidly growing mix of internet-based systems for sharing, swapping, lending or trading goods and services. The idea is that underused assets can be shared or rented to people who need them. Examples include swap sites for used books, clothes or films; schemes that allow people to lend or hire equipment to people in their community; lift-sharing and car-sharing schemes such as Zipcar; sites for selling or giving away second-hand goods such as eBay and Freecycle; garden-sharing schemes where people cultivate gardens in return for a share of the produce; time banks or skill-share sites; parking-space rental such as Park On My Drive; and schemes for swapping or renting private houses as holiday accommodation.[60] The spread of these systems has been made easier by the development of social networking, together with technology such as smartphones and apps which allow users to participate while on the move. User rating systems such as the one pioneered by eBay have helped to break down the barriers associated with trusting strangers (despite occasional instances of fraud), and the popularity of these schemes has increased the social acceptability of sharing or buying pre-owned goods. Although people are often attracted to the scheme because of the prospect of saving money or for environmental reasons, many find that the social aspects of belonging to the user community are a major benefit.[61]

Perhaps the ultimate example of collaborative consumption is communal housing projects. The co-housing movement began in Sweden, where there are currently around 40 projects.[62] They typically involve developments where small private apartments are clustered with larger communal facilities including cooking, living and outdoor areas. Residents often take turns to cook meals, and may share other activities such as gardening, cleaning and childcare. These communities tend to have low-carbon footprints because total living area is reduced (together with heating and lighting requirements), appliances such as kitchen and laundry equipment are shared and features such as renewable energy, recycling and car-pooling are often built in. By sacrificing some private space, residents gain the use of shared facilities such as children's play rooms, workshops, guest rooms for visitors and exercise equipment. Co-housing offers obvious opportunities to form social links, which may be particularly beneficial given the current trend towards smaller households and more solitary living, and it could be a useful way of helping elderly people to maintain independent yet socially integrated lifestyles.

Finally, as we will see in the next section, a shift to less materialistic lifestyles can leave more time available for contributing to local communities; for example, through voluntary work, socialising, taking part in local politics, joining sports or arts groups or caring for friends and relatives.

Beyond shopping: buy less, work less, be happy?

We have looked at the benefits of low-carbon transport and low-carbon diets – now we move on to the somewhat more controversial issue of low-carbon shopping. In practice, this mainly means *less* shopping, at least for affluent consumers who have already met their basic needs. The implications of this for the economy could eventually include less working – or at least, less paid work.

In Chapters 6 and 7, we saw that rising levels of income are linked to increasing consumption of material resources. As we get richer, we buy more 'stuff', which more than cancels out the greater efficiency and cleanliness of our factories. As a result, we are pushing vital ecosystem services such as climate, biodiversity and water supplies to the brink of collapse.

Hard as it may seem in a society accustomed to shopping as the driver of economic growth and higher living standards, a change is essential if we are to avoid ecological meltdown. We need to restructure our societies away from an economic and social dependence on the endless accumulation of material goods. Chapter 7 looked at possible strategies for this adjustment, including alternatives to GDP as a measure of progress, and policies such as ecological tax reform, job guarantee and basic income to help deal with any reduction in the size of the economy that might result from restraining material consumption.

It will certainly be a huge challenge to restructure the economy away from a dependence on consumption growth whilst protecting employment and funding public services; but it could be even more of a challenge to achieve the change in culture that is needed to persuade affluent consumers to stop shopping, or at least to limit shopping to a more sustainable level. Yet a growing body of research is finding that growth in material consumption does not necessarily bring happiness. Surveys show that happiness and life satisfaction are well correlated with income up to a level of around $15–20,000 *per capita*, after which further increases in income produce only small increases in satisfaction (Figure 8.4).[63]

A number of reasons have been proposed to explain this finding. One is to do with status competition: research shows that once basic needs are satisfied, people are more influenced by how much they earn compared to others than by their absolute level of income. A recent study of adults in Great Britain, for example, found that life satisfaction was strongly correlated with income rank, rather than with absolute levels of income. Interestingly, people were more influenced by

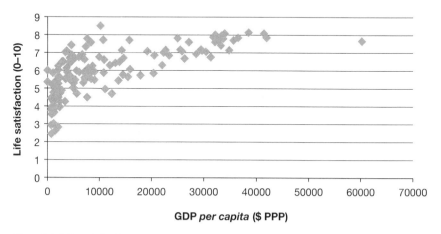

Figure 8.4 Life satisfaction against average GDP *per capita* for 179 countries

Source: Based on data from New Economics Foundation, 2009

Note: Life satisfaction is measured as the response to the question 'All things considered, how satisfied are you with your life as a whole these days?'

upwards comparisons (looking at people better-off than themselves) than by downwards comparisons (looking at people less well-off).[64] People tend to display their rank in society through 'conspicuous consumption' – using clothes, cars, houses and other possessions to show both their membership of a social group and their rank within that group – and this tends to be more pronounced in more unequal and more materialistic societies.

The problem is that this kind of status competition is a 'zero sum game' – people can only improve their happiness levels if they move up a rank in income, but this comes at the expense of someone else moving down a rank, so there is no overall increase in happiness levels. This helps to explain the 'Easterlin paradox', first identified in 1974, which states that although happiness is positively correlated with household income within a country at a given point in time, average happiness levels within a country do not increase significantly over time in the long term, even though incomes are rising in real terms due to economic growth. Even in developing countries with high growth rates, we do not see equivalent increases in satisfaction. Easterlin cites the cases of China, South Korea and Chile, where strong economic growth has led to a doubling of real incomes in 10 years, 13 years and 18 years respectively. Both China and Chile show slight declines in life satisfaction, and South Korea shows a slight increase at first, followed by a slight decline between 1990 and 2005.[65]

The law of diminishing returns applies to the relationship between happiness and income: an extra dollar of income for a rich person gives only a tenth of the happiness it would give to someone ten times poorer.[66] Recent studies appear to show that happiness can even decline at higher incomes. A 2011 study of household incomes in Germany and the UK identified a 'hump-shaped' relationship where happiness peaked at incomes of around $150,000 to $200,000 per year and declined after that. The researchers hypothesised that this was due to the 'psychological cost' of the gap between aspirational income and actual income – as incomes grow, people's expectations also grow so that they have to strive harder to achieve their goals.[67] Other researchers propose further explanations: people who are more materialistic are more likely to be anxious, unhappy, suffer low self-esteem and have difficulty forming relationships.[68]

There is evidence that the satisfaction gained from material possessions is fleeting – people rapidly become accustomed to the new item, and elevated feelings of happiness wear off, leaving them looking for their next 'fix' from another new purchase. This is known as the 'hedonic treadmill', where people become trapped in a cycle of working ever harder to buy more new goods, with little real long-term increase in satisfaction. Long working hours can be stressful and detrimental to health as well as placing a strain on relationships. On top of this, ownership of goods and property comes with responsibility to maintain those goods – cars need fixing, houses need cleaning and decorating, and appliances break, all creating stress and hassle. Even finding space to store possessions can be a problem, as demonstrated by a spate of television programmes on how to 'declutter' your home.

So if material goods and higher incomes do not make us happier in the long term, what does? Psychologists beginning with Abraham Maslow in 1943 have identified a range of needs, from basic physical needs to higher-order psychological needs:

- physiological needs (food, clothing, shelter, sleep);
- safety and security (including financial security and health);
- love and belonging (family, friendships, community, connectedness);
- esteem (status, recognition and respect from the community; feelings of self-worth and competence);
- self-actualisation (the ability to make the most of your own potential).

A large number of studies in the academic field of 'happiness research' have confirmed that successful relationships with family and friends are the most important indicator of happiness, followed by financial security, satisfying work, community relationships, good health and political freedom.[69]

Obviously, money can contribute to some of these needs – a certain amount of money is necessary for food, shelter, health and financial security. Money can also contribute to happiness for individuals when it is linked to an increase in social status – although this may be a 'zero sum' game at the level of the whole of society, as mentioned above. But there is evidence that too much focus on materialistic aims can start to detract from the likelihood of achieving other needs, especially the need for strong family and community relationships.[70] Studies published in 2009 and 2011 have linked a rise in consumerism and materialistic values in certain Western nations, including the United States and the UK, with increases in the rates of depression, mental illness and crime, and decreases in levels of trust.[71]

Interestingly, money can contribute more to happiness if you give it away, either as charitable donations or as gifts to others.[72] There is also evidence that spending money on experiences (such as a day out or a trip to the theatre) rather than on material goods creates more happiness. This is believed to be because experiences contribute to feelings of vitality and social connectedness as well as creating long-lasting pleasant memories. Spending on experiences rather than goods was also found to be less subject to social comparison.[73]

If material goods do not make us happier, why do so many people continue to shop to excess? It seems that personality and values are important in this respect. Studies suggest that around half of our happiness is determined by genetic factors – by inbuilt personality traits. But upbringing and cultural influences are also important. Some authors have suggested that people buy things to try to satisfy unmet emotional and psychological needs,[74] and there is evidence that once basic needs are met, our purchasing decisions are influenced more by desire and emotion than by rational decision making.[75] Companies exploit this phenomenon through advertising, by creating desire for things that we do not actually need. But people with life goals oriented towards family, relationships and social involvement tend to be happier than those with materialistic goals such as the accumulation of money, status and career success.[76]

Many people are already coming to believe that 'enough is enough' – there comes a point where more material goods do not improve quality of life, and may even start to detract from it. In the EU-27, for example, the material consumption of individual households with similar income levels varies by a factor of 10 or more, with lower-impact households tending to spend more of their income on leisure services, and looking for quality rather than quantity in the goods they buy.[77]

Perhaps the greatest benefits for health and happiness can be achieved by coupling 'buying less' with 'working less'. Chapter 7 warned that if we can no longer rely on economic growth to offset labour productivity improvements, we may face a future where there is less paid work available. A key part of adapting to this would be through voluntary reductions in working hours, with strategies such as:

- starting work later in life (more education);
- shorter hours per week (including more part-time working or job sharing);
- more holidays;
- opportunities for career breaks – e.g. for further training or voluntary work;
- earlier retirement;
- more support for unpaid carers, including parents of young children or those with sick or elderly relatives.

Working hours have been steadily falling since the 14–16-hour working days of the early Industrial Revolution. Productivity improvements, increased rates of pay (meaning that employees could afford to choose jobs offering shorter working hours) and the development of trade unions gradually brought annual work time in Britain, France, Germany and the United States down from 3,000 hours per year in 1856 to about 2,300 hours by the time of the First World War.[78] But in the 1920s and 1930s, it became apparent that productivity improvements were creating a labour surplus: a problem identified by John Maynard Keynes as 'technological unemployment'.[79]

At this stage, it was generally expected that future productivity improvements would continue to be taken as increases in leisure time. Keynes predicted that technical progress would soon lead to a situation where everyone had enough for their basic needs, and the main problem would be finding satisfying activities for people to pursue once they no longer needed to work as much. He proposed shorter working hours – 3 hours a day – to share out the available work and thus fulfil people's natural desire for constructive activity. And although he is responsible for the 'spend your way out of economic recession' strategy, this was only meant to be a temporary phase in the path towards a future in which basic needs were satisfied and we could all look forward to lives of leisure.

But it was not to be. A turning point was reached during the Great Depression of the 1930s. The US Senate voted to pass a law enacting a 30-hour working week, to combat unemployment, but the move was defeated.[80] Rather than reducing working hours, the government decided to adopt a strategy of encouraging increased consumer demand for the goods that industry was producing. Instead of taking productivity improvements as leisure time, workers were given wage increases, and the advertising and marketing industry was employed to create demand for an endless carousel of new products that people didn't know they needed.

In her book *Selling the Work Ethic*, Sharon Beder describes how this entailed a major cultural change, replacing traditional Protestant values of thrift and prudence with acceptance of waste, self-indulgence and artificial obsolescence. The 'culture of character', valuing hard work and moral strength, was replaced with a 'culture of personality' based on appearance, clothing and other markers of conspicuous consumption. She describes the beginnings of the 'work more to buy more' culture, in which leisure time became chiefly an opportunity for more consumption, fuelling the sale of goods such as radios, clothes and hi-fi equipment.[81]

After the Second World War, the release of a pent-up desire for consumer goods cemented consumerism as a way of life in industrialised countries, and the idea of tackling unemployment by reducing working hours fell out of favour. The 1950s saw a rise in materialistic attitudes in Britain, where ordinary people set about the acquisition of status symbols as a way of displacing the class system that had previously governed society. People began to see their chief role in society as consumers rather than producers, and to express their identity through clothing and possessions.[82] The spending strategy of Keynes was converted from a temporary solution to an economic crisis into a long-term strategy for growth, and this worked very well at first, with the global economy fuelled by cheap oil and electricity.[83] In the 1980s, neo-liberal economics became the established paradigm, with a reliance on the free market and deregulation to solve all problems, and any idea of respecting ecological limits or restraining consumerism became unthinkable.

Today, however, things are changing. Quite apart from the realisation that we are consuming the planet's resources at an unsustainable rate, many people are starting to feel the pressure of increasingly 'time-poor' lifestyles. Although working hours per person have continued a gradual decline in most countries, many households are facing new time pressures because both adults now work, so that the time available for activities such as childcare and housework is decreasing. A survey in Denmark showed that the proportion of people who would prefer to reduce their working hours instead of taking a pay rise has risen from around 43 per cent in 1964 to 73 per cent in 2007.[84]

This time deficit leads to a reduction in 'social capital', as people have less time to spend on family, friends, community and even on getting involved in political activities, thus potentially undermining democracy. Time shortages can also make environmentally sound behaviour harder – people who are pressed for time are more likely to take the car instead of walking, and are likely to grab processed and packaged meals from the nearest superstore rather than searching for organic vegetables at the farmers' market. But there is now a growing interest in 'slow food', such as growing your own and cooking from scratch, and 'slow travel', such as taking the train instead of flying, or walking instead of driving. Supporters point to benefits such as healthier, fresher food, more community interaction and greater appreciation of your surroundings. It has also been argued that 'slow' activities in general, such as chatting to friends, going for a country walk or reading a book, tend to be environmentally friendly because they create a 'time de-bound' (the opposite of rebound) – they take up time that cannot then be used for higher-impact activities such as driving, shopping, or earning money![85] By working less, we reduce our spending power, which makes it easier to resist materialistic urges to buy goods that we do not really need.

A number of countries have already implemented policies to allow shorter working hours:

- All EU countries except the UK have a maximum work time of 48 hours per week.
- In Belgium, a 'time credit' system allows people to work a 4-day week for up to 5 years and also to take a 1-year career break with a paid allowance from the state.[86]
- In the Netherlands, the working week was cut from 40 hours in the 1980s to 36 hours in the 1990s, in return for wage restraint. Part-time workers are entitled to the same pay, benefits and promotion opportunities as full-time workers, which has led to a major increase in the number of people choosing to work part-time. Since 2000, all workers are allowed to request part-time working, and part-time workers are allowed to request longer hours.[87]
- In Denmark, there is a 'job rotation' system: employees are allowed to request 6 months' paid leave for childcare, education or sabbaticals, and their position is temporarily filled by an unemployed person. Similar schemes exist in Belgium, Finland and Sweden.[88]
- In Germany, there is a 'short work' policy where the government encourages firms to cut work time rather than making employees redundant. The government pays for 60 per cent of the lost wages and the company pays for 20 per cent, so the workers still receive 80 per cent of the pay for their lost working hours. The government ends up paying out about the same amount as they would have paid in unemployment benefit, but the workers keep their jobs and the firm avoids the cost of losing skilled workers and having to retrain new people when demand picks up.[89]

In summary, there is a strong case to be made for shifting towards shorter working hours rather than more material consumption, as labour productivity increases. The benefits would be three-fold: a reduction in unemployment; easing of the environmental pressures caused by over-consumption; and redressing the work–life balance to reduce stress and increase social capital. To illustrate the potential impact of such a policy, one modelling study suggests that if labour productivity grows at 2 per cent a year, then half of this could be taken as an increase in leisure time and half as an increase in wealth. This policy would result in cutting consumption in 2050 by a third compared to the business-as-usual case (although it would still be 50 per cent bigger than today). The remaining 1 per cent annual increase in wealth could be used to help to overcome the problems of funding pensions and public services in a lower-growth future.[90]

Conflicts

It seems that there are remarkable benefits for health and well-being to be gained from climate-friendly behaviour such as active travel, low-carbon diets and a 'buy less, work less' strategy. But there are also significant areas of conflict. These include:

- *Travel.* Holidays, leisure activities and visiting friends and relatives make people happy, but travel (especially flying) can also be a major source of carbon emissions. There are also times when driving is the fastest, most comfortable and most convenient mode of travel. On the other hand, measures to avoid travel, such as tele-working and videoconferencing, can save time and stress for workers, and using trains instead of cars for business travel allows work to be done whilst travelling.

- *Processed food.* It cannot be denied that fat, salt and sugar taste nice, and so a move to a healthier low-carbon diet could make some people less happy. The use of processed food such as ready meals can also save a lot of time and stress for people with busy lifestyles, which is especially important with more women entering the workforce – fewer households are able to devote time to cooking healthy food from scratch.

- *Local food.* Restricting diets entirely to local food would reduce the pleasure of eating a diverse range of food from around the world, including healthy choices such as exotic and out-of-season fruit and vegetables. Exports of fruit and vegetables are also a particularly important source of income for a number of developing countries. There are also cases where imports from sunny countries can be less carbon-intensive than producing food locally using heated greenhouses.

- *Buying less.* There may be impacts on employment, especially in manufacturing and retail sectors. Chapter 7 discussed mitigation methods, including retraining, shorter working time, job guarantee, basic income and ecological tax reform.

- *Working less.* Any mandatory policy to reduce working hours needs to protect low-income households; for example, through a minimum wage and income support where necessary. Work is important for happiness and well-being – there is evidence that income received from the state (for example, through social security payments) does not generate the same degree of satisfaction as wages earned through work. However, it is also important to remember that the whole point of the 'work less' policy is to ensure that fewer people experience involuntary unemployment.

- *Economic and social concerns.* There are serious issues to be investigated over the implications of a lower-growth 'buy less, work less' economy. With earlier retirement ages and an ageing population, the ratio of workers to pensioners will decrease, potentially leading to difficulty in funding pensions. With lower rates of economic growth, there will be less tax revenue for funding public services and higher tax rates could be necessary. And although a 'work less' economy frees up more time for unpaid caring, it is also possible that this could have implications for gender equality and the career development of women.

The next section discusses some strategies that could help to maximise the synergies and reduce the conflicts between the goals of cutting carbon and improving health and well-being.

The way forward

In summary, a low-carbon society that encourages active transport, more sustainable eating habits and less materialistic lifestyles can significantly improve health, well-being and community cohesion. But a strong policy framework is needed to deliver these benefits and to minimise potential conflicts. It is very difficult to change long-established habits, culture and lifestyles, especially when efforts to promote change can be perceived as interference with personal freedoms.

This section first discusses the general problem of how to encourage more sustainable lifestyles. It then looks at some specific strategies for dealing with travel, diet and work–life balance.

Promoting lifestyle changes

There is a huge potential for behaviour change to reduce environmental impacts. European households with similar incomes can have ten-fold differences in their use of energy and materials because of different patterns of behaviour in diet, travel, shopping and home energy use.[91] But it can be difficult to persuade people to change their behaviour. Some are sceptical, or feel that individual actions cannot make a difference; but even those who want to help the environment often find that they are trapped in patterns of behaviour dictated by the available products and infrastructure, unfavourable cost structures, time constraints, working patterns, social norms and habits. Those who try to be green often find themselves 'swimming against the tide'. Rather than relying on a minority of committed green consumers to shoulder the burden, governments need to do more to help all citizens change their behaviour. Motivation can arise from the 'feel good factor', from awareness of health or cost benefits, from a feeling of 'being part of something', or if environmental behaviour becomes a social norm.[92]

Studies have shown that many people have an attitude of 'I will, if you will' – in other words, they are willing to make lifestyle changes if they can see that other people are also 'doing their bit' to help. They expect a clear vision and strong leadership from government, and they also need to see that policies are fair, with no free-riders, but with protection for vulnerable people such as those on low incomes.[93]

The key strategy is to make sustainable behaviour easy, cheap and attractive compared to the alternatives, and then to raise awareness of the benefits and build support for further action. It is important to put in place a wide range of consistent policies, including:

- *The right cost framework.* People need to see that they will be rewarded for 'good' behaviour and penalised for 'bad' behaviour; for example, through a co-ordinated system of taxes, charges, fees, grants and subsidies. Although higher taxes (e.g. on fuel) can be unpopular, this is often because people assume that government is imposing charges simply to raise revenue. Opposition tends to decrease if it can be seen that revenue is being used to subsidise 'good' behaviour; e.g. if fuel taxes or congestion charges are used to make public transport cheaper and better.

• *The right infrastructure, products and services.* This includes government support for efficient public transport systems, pedestrian and cycle-friendly town planning, bike-hire schemes and local food initiatives.

• *Government leadership.* If government does not set a strong example through its own actions as well as through consistent policies, the public will not take environmental threats seriously. Green public procurement has a key role in encouraging the market for sustainable goods and services – the public sector accounts for 16 per cent of all procurement in Europe.[94] But it is also important to focus on highly visible measures, such as installation of renewable energy technologies in public buildings, a switch to sustainable food in schools and hospitals, the use of cleaner vehicles by government employees and the pro-vision of infrastructure such as cycle lanes. This raises awareness of the need for action and builds confidence amongst the public in the viability of green choices. People also need to see that the government is making an effort to cut its own emissions and is receiving benefits such as cost savings in return.[95]

• *Choice editing.* Although labelling is useful, it can also be confusing if it leads to 'information overload'; and in any case, many people prefer to stick to their favourite brands. Green labelling can also be associated, rightly or wrongly, with higher product prices (e.g. for organic food). Instead, many people tend to assume and expect that there will be 'choice editing' by government and business, such as the use of regulations and voluntary agreements to phase out unhealthy food or inefficient vehicles and appliances. People expect govern-ments to make the choice easy and automatic by removing the most damaging products from the market.[96]

• *Community initiatives.* It can be easier to change behaviour as part of a group. Community groups such as schools, churches, charities and environmental organisations can help to stimulate change, provide practical and moral support to individuals, influence local government and set new social norms.[97] The Transition Towns network is a good example – from the first initiative in 2006 (Totnes in the UK), it has grown to over 450 official initiatives in 34 countries.[98] These groups provide a forum for individuals to get together and make their communities more resilient to the threats of climate change and Peak Oil; for example, by setting up local food initiatives and renewable energy projects. The 10:10 initiative is another example, encouraging individuals and businesses to try to cut their carbon emissions by 10 per cent.

• *Smart communication* can be used to encourage individual and community action. For example, Sutton Council in the UK used a pledge website to offer residents incentives such as discounts on cycle repair and public transport tickets, in return for a promise to install insulation, increase recycling or use the car less. A pilot scheme in the UK also increased the number of people recycling by

39 per cent by showing them how their street's recycling rate compared with that of nearby streets.[99]

- *Debate and engagement.* Governments need to show leadership and vision in tackling environmental problems, but they also need public support. Although studies in the UK show that the existing mandate for action may not be fully exploited – with people expecting the government to take a stronger lead than it is currently doing – there is also a need to extend the mandate to allow stronger action. A crucial part of this process is to engage the public in wider debate, and involve them in decision making and policy design. In Japan, for example, public debate and engagement helped to build support for a previously unpopular system of charging for waste disposal.[100] Incorporating sustainability issues into educational curricula is also important.

Encouraging low-impact diets

It can be very difficult to change eating habits – the size of the diet industry bears witness to the number of people around the world who are struggling to lose weight. Nevertheless, there is evidence that a well-co-ordinated and consistent approach can work. In Finland, for example, a health education programme involving national and local government, schools, health professionals, the private sector and non-governmental organisations managed to cut heart disease by 80 per cent from very high levels in the 1970s, mainly through persuading people to eat less saturated animal fat.[101]

A co-ordinated approach would include measures such as:

- Advice from national governments on the need to eat less animal produce. For example, Sweden produced guidance in 2009 advising the public to eat less meat and to choose locally produced, grass-reared and organic produce where possible.[102] The government of South Korea even runs training courses teaching newly married women how to cook healthy traditional dishes.[103]
- Public procurement of sustainable, healthy, locally sourced food for schools, hospitals and government buildings.
- Clear labelling of the health impacts of food; e.g. fat, salt and sugar content, including in restaurants.
- Choice editing by government and business to eliminate or discourage the worst food choices; e.g. by regulations to limit the fat, salt and sugar content of certain foods; by placing local, seasonal or vegetarian food choices in prominent positions on supermarket shelves and restaurant menus; and by discouraging large portion sizes in restaurants and fast-food outlets.
- Public education and debate, in schools and via the media, about the health and environmental impacts of different food choices.
- Community initiatives; e.g. awareness campaigns and support services run by local authorities, health professionals or other community organisations.
- Restrictions on advertising and marketing of unhealthy foods, especially to children.

- Consideration of the potential for taxing unhealthy foods and subsidising healthy foods, although research suggests that the price differential would have to be large to make a significant impact.[104]

A balanced approach is important. Processed food is not universally a bad thing – it can save time for those with busy lifestyles, it lasts longer than fresh food and it can provide variety in the diet. Similarly, a diet based solely on locally produced seasonal food could be very restricted in certain parts of the world. But by cutting out excessive consumption of animal produce and processed food and eating more fresh fruit and vegetables, seasonal and local where possible, we could cut climate emissions and gain very significant co-benefits for health and the environment.

Enabling sustainable travel choices

Active transport can improve health and well-being, but it also presents two obvious conflicts: people like their cars because they are fast and convenient, and people like to fly because they can visit far-away places.

It is therefore important to put in place attractive alternatives to car use, including a fast, cheap, convenient and efficient public transport system; safe and pleasant walking and cycling routes; secure cycle parking; bike-hire schemes and better urban planning to minimise trip distances. But it will also be necessary to cut traffic levels, through measures such as traffic calming, fuel taxes and parking fees. Congestion charges are particularly effective – those in London and Stockholm cut traffic by 18 per cent and 25 per cent respectively.[105] Although these restrictions can be unpopular, their acceptability can be improved by demonstrating a clear commitment to investing in better active transport options, and by emphasising that cutting and calming traffic makes walking and cycling safer and cleaner. It is also important to ensure that those on lower incomes still have affordable transport options – perhaps by providing subsidised public transport passes. To achieve a lasting cultural change, there needs to be a clear message that the needs of pedestrians, cyclists and public transport users will take priority over those of motorists.

Flying is perhaps a more intractable problem. There is little that can be done in the way of technical fixes to reduce emissions from aircraft, and so the need to reduce journeys is paramount. For business travel, this might not be too painful – videoconferencing and tele-working can reduce trips whilst saving time and money for businesses, and reducing the time employees spend away from home. But leisure travel is another matter – cheap flights have enabled many people to enjoy foreign holidays or take long-distance trips to see friends and relatives. There is some scope for shifting short-haul flights onto rail or long-distance coach services. But for longer flights, the only feasible option is simply to cut down on foreign trips. Of all the options for reducing carbon emissions, this is perhaps the only case where a genuine and unavoidable sacrifice of well-being may be involved.

The main tool for discouraging air travel would be to impose emission charges on flights. This would help to ensure that rail travel is cheaper than short-haul flights, which is rarely the case at present. Although this measure is likely to be unpopular because it would be unfair to less affluent consumers, public acceptance could be increased to some extent if it was clear that the revenue was being reinvested to make long-distance rail travel cheaper and faster.[106]

There is one further option available – the carbon offset, whereby travellers pay a charge for the carbon emissions generated by their flight, and the money is used to offset the emissions by investing in a carbon-reducing project such as energy efficiency or tree planting (usually in developing countries where emission reductions can be achieved at lower cost). Offsetting is controversial: many argue that it simply absolves the traveller of responsibility, and it is difficult to verify that emission reductions paid for under the offset scheme are genuine and would not have taken place anyway. But offsetting is probably better than nothing, and it has been suggested that airlines could be encouraged to add offset payments to ticket prices on an 'opt-out' basis, which could increase the proportion of passengers offsetting their flights to around 60 per cent. Indeed, a forum participant in a UK study even suggested that those who had opted out of offsetting could be sent to sit at the back of the plane![107] If nothing else, at least the practice of opt-out offsetting would raise awareness of the climate impacts of air travel, and perhaps help to make excessive flying less socially acceptable.

Finally, there is considerable potential for encouraging more tele-working. A recent study found that 64 million US workers could work from home at least part of the time, and half of these would like to tele-work (in fact 37 per cent would take a pay cut in order to do so), yet only 3 million get the chance.[108]

Achieving a better work–life balance

We have seen how over-consumption of material goods, far beyond the level needed to satisfy basic needs, is endangering our ecosystems whilst giving little real benefit in terms of well-being. At the same time, many workers feel under pressure to work long hours in order to afford more consumer goods, leaving little time for family life and community participation. Yet governments around the world actively encourage this consumerism. With economic growth as the be-all and end-all of government policy, any dip in monthly growth rates inevitably triggers policies such as cuts in VAT together with media exhortations aimed at getting consumers out to the shops. The message to consumers is clear – it is your duty to shop, or the economy will suffer. Ironically, this is in direct opposition to the aim of other government departments and international organisations that are trying to minimise the environmental damage caused by over-consumption, but most governments seem oblivious to this point.

On top of this, some governments (e.g in the UK) are also engaged in a drive to increase working hours; for example, by raising retirement ages, withholding welfare from single parents to encourage them to seek work, and encouraging schools to provide 'dawn to dusk' care for children of working parents. The Social Market Foundation concludes that 'rather than creating a family-friendly economy, the goal is the creation of economy-friendly families'.[109]

It is unlikely that we will be able to cut consumption to sustainable levels without a radical change in government policy. Unless governments recognise that our economies need fundamental restructuring, the public will continue to receive conflicting messages, and inevitably the 'go shopping' message will continue to dominate, undermining efforts at voluntary restraint by concerned citizens. Fortunately, as we saw in Chapter 7, there is now some recognition that GDP should not be the sole measure of progress and that other measures of well-being should be taken into account. But it will take a considerable effort to overturn decades of economic orthodoxy and to find ways of making sustainable lower-growth economies work.

If governments can be persuaded to begin this process of restructuring, one of the first steps could be to discourage excessive advertising of the type that promotes over-consumption. Some countries have already taken steps to limit advertising in specific cases. In Sweden, for example, there is a ban on advertisements aimed at children under 12.[110] Other countries have restrictions on the siting of billboards, or bans on advertising cigarettes or junk food. Education is important – children can be made aware of the techniques used by advertisers and encouraged to develop a healthy resistance. Governments could also place higher taxes on 'status goods', and could introduce product durability and upgradeability standards to discourage 'planned obsolescence' (see Chapter 6). And although 'easy credit' is not quite the problem it was before the 2008 financial crash, further restrictions could help to discourage consumers from spending more than they can afford.

Next we come to the issue of the work–life balance. It is important to note that a 'work less' policy does not seek to reduce working hours because work is bad for well-being – quite the opposite. Satisfying work can be a vital source of status, creativity, learning and social interaction, and provides a purpose in life and a means of contributing to society, as well as income. But we may need to consider reductions in working hours in future, in order to maintain high employment in a lower-growth economy.

Some people work long hours because they enjoy their jobs, but others feel compelled to work longer than they would like for a variety of reasons: low wages – meaning that they must work long hours to satisfy basic needs; difficulty in persuading an employer to allow part-time working; concern that career progression will suffer if hours are reduced; or peer pressure to conform to a social norm that involves buying status goods such as expensive cars or clothing.

There are many ways in which governments can make it easier for people to voluntarily reduce their working hours. Some of these, such as the possibility of taking long periods of maternity and paternity leave, are already in operation in various northern European countries, as is legislation allowing workers the right to ask for part-time work or flexible hours. The legislation in the Netherlands which guarantees part-time workers the same rights and benefits as full-time workers is particularly important, because concerns about career progression and status can deter employees from choosing to work part-time. Long working hours could also be discouraged by promoting awareness that working too much, like drinking or smoking too much, could be bad for your health.

Naturally people are unlikely to voluntarily choose to work less if their incomes would fall below the level needed for a decent quality of life. Several points come into play here. First, the policy of a 'basic income' paid to every citizen, as discussed in Chapter 7, would help to top up incomes and allow more people to choose to work less. Second, a higher minimum wage can help to ensure that people are not forced to work excessively long hours to make ends meet – this is the phenomenon of the 'working poor', who have jobs but still live in poverty. Third, progressive taxation – higher tax rates for high earners; lower rates for low earners – would reduce income inequality, which is likely to reduce the status competition that drives conspicuous consumption. In their book *The Spirit Level*, Richard Wilkinson and Kate Pickett cite evidence that more equal societies tend to be more successful in many ways, with lower homicide rates, better mental and physical health, lower rates of teenage pregnancy, less drug abuse, higher literacy rates and higher levels of trust.[111] Finally, there is a possibility that hourly wages could be increased through a shift towards making higher-quality, more durable goods.

Although there are already mandatory limits to working hours in the EU (except in the UK), it is important to ensure that any legal limit protects those on low incomes through measures such as minimum wages and supplementary incomes where necessary. On the whole, it seems preferable to focus efforts first on enabling voluntary reductions in working hours, rather than imposing mandatory limits.

A mass shift to a 'buy less, work less' ethic will not happen without a deep culture change – especially in countries such as the United States where long working hours and material acquisition are now seen as badges of honour, in line with the Protestant work ethic. To counteract this, over-consumption will have to become socially unacceptable. The obvious difficulty in spreading this message is that most people do not want to hear about the dangers facing the planet – resource shortages, pollution, climate change and biodiversity loss. A message of 'doom and gloom' tends to alienate people and can be counterproductive – fear and guilt make people feel more insecure which tends to make them even more materialistic.[112]

A more promising approach could be to highlight the benefits for well-being that can come from downshifting to a less materialistic lifestyle – reduced stress and more time for family, community and leisure. Of course this can lead to over-simplifications: many enjoyable leisure activities such as travel or adventure sports are not environmentally benign, and some (such as computer gaming) may not be good for health or social interaction if pursued to excess! There is also a danger that people may quite rightly reject an authoritarian approach that attempts to force them to be happy in certain ways – for many, material goods such as stylish clothes and electronic gadgets are an important source of happiness and this is unlikely to change. The challenge is to encourage moderation and more thoughtful consumption, so that society has a chance of satisfying everyone's basic needs both today and in the future, without exceeding ecological limits. This could lead to a culture where over-consumption is acknowledged as a problem, and frugality is applauded rather than derided – one where people compete to have the lowest energy bill or the oldest mobile phone. There could also be more recognition that the work ethic can also apply to unpaid work or other activities that benefit the community.

This book has stressed the many co-benefits of well-designed climate change policies, such as cleaner air, more secure supplies of energy and resources, economic stability, cost savings and (in this chapter) personal health and well-being. Yet a recent line of argument developed by the wildlife charity WWF suggests that perhaps we should move away from stressing the value of these benefits to individuals, because that simply reinforces the cultural belief that personal gains are what matters, as well as sending the message that only simple and painless actions are needed. Instead, the charity argues, we should work to reinforce values based on empathy with others, both humans and other species, on self-development and on co-operatively working towards the greater good.[113]

The WWF also argues that asking 'what can nature do for humans?' tends to reinforce our perception that nature is something separate to humans, or 'not in our group', serving mainly as a source of resources for human exploitation. Excluding humans from 'conservation areas', which are reserved for wildlife, also has this alienating effect. Instead a strategy that encourages 'optimal contact with nature', such as enjoyable countryside holidays, walks, wildlife viewing and other nature-based activities, is far more likely to foster a set of values in which nature is valued in its own right and humans are seen as a part of nature. Making these activities part of the school curriculum could be helpful in this respect.

9 Joining the dots

Many people still think of climate policy as being a burden on society – a significant cost, with few immediate or visible results. But in fact, many of the actions we need to take to cut greenhouse gas emissions are necessary for other reasons. We need to save our disappearing tropical forests; we need cleaner, safer energy for the future; we need to cut waste and conserve precious materials. Strong action on climate change can deliver a host of other benefits for health, ecosystems and economies – the Climate Bonus. But this bonus is not guaranteed. Conflicts can arise, and careful planning will be necessary to reap the full potential benefits. This concluding chapter will draw together the key findings of the previous chapters, summarising the main co-benefits of climate policy and discussing how we can use joined-up policies to maximise these benefits and minimise potential conflicts.

The Climate Bonus

A well-planned climate strategy can deliver a multitude of additional benefits.

- *Cleaner air* (Chapter 3). Climate change and air pollution go hand in hand – both are caused largely by fossil fuels. Climate policy to halve global carbon emissions by 2050 could result in a 42 per cent reduction in premature deaths from fine particle pollution, avoiding over 5 million early deaths. There can also be considerable reductions in acidification, eutrophication and eco-toxicity in the environment, with benefits for ecosystems, and a reduction in ozone damage to crops. In the long term, the costs of climate policy can be completely offset by the benefits of reducing air pollution.
- *Greener land* (Chapter 4). Climate policy can be a major driver for reducing deforestation, with benefits for biodiversity, rainfall regulation, flood and erosion control, provision of medicinal plants and other forest products, aesthetic pleasure, recreational opportunities and the preservation of the culture and lifestyles of indigenous forest communities. Climate-smart sustainable agriculture – including measures to reduce soil erosion, add soil carbon through greater use of organic fertilisers, plant trees and reduce over-application of inorganic fertilisers – also has a wide range of co-benefits including reduced

water and air pollution, improved soil fertility and water retention, improved farm incomes and enhanced biodiversity.

- *Safe and secure energy* (Chapter 5). Reducing our reliance on fossil fuels can improve energy security and price stability, provide safer and more affordable energy in the long term and reduce the environmental impacts of fossil fuel extraction. A scenario to halve global carbon emissions by 2050 can halve our use of oil and gas, and cut coal use to a quarter of the baseline value, with major benefits for energy security. Even more ambitious scenarios are possible, with a diverse mix of renewable energy sources supplying up to 95 per cent of our energy by 2050.

- *Less waste* (Chapter 6). We are accustomed to living in a throwaway society, but there is tremendous scope to reduce both climate impacts and a wide range of other pressures on the environment by moving to a zero-waste economy where products are designed to be durable and upgradeable, with all materials easily reused or recycled. This has numerous benefits, including conservation of scarce resources, financial savings for businesses and households, and reduction of the environmental impacts of resource extraction and waste disposal. In the UK alone, cheap and simple resource efficiency improvements could save businesses £55 billion every year, and households could save £47 billion each year by cutting food waste and using products to the end of their useful life.

- *A stronger economy* (Chapter 7). Many people still believe that moving to a low-carbon economy will impose unacceptable costs on businesses and households. There will certainly be winners and losers, but most studies show that in the long term, the new jobs provided by a low-carbon economy are likely to exceed those lost in high-carbon sectors, and that fuel cost and healthcare savings are likely to far outweigh the costs of investing in low-carbon technology (which are estimated to be around 2 per cent of GDP). The IEA, for example, estimates that cumulative low-carbon investments of $45 trillion over the period from 2010 to 2050 would yield fuel cost savings of $110 trillion. The UN report *Towards a Green Economy* illustrates how investments in green sectors can result in a more prosperous global economy in the long term, with over 200 million more jobs, less poverty, better crop yields and less water stress than the business-as-usual case. A green economy can be more stable and productive than one which is dependent on high levels of material consumption and increasingly scarce, expensive and environmentally damaging fossil fuels, even before taking into account the damaging effects of climate instability on the economy.

- *Health and well-being* (Chapter 8). Low-carbon lifestyles can be associated with surprisingly large benefits for health and well-being – even outweighing the substantial benefits from reduced air pollution. Two-thirds of all attributable deaths – 21 million deaths globally – are due to lack of exercise or unhealthy diet, and the problem is growing. Walking and cycling instead of driving, and eating less meat and dairy produce, can dramatically improve health and fitness, reducing the risk of obesity, heart disease, diabetes, strokes and cancer.

There are financial benefits for society too, with fewer days lost from work due to sickness, and reduced healthcare costs. Cycle-friendly towns and cities can also be more pleasant places to live, with less congestion, noise, air pollution and accidents, and more social interaction. More controversially, a low consumption 'buy less, work less' lifestyle can promote well-being by reducing stress levels and giving more time for family, friends, art, sport and community activities.

Health and well-being: the heart of sustainability

With such a wide array of co-benefits, why has the general public still not united to push policy makers into stronger action on climate change? Partly this may be because most people are unaware of the co-benefits – it is hoped that this book may raise awareness, as well as dispelling some of the fear and confusion over the economic costs of climate policy. Also, conflicts such as concerns over nuclear power, biofuels, wind farms and fuel poverty undoubtedly cause many people to reject climate policy. The final section of this chapter will draw together a set of policies that can help to mitigate these conflicts, and present a joined-up approach to climate policy that brings it into line with other environmental and social concerns.

But one more factor may be important. Climate policy is generally perceived as being an abstract political process, far removed from our daily lives. From the point of view of Western consumers, who are responsible for the greatest *per capita* emissions, the worst consequences of climate change are practically invisible, as they take place in distant countries, several years into the future and with a high degree of scientific uncertainty.

When we start to look at the co-benefits of climate policy, however, most of the impacts become far more relevant to the general public. We can start to see how climate policy will affect our daily lives, with issues such as local air pollution, energy security and jobs rising to the fore. Global environmental issues that may have more widespread and immediate appeal than climate change, such as rainforest protection and avoidance of oil spills, also become part of the argument for climate action. But perhaps the most striking, relevant and pressing argument for climate action is the impact on health and well-being – an impact that has been largely ignored in the debate so far.

Climate change itself will have severe health impacts for many people, with an increase in floods, droughts, storms and heat waves, leading to food shortages and mass migration, as well as the spread of diseases such as malaria. The problem is that the worst of these impacts will hit people with little control over the cause of the crisis: *per capita* emissions in drought-struck Africa and flood-prone Bangladesh are very low. But when we start to look at climate co-benefits, the picture changes. Table 9.1 summarises the co-benefits identified in the preceding chapters, classifying them broadly into four groups: health, ecosystems, economy and society. It is immediately obvious that there is a wide range of benefits that will apply to people in developed countries. What is more, many of these benefits are related

Table 9.1 Co-benefits for health, ecosystems, economy and society

Chapter	Health	Ecosystem	Economy	Society
3: Cleaner air	Less heart and lung disease, cancer, asthma	Reduced acidification, eutrophication and eco-toxicity	Savings in health costs; less crop damage from ozone	
4: Greener land	Rainforests protected as source of medicinal plants	Less deforestation – more biodiversity	Value of forests for flood protection, soil stability and forest products	Protected forests for indigenous societies
	Amenity value of forests for health and happiness	Sustainable agriculture – less water pollution and soil erosion, more biodiversity	Value of biodiversity for pest and disease resistance and useful plants	Safeguarding future food supply
5: Safe and secure energy	Reduced fuel extraction impacts (coal mine and oil rig accidents, coal miners' lung disease, water pollution from shale gas and tar sands)	Reduced fuel extraction impacts (water pollution, oil spills, habitat destruction, land use)	Secure energy supplies; improved price stability; cheaper energy in long term	Renewables give energy access in remote rural areas
6: Less waste	Reduced material extraction impacts (mining accidents, air and water pollution)	Reduced material extraction impacts (water pollution, habitat destruction, land use)	Eco-efficient economy – cost savings; competitiveness	Less resource-related conflict
	Fewer deaths, injuries and refugees from resource wars			
7: Stronger economy	Health benefit of warmer (better insulated) homes		Green jobs in energy, recycling, eco-design, sustainable agriculture, etc.	Less fuel poverty from household energy efficiency
			Cost savings from energy and resource efficiency	
8: Lifestyle	More walking and cycling –		Savings in healthcare costs	Cycle-friendly towns – more

Chapter	Health	Ecosystem	Economy	Society
	less obesity and heart disease; less traffic noise; fewer accidents			opportunities for social interaction and outdoor play
	Lower meat and dairy diet – less heart disease, obesity, diabetes	Less deforestation for meat production		Lower food prices from reduced land use for meat production
	Less stress from better work–life balance		More stable economy less dependent on growth	Stronger communities from better work–life balance

to health and well-being – and are therefore likely to have immediate relevance for many people in their daily life.

Sustainability is usually thought of as a three-cornered triangle, with economy, society and environment at the corners. Yet the reason why we are concerned about these three issues is because they all have an impact on our well-being, which could be summed up as a combination of health and happiness. Damage to ecosystems has adverse impacts on health in a number of ways (Figure 9.2). A strong economy contributes indirectly to both physical and mental health, by providing employment, income and funding for public services such as schools, hospitals and social care. And a healthy society with few conflicts and good community interaction is important for personal safety and emotional well-being. Health and happiness should lie at the centre of the sustainability triangle (Figure 9.1) and should play a

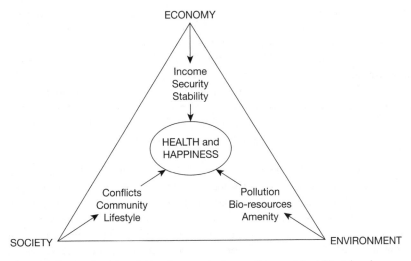

Figure 9.1 Putting health and happiness at the heart of the sustainability triangle

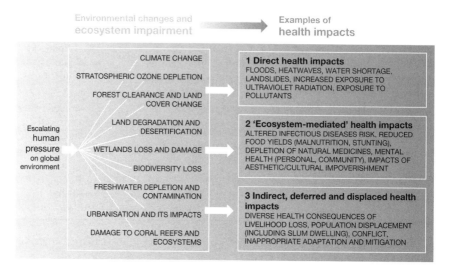

Figure 9.2 Effects of damage to ecosystems on human health

Source: Corvalan *et al.*, 2005, p. 1, Figure SDM1. © World Health Organization 2005; reprinted with permission

central role in the whole sustainability debate. The issue of health appeals directly to the general public in the way that more abstract words like 'sustainability', 'environment', 'ecosystem' and 'social impacts' do not.

Looking at the big picture

For too long, governments have looked at climate change in isolation, failing to consider the impacts of climate policy on other areas and missing the co-benefits and conflicts. Only by looking at the big picture, viewing climate policy as part of a wider strategy and making sure that it does not conflict with other goals, can we reap the full benefits of the Climate Bonus. Figure 9.3 (see also plate section) summarises the main co-benefits and conflicts of key climate policies.

Although Figure 9.3 is only schematic, and is guilty of grossly simplifying some very complex issues, it does show how certain policies stand out as having a wide range of benefits, whereas others tend to be associated with conflicts. In fact, it is possible to identify four groups of policies from the table:

- *Cutting waste*: energy efficiency, resource efficiency and lifestyle change. Cutting waste avoids a wide range of impacts from resource extraction and use, including deforestation, pollution, landscape damage and the impacts of waste disposal. All of these options will cut fossil fuel use, thus reducing the damaging effects of air pollution; avoiding the environmental and safety impacts of oil, gas and coal extraction; and improving energy security, at the same time as saving money for households and businesses. Material efficiency also saves

Joining the dots 323

	Air	Water	Biodiversity	Energy security	Resources and waste	Cost
Energy efficiency	☺☺	☺☺	☺	☺☺	☺	☺☺
Material efficiency	☺	☺☺	☺	☺	☺☺	☺☺
1st-generation biofuels	–/X	XX	X	☺☺	X	–
2nd-generation biofuels	–	–	–	☺☺	☺	?
Renewable energy	☺☺	☺	–	☺☺	–/?	?
Fuel switching (coal to gas)	☺	–	–	X	–	–
Nuclear	☺☺	X	–	☺	X	–
CCS	X	X	–	X	X	X
Forest protection	–	☺	☺☺	–	–	X
Sustainable agriculture	☺	☺	☺	☺	☺	–
Geo-engineering	XX	X	XX	X	–	X
Green lifestyle	☺☺	☺☺	☺☺	☺☺	☺☺	☺

Figure 9.3 Summary of climate policies and their co-benefits and conflicts

Notes: The ratings refer to the potential change if the deployment of this technology or policy was significantly increased compared to the current situation. ☺☺ = large improvement; ☺ = improvement; [-] = no significant change; X = deterioration; XX = large deterioration; ▨ = variable or uncertain.

'Water' refers to both water demand and water pollution.

'Geo-engineering' refers to sulphate aerosols and iron fertilisation of the oceans.

other scarce resources such as metal ores, water, phosphorous (for fertilisers) and land, thus protecting biodiversity and reducing the potential for international conflict, as well as cutting the costs and impacts of waste disposal. Lifestyle change is particularly important within this group of policies, because it can help to offset the rebound effect. Without behaviour change, improved energy and material efficiency is likely to continue to be outweighed by increasing consumption.

- *Low-carbon energy*: biofuels, other renewables, nuclear energy and switching from coal to gas. Although these policies cut fossil fuel use, and so generally have the co-benefits of cutting air pollution, reducing fossil fuel extraction impacts and improving energy security, they also present a range of negative impacts, such as the landscape impacts of wind farms; the waste, proliferation and safety implications of nuclear power; and the land- and water-use problems of biofuels. It is important to weigh these impacts carefully in policy decisions, and not just to look at climate benefits in isolation.

- *Agriculture and forest protection.* Policies in these sectors have a very wide range of benefits and conflicts, and complex interactions between economic, social and environmental issues. There is tremendous potential to protect and enhance biodiversity through forest protection schemes, but there can be economic and social conflicts attached to funding the schemes and ensuring that communities are not displaced from their land. Similarly, climate-smart agriculture based on increasing soil carbon levels and reducing emissions from fertilisers could dramatically reduce the large environmental impacts of food production, but there are trade-offs between the use of agrochemicals, which can increase yields and thus reduce deforestation from cropland expansion, or less intensive organic methods that reduce pollution and protect farmland biodiversity, but may have lower yields, with implications for food security and food prices.

- *End of pipe and beyond*: carbon capture and storage (CCS) and geo-engineering. Neither of these techniques harnesses any of the co-benefits of the other options discussed, and both entail extra economic, environmental and energy costs. Carbon capture and storage carries an energy penalty of around 25 per cent, increasing the impacts from fossil fuel extraction and decreasing energy security, though emissions of some air pollutants could be reduced. There could, however, be a role for CCS in conjunction with sustainable biofuels, where these negative fossil-fuel-related effects would be absent, although the energy penalty would increase the biofuel feedstock required. Geo-engineering could be even more damaging. The two most commonly advocated forms of geo-engineering – sulphate aerosols and iron fertilisation of the ocean – pose major risks to biodiversity. It is almost impossible to predict the impacts of such major intervention in the environment. Sulphate aerosol injection threatens to increase acid rain and have unpredictable impacts on rainfall patterns, and iron fertilisation could have drastic impacts on marine ecosystems.

By looking at the big picture, taking into account not just climate change but also all the other benefits and impacts of policy options, we can see that some policies are more promising than others. Cutting resource use – through energy and material efficiency coupled with demand reduction – tackles a multitude of different environmental problems at source – effectively 'turning off the tap' of environmental damage. Carbon capture and storage and geo-engineering, on the other hand, only try to tackle the problem of carbon emissions after they have already been created – just 'mopping up the water'. They therefore fail to produce any co-benefits, and they also have serious negative impacts. Agriculture and forest policies have the potential to deliver major benefits for biodiversity and the environment, provided that the right safeguards are put in place to protect against economic and social conflicts. That leaves the diverse group of low-carbon energy technologies – nuclear, biofuels, renewables and fuel switching. No energy source is impact-free: the policy challenge here is to ensure that the market chooses the 'least bad' technologies, and to look for ways of mitigating the impacts by good policy design.

But how can we ensure that the full range of benefits and impacts are taken into account in policy decisions? One approach, as described in Chapter 7, is to try to include external costs (such as health impacts) in energy prices through levying pollution taxes; another is to impose emission caps and then let companies trade emission permits between themselves. However, neither of these approaches can take into account the full range of impacts, including issues such as energy security or biodiversity loss. Many governments, therefore, supplement these approaches with other policy options such as regulation to discourage the most polluting technologies, or support for specific technologies that are seen as being desirable.

A coherent policy requires some kind of assessment of the relative impacts of different technologies. This often involves comparing 'apples and pears' – weighing up the small chance of a serious nuclear accident, for example, against the visual impact of a wind farm. One important aspect that can become obscured in economic analysis is that some impacts are relatively short term and local, whereas others are long term and global. Table 9.2 summarises this for a range of energy sources.

The first thing to note is that greenhouse gas emissions have a long-term, global impact. If we fail to deal with climate change today, the consequences will affect the world for many generations. Other impacts of fossil fuels tend to be slightly less extensive – air pollution has a medium-term impact on a regional to global scale, and extraction impacts tend to be local. Of course, as has been pointed out already, the fact that these impacts tend to be shorter term and more local can, ironically, mean that they offer stronger motivation for cutting fossil fuel use, as this will produce immediate local co-benefits.

The most interesting contrasts, however, arise when comparing the main alternatives to fossil fuels. Nuclear energy generates uranium and plutonium radionuclides with very long half-lives, so that pollution from accidents, waste disposal and uranium mining can remain a hazard for millions of years. The key impact of wind power, by contrast, is the landscape impact, which is localised and is only a problem for as long as the turbine remains standing.

Although we can draw some general conclusions on the relative benefits and drawbacks of these different technologies, it is important to remember that in many cases the situation will vary between countries. A country that is well endowed with coal, for example, might not be concerned about the effect of CCS on energy security. Similarly, countries with abundant rainfall and a favourable climate might find biofuel cultivation less of a problem than others where water and arable land are scarce. And countries with good grid interconnections and large hydropower resources might find it easier and cheaper to increase the share of intermittent renewable energy.

A tale of two strategies

This book has tried to emphasise the importance of looking at the big picture. Rather than drawing up climate policy in isolation, we need to integrate it fully with other concerns such as energy security and air quality. We need to take all the co-benefits and conflicts of different policy options into account, and set up an integrated policy framework. Perhaps we can illustrate this with a look at two visions of the future: one in which we continue with isolated policy initiatives, and one in which we move to a fully joined-up approach.

Table 9.2 Significant health and environmental impacts of various energy sources

Fuel	Impact	Time horizon	Geographic scale
Coal and lignite	Coal mining accidents	Medium*	Local
	Occupational illness	Medium	Local
	Subsidence	Long	Local
	Regional air pollution (SO_2, NO_x, PM)	Medium	Regional to global
	Emissions of heavy metals	Long	Regional to global
	Greenhouse gas emissions	Long	Global
	Generation of solid wastes	Long	Local
	Water pollution from mines	Medium to long	Local
Oil-based fuels	Accidents at drill sites	Medium*	Local
	Oil spills	Short to medium	Local to regional
	Air pollution	Medium	Regional to global
	Emissions of heavy metals	Long	Regional to global
	Greenhouse gas emissions	Long	Global
	Water pollution from effluents	Medium	Local
Natural gas	Accidents at drill sites	Medium*	Local
	Air pollution	Medium	Regional to global
	Greenhouse gas emissions	Long	Global
Nuclear	Uranium mining impacts	Long	Local
	Routine discharge of radiation	Medium to long	Local to regional
	Major accidents	Very long	Regional to global
	Nuclear proliferation	Very long	Global
	Risks associated with theft or misuse of radioactive material	Very long	Global
Wind	Land use	Short	Local
	Noise	Short	Local
	Visual intrusion	Short	Local
	Intermittency of supply	Medium**	National
Large-scale hydro	Dam failure	Long	Local
	Disruption to river systems; water rights	Long	Local to regional
	Population displacement	Long	Local
Large solar plants	Visual intrusion	Short	Local
	Land use	Short	Local
Biofuels (1st-generation)	Land use	Medium	Local to regional
	Water, energy, fertiliser use	Medium	Local to global

Source: Adapted from various ExternE reports (http://www.externe.info)

Notes: Time horizon refers to length of time for which impacts would persist if technology was no longer used.

Indicative horizons: short term = 5 years or less; medium term = 5–50 years; long term = 50–500 years; very long term = 500 years+.

*Impacts of accidents are considered medium term because of loss of life-years.

**Intermittency considered medium term because it necessitates restructuring of electricity supply infrastructure.

Isolated policies: fossil fuel lock-in

Business continues pretty much as usual. While paying lip service to environmental concerns, governments continue to focus on maximising economic growth. In their view, this means minimal environmental regulation for business, and minimal taxation. In governments across the world, proposals for carbon taxes or renewable energy subsidies put forward by the Environment Department are overruled by the Treasury. Economies need energy in order to grow, so restrictions on drilling for oil in sensitive regions are relaxed, and subsidies are pumped into tar sands and shale gas. New coal-fired power stations are built, with vague plans to deal with the climate impacts at a later stage by installing CCS systems. There is a 'dash for gas', with new power stations being built in the expectation that abundant supplies of shale gas will be forthcoming. There is continued support for energy efficiency, but no restrictions on total carbon emissions or fuel use. Global climate change negotiations continue at a leisurely pace, following the Durban plans for a global agreement by 2015, which will not come into force until 2020.

With rapid acceleration in global energy use, driven by growth in developing countries, reserves of conventional oil and gas dwindle. As production shifts to deeper, lower-quality and more remote sources of oil, prices rise steeply. Faced with angry voters, governments respond by cutting fuel taxes. This ensures that there is no major shift to alternative energy sources – we remain locked into a fossil fuel economy. Despite the cut in fuel taxes, prices continue to rise as demand exceeds supply. The loss of fuel-tax revenue means that there is less money available to fund a transition to alternative low-carbon technologies. More households slip into fuel poverty. Food production is still highly dependent on the input of fossil fuels for mechanisation and fertiliser manufacture, so food prices rise. With the development of shale gas in Europe slower than expected, due to the difficulty of sinking multiple wells in densely populated areas, there is a shift towards coal for power generation, leaving the newly built gas power stations as 'stranded assets'. The increased demand for coal causes coal prices to rise as well, making CCS plants (which increase coal use by a quarter) uneconomic.

As car use continues to increase across the world, air pollution rises. Environment departments respond by requiring manufacturers to fit more end-of-pipe controls to their vehicles. Unfortunately this increases fuel use, thus exacerbating energy security concerns and fuel price rises. Meanwhile, health departments are struggling to cope with an increase in pollution-related illnesses such as asthma, cancer and heart and lung diseases, as well as a rising wave of lifestyle diseases caused by inactivity and poor diet. And planners faced with rising levels of traffic congestion, noise and road accidents are also struggling: it seems that as fast as they build new roads, extra traffic appears to fill them up.

The impacts of a changing climate start to be felt more widely. More frequent storms, droughts, floods and heat waves cause loss of life, damage property and destroy crops and livelihoods, pushing food prices up further. Water scarcity increases, and more species are lost. Other resources are under stress as well: prices for certain rare metals are increasing sharply, and phosphorous supplies are running out. There is a noticeable increase in 'bad news' events associated with resource extraction – more oil spills, due to the difficulties of drilling in deep water, and more mining accidents as the hunt for scarce metals intensifies. The global economy becomes ever more unstable, with volatile commodity prices (worsened by financial speculation), mass migration of climate refugees, frequent energy supply disruptions and more resource wars. Growth falters and unemployment rises.

Too late, policy makers realise that it is now impossible to avoid dangerous levels of climate change. The world no longer has the economic or physical resources or the political stability needed for the herculean effort of replacing all our fossil-fuel-powered infrastructure, cars, homes and factories with low-carbon alternatives. In desperation, we turn to geo-engineering. Aircraft spray the stratosphere with sulphate aerosols, and fleets of boats pump iron into the oceans to stimulate uptake of carbon dioxide by algae. These measures have serious side effects: an increase in acid rain; disruption of rainfall patterns and a collapse of marine life that signals the death knell for already over-stressed fish stocks. The limited impact on global warming is partly offset by additional emissions from the aircraft and ships used in the operation. The geo-engineering techniques also fail to address another problem: rising levels of dissolved carbon dioxide that are acidifying the oceans.

Bizarrely, GDP continues to grow at first. Poverty and starvation are increasing; water, food and energy are in short supply; and environmental devastation is widespread, but at the same time governments are spending more than ever on expensive geo-engineering projects, healthcare costs, resource wars and reconstruction of infrastructure after floods and storms. It is not until resource limits really begin to bite – when fuel shortages restrict supply of the concrete and steel needed for reconstruction, for example – that GDP finally starts to fall . . .

Joined-up policies: a green economy

Policy makers experience an epiphany – they realise that climate change can no longer be viewed in isolation, but must form part of an integrated policy framework that takes account of all economic, social and environmental concerns. A new style of policy making sees ministers from all government departments sitting down together to work out how to maximise well-being and achieve a decent quality of life for 9 billion people without trashing the planet.

First and foremost, world leaders decide to listen to scientific advice on safe levels of greenhouse gas emissions. They will then try to plan policy so as to keep within those limits. A global cap is set to reduce greenhouse gas emissions steadily to one tenth of current values by 2050. An agreement is thrashed out that recognises the greater responsibility of industrialised countries for historic emissions, and allows all countries to converge on an equal *per capita* emission limit by 2050. With firm promises of financial and technical help to shift to a low-carbon economy, developing countries happily sign up to this agreement. Action begins immediately.

All countries recognise the co-benefits of a strong climate policy. To cut air pollution, reduce waste and increase energy security, there is a massive drive to increase energy and resource efficiency. This is backed up with information campaigns to promote behaviour change – cutting waste and over-consumption. Governments no longer exhort consumers to shop until they drop, in order to save the economy. Instead, there is urgent research into how to shift to a lower-consumption economy. Alternatives to GDP are put in place, in recognition of the fact that blind pursuit of GDP growth ignores environmental limits and does not necessarily improve well-being. Economists and ecologists work together to develop new approaches to finance, taxation and social policy that allow consumption to be reduced without endangering employment and social spending. This may include measures such as ecological tax reform, job guarantee schemes, basic income and encouragement of voluntary reductions in working hours. Financial speculation is curbed, and the role of banks in creating money as loans is reviewed in the light of concerns that this may be a barrier to moving successfully to a lower-consumption economy.

Production systems are completely transformed. Instead of adding bolt-on recycling schemes to existing systems, all products are now designed from scratch to be part of a zero-waste system. Products are designed to be durable, with modular parts that can easily be upgraded or reconfigured. A skilled repair industry springs up. Because of the carbon cap and ecological taxation, material prices are higher, so that it is now cheaper to repair or upgrade a product than to throw it away and buy a new one. Surprisingly, most people find this far more satisfying – it turns out that many felt annoyed by the poor quality and frequent break-downs of goods in the throwaway society, and felt uncomfortable with generating mountains of waste. Even buildings are now modular – internal walls can be moved around to adapt to changing needs. Products that cannot be repaired or re-manufactured are dismantled for recycling – a simple task, given that products are now designed so that different materials can be easily separated for recycling at the end of the product life.

Our energy needs are slashed by improved efficiency. All new buildings are super-insulated and built to harness passive solar energy, with large south-facing windows, for example, to maximise solar heat gains in cold countries. Town planners make sure that journeys between homes, shops, schools and businesses can be served by high-quality public transport systems, and by safe, pleasant cycling and walking routes. This encourages far more people to leave their cars at home, cutting accidents, traffic noise and air pollution, and improving public health. Encouraged by the new taxation systems that clearly penalise over-consumption of energy and materials, businesses find ever more innovative ways to reduce their resource use. Productivity increases, with a flood of new, green products and services that improve living standards, but with less cost to the environment. Air and water pollution falls, and there is no need to produce energy from high-impact sources such as tar sands and Arctic oil.

Governments set a policy framework that encourages choice of the lowest-impact energy technologies. Renewable energy is high on the agenda, in recognition of the relatively low environmental and health impacts compared to other energy sources. However, there are strict criteria to limit the negative impacts of renewables. Planning policy prevents the installation of wind farms in the most sensitive landscapes, but encourages local communities to be involved in co-operative schemes where they share in the benefits of wind farm development and other local renewable schemes. Biofuels must pass strict criteria to show that they come from certified sources, not involving destruction of natural forests, for example. This encourages the development of more second-generation biofuels derived from waste and algae. Governments encourage a diversity of supply from different renewable technologies, and put in place the supporting infrastructure needed to enable a high contribution of intermittent renewable energy: smart grids, more pumped-storage schemes, smart supply and demand management and good interconnections with other countries.

Nuclear power must demonstrate that it has credible plans for safe, permanent disposal of radioactive waste and prevention of theft of radioactive material. This results in a shift away from construction of existing Generation II and III reactor designs, but a renewed focus on the development of Generation IV designs, which are capable of reducing the existing stockpiles of waste and have intrinsic passive safety features. Governments also resist the temptation to 'dash for gas', wisely recognising that we need to avoid lock-in to fossil fuel technologies, especially in view of the strict limits on total carbon emissions.

Recognising the importance of preserving our remaining natural forests, governments set up a well-monitored REDD scheme so that communities find it is more profitable to leave their forests standing than to cut them down. At the same time, pressure to exploit forests decreases due to higher resource efficiency, cutting the demand for paper, timber and minerals. Also, people are eating a healthier diet involving less meat and dairy produce, which cuts the incentive to convert forest to grazing land or animal feed crops. Agriculture on the whole is becoming more sustainable. Compost is replacing synthetic fertilisers to a large extent, improving soil structure and cutting water and air pollution from nitrogen fertilisers, as well as improving the amount of carbon stored in soils and improving water retention.

Although this strong and co-ordinated climate policy requires considerable upfront investment, the new approach to policy making means that governments can see that they are saving money on healthcare costs, environmental clean-up costs, and the costs of importing fuel and resources. The total costs are affordable – around 2 per cent of GDP – and are outweighed by the co-benefits. A clear long-term policy framework gives businesses the confidence to invest in new low-carbon technologies, and governments back this up by doubling their support for appropriate R&D. Although some businesses fail to adapt, the most dynamic and forward-thinking businesses thrive, becoming smarter, cleaner, more efficient and more productive. Governments soften the social pain of the low-carbon transition by providing retraining for redundant workers.

The new carbon cap means that energy prices remain high, although at least they are stable and predictable, and governments have well-co-ordinated policies in place to protect low-income households from fuel poverty. There is a well-targeted programme of building renovations, so that all homes are super-insulated and free from draughts and damp, requiring little fuel to stay warm and comfortable. Public transport is cheap and accessible, and appliances and lighting are more efficient so that electricity bills are affordable.

In developing countries, an international climate fund helps to provide clean, renewable energy for remote communities. Deaths from indoor air pollution plummet as households are provided with more efficient stoves. Exports of raw commodities such as metal ores have declined, but reform of international trade laws and cancellation of unpayable debt have enabled developing countries to shift to higher-value-added production, giving them a more sustainable income source for the future. Free of the threat of escalating climate-related floods and droughts, developing countries are able to increase their living standards to the point where family sizes begin to fall voluntarily.

By 2050, we have managed to cut carbon emissions to the point where the worst impacts of climate change have been avoided. Human ingenuity has been channelled into getting the most benefit and use out of our natural resources, whilst living within our means. There has been a gradual shift in culture, with less emphasis on conspicuous consumption and more time for family, friends, leisure and community involvement. The natural world is thriving, with vibrant forests, clean air and water and abundant wildlife. We look forward to a stable and prosperous future.

An integrated policy framework

These two visions of the future will, it is hoped, help to illustrate the benefits that we can gain by moving to a low-carbon economy, provided that we design policies carefully to take into account all the co-benefits and conflicts of the different climate policy options. The choice is stark – stay with the business-as-usual path, based on the use of polluting fossil fuels; or switch to a clean and sustainable future in which strong action on climate change brings a host of other benefits for health and the environment. But time for action is running out. As the IEA asks in its *World Energy Outlook 2011*, 'The door to 2°C is closing but will we be locked in?'. Figure 9.4 shows that we are already locked in to 80 per cent of the energy-related carbon emissions that we are allowed to produce by 2035 if we are to achieve the 450ppm target. Room for manoeuvre is extremely limited.

One thing is certain – we cannot just leave everything to market forces. The speed and scale of the transformation that is needed to meet climate targets requires a radical shift to low-carbon technologies, including major changes to infrastructure. This will not happen without strong government support, including a clear regulatory and cost framework; ambitious long-term targets; and investment in education, research and deployment. A comprehensive policy framework is needed so that we can balance all the different benefits and impacts. This final section lists the main recommendations for an integrated policy framework that will allow us to reap the full benefits of the Climate Bonus.

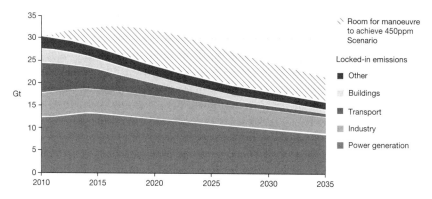

Figure 9.4 World energy-related CO_2 emissions from locked-in infrastructure in 2010 and room for manoeuvre to achieve the 450ppm Scenario

Source: IEA, 2011a. © OECD/IEA, 2011

General policies

- *Take co-benefits into account in policy decisions.* For example, air quality benefits should be included in cost–benefit analysis of climate policy options.
- *Focus support on win-win technologies and policies.* These include energy efficiency, material efficiency and behaviour change. Beware of policies with few co-benefits and major conflicts, such as CCS and geo-engineering.
- *Avoid lock-in to fossil fuels.* Set ambitious targets for a rapid shift to low-carbon energy sources. End fossil fuel subsidies. Place a moratorium on exploitation of environmentally damaging sources such as tar sands and Arctic oil.
- *Address the rebound effect.* Control total resource use and carbon emissions with caps, taxes and policies to encourage behaviour change.
- *Compensate losers.* Help businesses, households and developing countries to adapt to a low-carbon economy, with appropriate financial and technical support. Provide support and retraining for workers who lose their jobs.
- *Rethink the economy.* Investigate ways of coping with the potential impact of a low-consumption economy on jobs, pensions and social spending. These might include encouraging shorter working hours, ecological tax reform, job guarantee schemes, basic income schemes and changes to the monetary system.
- *Lifestyle and culture.* Encourage a shift to low-carbon lifestyles, with incentives, rewards and information campaigns to emphasise the health and well-being benefits of walking and cycling, eating less meat and adopting a less materialistic lifestyle. Government should lead by example, fostering an 'I will if you will' approach with low-carbon public procurement policies.

Specific areas

- *Air quality.* There should be more emphasis on reducing emissions of black carbon, ozone precursors and methane, which are greenhouse gases as well as causing health and ecosystem impacts. Key policies include addressing emissions from vehicles, and providing more efficient stoves in developing countries.
- *Nuclear.* Problems that need to be solved include waste disposal, safety risks, routine releases of radionuclides into air and water, the impacts of uranium mining, and the risks of proliferation and terrorism. Generation IV reactor designs offer potential to mitigate these problems.
- *Intermittent renewables.* Enable a maximum contribution from renewable energy by encouraging a diverse mix of renewable energy sources, setting up good grid interconnections with neighbouring regions, building more energy storage facilities and investing in smart grids and smart supply and demand management technologies.
- *Landscape impacts of renewables.* Address landscape and other impacts of large wind, solar and hydropower schemes through sensitive siting, full environmental assessments, involving local communities at an early stage, providing transparent information and encouraging co-operative schemes that share the benefits with local communities.
- *Biofuels.* Safeguards such as certification and regulation are needed to ensure that biofuel production does not entail destruction of valuable habitats, such as rainforests, or displacement of local communities. The impacts of biofuels on land use, water demand and food prices should also be taken into account. Second-generation biofuels can mitigate these impacts and deserve more support.
- *REDD and offsetting schemes.* If monitoring and verification can be resolved and the rights of local communities can be protected, REDD schemes offer tremendous potential for climate and biodiversity protection, and international funding should be made available. There is an urgent need to protect areas of high biodiversity in the short term, while plans for REDD are being finalised.
- *Agriculture.* Provide financial and technical support to promote the uptake of climate-smart agricultural techniques such as agroforestry, low-till agriculture, cover crops and addition of organic matter to the soil, and encourage more research into methods of improving yields from organic farming.
- *Resource efficiency.* A zero-waste economy offers huge co-benefits for climate, economy and environment. Priority areas are reuse and recycling of electronic waste and construction waste, and avoidance of food waste. The emphasis should be on designing out waste from the start of the production process, with eco-design of products to be durable, upgradeable, reusable and recyclable. Regulations such as landfill bans, recycling targets and product take-back have a key role.
- *Transport sector.* The transport sector offers the largest co-benefits for air quality and energy security. Important measures include fuel-efficiency standards for

vehicles; promotion of public transport, walking and cycling; control of traffic levels through congestion charges, fuel and aviation taxes, traffic calming or parking restrictions; better urban planning; and encouragement of demand reduction measures such as videoconferencing or tele-working.

Far from being a burden on society, tackling climate change presents us with an opportunity to move to a cleaner, safer and healthier world. Even without the threat of climate change, there are many good reasons for improving energy efficiency, investing in clean energy, protecting forests and reducing material consumption. By looking at the big picture, seizing the opportunities to tackle multiple problems at once and avoiding false solutions, we can deliver the Climate Bonus.

Notes

1 Introduction

1 The two degree target refers to the increase since pre-industrial times, generally taken to be around the year 1750. Since then, temperatures have already risen by 0.6°C.

2 The IPCC *Fourth Assessment Report* (IPCC, 2007) predicts that at 450ppm, there would be a 50 per cent chance of avoiding a 2°C rise; and at 400ppm, there would be a 66–90 per cent chance (for an explanation, see EU Climate Change Expert Group, 2008, p. 28). But Hansen *et al.*, 2008 call for the 450ppm target to be revised to 350ppm, based on new analysis of climate feedback effects such as release of methane from melting permafrost. For comparison, pre-industrial concentrations were around 280ppm.

3 The average atmospheric concentration of CO_2 during 2009 was 386ppm. Including the other greenhouse gases covered by the Kyoto Protocol (methane, nitrous oxide and fluorinated gases) brings the total up to the equivalent of 439ppm CO_2. However, including the effect of other climate-altering gases such as sulphate aerosols (which have a cooling effect) brings the total back down to 399ppm (EEA, 2012a).

4 EU Climate Change Expert Group, 2008.

2 The big picture: how climate policies are linked to co-benefits

1 The fluorinated gases are HFCs (hydrofluorocarbons) used mainly as coolants in refrigeration and air conditioning, PFCs (perfluorocarbons) used in electronics manufacturing and SF_6 (sulphur hexafluoride) used in electrical switchgear. Although accidental releases are small, these are all very powerful greenhouse gases, with global warming potentials hundreds or even thousands of times larger than carbon dioxide.

3 Cleaner air: cutting fossil fuel pollution

1 During the Great Smog of London, visibility was reduced to a few metres by a thick yellow-black fog caused by pollution from domestic coal burning, power stations and diesel vehicles, all trapped at ground level by a temperature inversion in windless conditions. The smog lasted from 5 to 9 December 1952. In the week ending 13 December, there were around 4,500 extra deaths compared to the same period in other years. There were 13,500 excess deaths between December 1952 and February 1953; of these, up to 12,000 may have been due to pollution (with the rest being attributed to influenza) according to Bell *et al.* (2004). See also Mayor of London (2002).

2 Newman *et al.*, 2009.

3 EEA, 2010a; Karlsson *et al.*, 2011.

4 US EPA at http://www.epa.gov/mats/powerplants.html (emissions); Newman *et al.*, 2009 (accumulation).

5 National Academy of Engineering and National Research Council, 2008.

6 Sundseth *et al.*, 2010.

7 Vlachokostas *et al.*, 2010.

8 Pleijel, 2009, pp. 61, 125. Modelling studies used data based on the European and North American dose-response functions – recent studies suggest that this could underestimate the damage in Asia.

9 Sitch *et al.*, 2007.

10 Pleijel, 2009.

11 Nitrogen oxide emissions from EU shipping were equivalent to 30 per cent of land-based emissions in 2000, and are still growing despite legislation to control emissions through Annex VI of the International Convention for the Prevention of Pollution from Ships (MARPOL), introduced in 2005 and 2008. Land-based emissions are decreasing, but are projected to fail to reach the target of the EU Thematic Strategy on Air Pollution by 2020. See Pleijel (2009, p. 75).

12 EEA, 2011a. Applies to the 32 countries in the European Economic Area.

13 EEA, 2011a. Targets under the National Emission Ceilings (NEC) Directive of the European Parliament and the Council on National Emissions Ceilings (Directive 2001/81/EC). Evaluation was not complete at the time of writing (2012), but ten countries were expected to miss the target for NO_x, four the target for NMVOCs (non-methane volatile organic compunds) and two the ammonia target.

14 EEA, 2010a. A number of countries applied for time extensions to 2011 for PM_{10} and 2015 for NO_x.

15 EEA, 2011b, p. 7, referring to the period from 2006 to 2008. The WHO guideline for PM_{10} is an average yearly value of 20 micrograms per cubic metre. In addition, 18–40 per cent are exposed to levels above the EU limit (a daily average of 50 micrograms PM_{10} per cubic metre not to be exceeded more than 35 times per year).

16 Brandt *et al.*, 2011, using data from Table 9 and dividing the years of life lost by a factor of 10.6 as stated in the paper on p. 33. The definition of Europe in this paper includes parts of Russia and Ukraine.

17 EEA, 2005.

18 American Lung Association, 2011.

19 UNEP, 2011a, Chapter 7.

20 EEA, 2011b.

21 AEA, 2008.

22 Developing countries consume on average 472kg oil equivalent *per capita* compared to 976kg for developed countries. World Resources Institute (WRI, 2010) based on OECD/IEA energy balances.

23 SEPA, 2007; Pleijel, 2009, p. 119.

24 World Health Organization, 2011a – deaths from particulate pollution in towns and cities with over 100,000 inhabitants in 2008.

25 EEA, 2010a, for the 32 countries of the European Economic Area (EEA-32) (a further 20,000 deaths are attributed to ozone pollution). World Health Organization estimates of deaths in the EEA-32 are obtained by summing data for individual countries from the WHO Global Health Observatory database at http://apps.who.int/ghodata/?vid=34300 (accessed 28 January 2012). The OECD paper is Bollen *et al.* (2009a) (includes both ozone and particulate pollution).

26 Selin *et al.*, 2009.

27 OECD, 2007a; World Bank and SEPA, 2007.

28 Megacities have over 10 million inhabitants. There are currently (2012) 25, with Tokyo being the largest (33 million inhabitants) followed by Seoul, Mexico City, Delhi, Mumbai and New York. Many megacities in developing countries are growing at rates of 3–4 per cent per year. See Brinkhoff (2009).

29 For example, Cooper *et al.* (2010) found that pollution from South and East Asia and from shipping in the Pacific Ocean was contributing to rising background levels

of ozone above the west coast of the United States between 1995 and 2008, although US emissions were falling. Similarly Wenig *et al.* (2002) detected a plume of nitrogen oxide pollution from power stations moving from South Africa to Australia. See also Committee on the Significance of International Transport of Air Pollutants (2009, Figure 2.5, p. 46), which shows ozone plumes extending eastwards from the eastern United States, Europe and South and East Asia.

30 Studies vary, but converge on a best estimate of about a 33 per cent increase in global tropospheric ozone over pre-industrial levels, with an increase of 40–100 per cent in the northern hemisphere. Ozone concentrations have been increasing at around 0.5–2 per cent per year (approximately 0.2–0.8 ppb per year) (Committee on the Significance of International Transport of Air Pollutants, 2009, pp. 40–41). At Mace Head on the west coast of Ireland, the background concentration of ozone increased by approximately 0.5 ppb per year (1.0 microgram per cubic metre per year) between 1987 and 2003 (AQEG, 2009, p. 58).

31 Pilling, 2005.

32 AQEG, 2007.

33 Tan, 2008.

34 Pleijel, 2009.

35 IEA, 2010a, p. 115. Supercritical plants operate at 21 bars pressure or above; those that operate above 600°C are deemed to be USC (though there is no precise definition). Average efficiency for all existing coal plants is 34 per cent.

36 IEA, 2010a, pp. 113–114.

37 Cooling is provided using heat-fired absorption chillers, in which the waste heat from the CHP plant is used to drive an absorption cooling system which produces chilled water for air conditioning or refrigeration systems. The heat can also be used for industrial applications.

38 IEA, 2010a, p. 117.

39 DEFRA, 2010.

40 Comparing a combined-cycle gas turbine plant with low-NO_x burners and a CSC pulverised coal plant fitted with flue gas desulphurisation and selective catalytic reduction (WEC, 2004).

41 EEA, 2010a, Figure 2.10.

42 IPCC, 2005; EEA, 2011c.

43 EEA, 2011c, p. 44.

44 Fischedick *et al.*, 2008; EEA, 2011c.

45 Campbell *et al.*, 2009, p. 107.

46 EEA, 2011c.

47 Simms *et al.*, 2010.

48 IPCC, 2011.

49 IPCC, 2007, Chapter 10, p. 600. This is over the lifetime of the landfill site, allowing for the time before and after the collection system is operating. The US Environmental Protection Agency (EPA) provides a higher estimate of 62 per cent for best-practice landfill sites: see http://www.epa.gov/lmop/faq/lfg.html#17 (accessed 21 February 2012).

50 Fthenakis *et al.*,, 2008.

51 WEC, 2004.

52 Pehnt (2006) presents a 'dynamic' life-cycle assessment of renewable energy technologies which demonstrates the effect of future technological changes on the relative impacts of renewables and conventional fossil sources.

53 Storm van Leeuwen and Smith, 2005.

54 IEA, 2010a, Figure 7.19.

55 Brinkman *et al.*, 2005.

56 See Pleijel (2009, p. 57).

338 *Notes*

57 Ramanathan and Carmichael, 2008.
58 World Health Organization, 2003a.
59 UNEP, 2011b, p. 21.
60 Ramanathan and Carmichael, 2008.
61 Jacobson, 2002.
62 Nowak *et al.*, 2006.
63 Tallis *et al.*, 2011.
64 Nowak *et al.*, 2006.
65 Clancy *et al.*, 2002.
66 Holland *et al.* (2011a), citing Pope (2009).
67 Crouse *et al.*, 2012.
68 DEFRA, 2010.
69 Williams, 2007.
70 Williams, 2009.
71 DEFRA, 2010. Net present value of cumulative benefits 2000–2050.
72 European Commission (2008a) analysed a 20 per cent cut in greenhouse gases; Holland (2009) extended analysis to a 30 per cent cut.
73 Holland *et al.*, 2011a, 2011b. Costs are based on a value of €60,000 for a life-year. Using an alternative valuation approach based on the value of a statistical life, assumed to be €3.8 million, the costs could be as high as €569 billion. However, the study authors advocate the first approach because they do not believe that air pollution is generally the sole cause of death.
74 Pleijel, 2009, pp. 103–104. Based on GAINS model by the International Institute for Applied Systems Analysis (IIASA). The life expectancy target is part of the EU Thematic Strategy on Air Pollution (2005).
75 EEA, 2006. This is consistent with the aim to limit the global temperature rise to 2ºC.
76 Pleijel, 2009, p. 102, Figure 8.2. Based on GAINS model by IIASA.
77 IEA, 2010b. Modelling by IIASA, see http://www.worldenergyoutlook.org/docs/weo2010/IIASA_Emissions_Impacts.pdf
78 Bollen *et al.*, 2009a, p. 74.
79 Bollen *et al.*, 2009a, Figure 3.1.
80 See IPCC (2007) and Bollen *et al.* (2009b) for reviews.
81 Pleijel, 2009, p. 105.
82 Nemet *et al.*, 2010. The study surveyed 37 peer-reviewed studies of air quality co-benefits, containing 48 estimates of co-benefits covering different areas, time horizons and valuation techniques, and involving different mitigation policies.
83 Pleijel, 2009, p. 108. Based on GAINS model by IIASA.
84 Pleijel, 2009, p. 102.
85 Bollen *et al.*, 2010.
86 West *et al.*, 2006.
87 O'Connor *et al.* (2003), described in IPCC (2007, Working Group III, Chapter 11, p. 673).
88 Pleijel, 2009, pp. 105–106.
89 UNEP, 2011b and UNEP/WMO, 2011.
90 Shindell *et al.*, 2012.
91 IEA, 2011a. Of these, 2.7 billion use traditional biomass; 0.4 billion use coal.
92 World Health Organization, 2011a.
93 Partnership for Clean Indoor Air, 2011, p. 55; costs from Wilkinson *et al.*, 2009.
94 Wilkinson *et al.*, 2009.
95 EEA, 2011c, Table 6.2.
96 AQEG, 2007.
97 The direct cooling effect of sulphate aerosols is estimated to be –0.5 watts per square metre (w/m^2), compared to a positive warming effect of 1.7 w/m^2 for anthropogenic

carbon dioxide. Aerosols also affect the size and number of water droplets in clouds, which in turn affect the amount of solar radiation reflected by the cloud (the Twomey effect) and the lifetime of the cloud (the Albrecht effect). The current estimate of total forcing effects from aerosols is -1.7 w/m^2. This masks the warming effect of greenhouse gases, currently estimated as 3.0 w/m^2 (for all greenhouse gases including ozone). See Pleijel (2009, Chapter 3).

98 Bollen *et al.*, 2009a, p. 15. In the long run (2100–2150), global temperature does eventually stabilise at a level well below the baseline projection.

99 Soot and ozone were estimated to have a warming effect of 0.8 watts per square metre (w/m^2), whereas other aerosol particles have a cooling effect of about -1.6 w/m^2 (IPCC, 2007). Ramanathan and Carmichael (2008) indicate that the warming effect of soot could be higher than the 0.2–0.4 w/m^2 assumed by the IPCC.

100 Shindell *et al.*, 2011.

101 Williams, 2009.

102 HEI, 1999.

103 Williams (2009), citing a study by Mazzi and Dowlatabadi (2007).

104 IEA, 2010a, p. 639.

105 'China quietly shelves new diesel emission standards', *The Guardian*. Online at http://www.guardian.co.uk/environment/blog/2012/feb/01/china-shelves-plan-diesel-emissions

106 Scottish Executive (2006) reports that biofuels derived from energy crops cause more eutrophication and acidification emissions than oil- and gas-based systems.

107 AQEG, 2011.

108 Environmental Protection UK and LACORS, 2009.

109 Scottish Executive, 2006.

110 Sustainable Development Commission, 2006.

111 This led Norway and Ireland to complain to the European Commission about the cost to their lobster fishing industry. See http://www.guardian.co.uk/uk/2003/apr/29/nuclear.fishing

112 Ghirga, 2010; Sermage-Faure *et al.*, 2012.

113 The half-life is the time taken for a given amount of the material to decay to half of the initial radioactivity. The half-life of iodine-131 is 8 days, so this radionuclide is only of concern in the short term following an accident.

114 WEC, 2012.

115 Von Hippel, 2011.

116 Sovacool, 2010.

117 *Daily Yomiuri Online*, 2011. The overall estimate of damage is 3.89 trillion yen (around €36 billion) for a 1.2 GW reactor. Japan's nuclear reactors have operated for over 1,400 reactor years and three of the reactors at Fukushima failed; so it was estimated that, on average, an accident occurs every 500 years, giving an estimated external cost of 1.1 yen/kWh (c€1/kWh) (assuming a capacity factor of 70 per cent). The ExternE estimates for Europe multiplied total damage cost (€83,252 million) by a much lower probability of core melt (0.00005) and containment failure on core melt (0.19) and then divided by the annual output of electricity (7.6 TWh/year). Even taking the upper bound damage cost, the result is only c€0.01/kWh.

118 Stohl *et al.*, 2011.

119 See http://www.energyfair.org.uk/home and http://www.mng.org.uk/gh/private/nuclear_subsidies1.pdf

120 IEA, 2007a.

121 Kravitz *et al.*, 2009.

122 Pleijel, 2009, p. 48.

123 Robock, 2008.

124 Campbell *et al.*, 2009.

125 Nemet *et al.*, 2010. The worst impacts of climate change will occur in the future, so the standard method of discounting future costs (which makes future damage appear smaller) produces relatively low estimates of climate policy benefits. Air quality benefits, on the other hand, are immediate, so including them in climate policy assessments would reduce the cost–benefit ratio and encourage stronger action in the short term.

4 Greener land: forests, food and farming

1 FAO (2001, Chapter 1) states that forest cover has declined from 50 per cent to 30 per cent over the last 8,000 years, which is a 40 per cent decline; wetland and mangrove figures from TEEB, 2011b, p. 10.

2 Millennium Ecosystem Assessment, 2005.

3 Figures from EDGAR database (EC/JRC-PBL, 2011). Note: IPCC (2007) has 17 per cent from forests; 13 per cent from agriculture.

4 Campbell *et al.*, 2009.

5 Campbell *et al.*, 2009; Canadell *et al.*, 2007.

6 Permafrost is soil that has remained frozen for more than 2 years. It covers around 25 per cent of the earth's land surface, mainly in Siberia, Canada and Alaska.

7 Campbell *et al.*, 2009, p. 93; fossil fuel emissions from EDGAR database (EC/JRC-PBL, 2011). See also http://www.tyndall.ac.uk/global-carbon-budget-2010 and http://co2now.org/

8 Luyssaert *et al.*, 2008; Lewis *et al.*, 2009. Lal (2008) says that this 'carbon fertilisation' effect could persist up to concentrations of 800–1,000ppm; but in practice, the adverse impacts of climate change (and possible limits in supply of other nutrients) would halt growth before then.

9 Pan *et al.*, 2011.

10 Trumper *et al.*, 2009, p. 29.

11 Minnemeyer *et al.*, 2011.

12 Trumper *et al.*, 2009.

13 Eliasch, 2008. The cost estimate is compatible with UNEP (2011a) which estimates $29 billion per year to halve global deforestation. Net benefits are the total net present value accumulated over the period from 2010 to 2200, taking into account both costs (payments to preserve forests) and benefits (avoided climate change damage). Note that claims of cost-effectiveness have been disputed: see Llanos and Feather (2011) who point out that there are high transaction costs and also that the level of payments specified may not be enough to compete with alternative uses of the forests for agriculture, timber or oil production.

14 Lal, 2008, p. 821.

15 Lal, 2004.

16 It is also worth noting that no-till agriculture has always been widely practised by traditional subsistence farmers around the world, who would simply sow seed directly into a hole without disturbing the surrounding soil. The concept of ploughing spread from Europe to the Americas and elsewhere at the end of the eighteenth century. Although successful in Europe, ploughing proved disastrous in different soils and climates, culminating in the 'dustbowl' of the US Great Plains in the 1930s. But no-till agriculture did not become a viable large-scale alternative until herbicides for weed control became widely available in the 1960s and 1970s. See http://www.rolf-derpsch.com/notill.htm

17 IAASTD, 2008.

18 Lal, 2008, p. 821 (Lal's figures of 0.4–1.2 Gt C have been converted to Gt CO_2 by multiplying by 44 ÷ 12).

19 Lal, 2008, p. 822.

20 Lal, 2008, p. 822.

21 Liddicoat *et al.*, 2010, p. 22.
22 Lal, 2008, p. 822.
23 Shackley and Sohi, 2011.
24 Shackley and Sohi, 2011.
25 Woolf *et al.*, 2010.
26 Trumper *et al.*, 2009, p. 38.
27 Cox *et al.*, 2006. Net emissions from annual crops, including fertiliser emissions, are 400–1,100 $kgCO_2e$ per hectare per year, whereas perennials act as a net sink for 200–1,000 $kgCO_2e/ha/yr$.
28 Nitrous oxide emissions depend on a number of factors including temperature, soil type, and storage or handling of manure before it is applied to land. Sutton *et al.* (2011, p. xxix) conclude that the percentage of nitrogen lost from manure is around twice that from inorganic fertilisers; but Davidson (2012) assumes that 2 per cent of the nitrogen in manure is lost as nitrous oxide compared to 2.5 per cent from synthetic fertilisers. There is evidence that inorganic fertilisers are taken up more readily by plants, and degrade less rapidly to nitrous oxide than organic fertilisers (FAO, 2006; Sutton *et al.*, 2011), but others argue that overall emissions from organic systems are lower because of the use of catch crops and cover crops to take up surplus nitrogen, and better soil structure resulting in more aerobic conditions (Scialabba and Müller-Lindenlauf, 2010).
29 FAO, 2006, p. 108.
30 IEA (2011b) gives a 13 per cent share of biofuels in final energy (this excludes biofuels used for electricity and district heating); REN21 (2011) estimates that 10 per cent of final energy is traditional biofuels.
31 REN21, 2011.
32 IEA, 2011c; see also the US Environmental Protection Agency website on the Renewable Fuel Standard: http://www.epa.gov/otaq/fuels/renewablefuels/
33 European Commission Directive 2009/28/EC on the promotion of the use of energy from renewable sources.
34 FAO, 2009a, p. 16.
35 IEA, 2010b, p. 282; IEA, 2010a, p. 92; IEA, 2011c.
36 IEA Bioenergy, 2011.
37 Carrington, 2012.
38 WWF *et al.*, 2011, p. 187.
39 IEA Bioenergy, 2011.
40 IEA, 2011c, p. 23.
41 The estimate of 18 per cent is from FAO (2006, p. 113): a third of the 18 per cent is from deforestation in the Amazon. This has been criticised by Fairlie (2010) who asserts, for example, that not all Amazonian deforestation is primarily driven by agriculture, even though logged land usually ends up as pasture. He estimates the true figure to be 10 per cent. For comparison, the EDGAR database (EC/JRC-PBL, 2011) estimates that 24 per cent of all man-made greenhouse gases originate from land-use change and agriculture combined. The FAO estimate includes processing and refrigerated transport of animal produce.
42 De Schutter, 2010.
43 Davidson, 2012. This would enable nitrous oxide emissions to be stabilised by 2050, which is consistent with the IPCC's emission reduction scenario that could meet a 450ppm target.
44 IPCC (2007) estimates that the agriculture sector could absorb 5.5–6 $GtCO_2e$ by 2030, similar to total emissions from the sector, with a net absorption of 0.6 $GtCO_2e$ per year.
45 FAO, 2011b. Gibson *et al.* (2011) compare the biodiversity of natural forest and plantations.
46 TEEB, 2011a.

47 TEEB, 2011b.
48 Benyus, 1997. See the website of the Biomimicry Institute (www.biomimicryinstitute. org) for a wide range of fascinating examples.
49 IUCN (International Union for Conservation of Nature) Red List Summary Statistics, Figure 2. Online at http://www.iucnredlist.org/about/summary-statistics (accessed 22 March 2012).
50 Secretariat of the Convention on Biological Diversity, 2010.
51 Braat and ten Brink, 2008.
52 TEEB, 2011b.
53 Secretariat of the Convention on Biological Diversity, 2010. Measured as the Living Planet Index, which monitors over 7,100 populations of over 2,300 species around the globe.
54 IUCN Red List website at http://www.iucnredlist.org/apps/redlist/search
55 UNEP-WCMC, 2008.
56 TEEB, 2011b.
57 Venter *et al.*, 2009. Current prices under the European Union Emissions Trading Scheme (April 2012) are €7.7/t ($10.2/t), having fallen from a high of €20/t in 2008.
58 Miles *et al.*, 2010.
59 Trivedi *et al.* (2008), citing a study by Bradshaw *et al.* (2007).
60 TEEB, 2011b. The programme is also known as 'Grain for Green'. Farmers are paid for 8 years for converting cropland to 'ecological forest', for 5 years for 'economic forest' (such as fruit or nut trees) and for 2 years for grassland.
61 TEEB case study: Enhancing agriculture by ecosystem management, India. Online at http://www.eea.europa.eu/atlas/teeb/enhancing-agriculture-by-ecosystem-management/view. See also TEEB, 2011a, p. 84.
62 TEEB, 2011a.
63 UNEP, 2011a, Chapter 5.
64 Trivedi *et al.*, 2008.
65 TEEB, 2011a. The programme has more than halved deforestation, from 1.6 per cent to 0.6 per cent of forest cover lost each year.
66 Balmford and Bond, 2005. Costs in 2005 dollars.
67 Trivedi *et al.*, 2008.
68 TEEB, 2010, Chapter 1, Appendix 3 (Amazon case study).
69 Trivedi *et al.*, 2008.
70 TEEB, 2011a.
71 World Bank, 2012a. Values in year 2000 dollars.
72 World Watch Institute, 2011: supplement: 'Innovations in action'.
73 See http://www.guardian.co.uk/global-development/2011/feb/25/great-green-wall-sahel-desertification
74 Ellison *et al.*, 2012.
75 TEEB (2011a), citing an estimate by the FAO that 18,000–25,000 tropical plants are used as food.
76 World Bank, 2004.
77 TEEB, 2011a.
78 Trivedi *et al.*, 2008.
79 UNEP (2011a, Chapter 5) cites the finding of Vedeld *et al.* (2004) that forests typically provide 22 per cent of household income, but TEEB (2011b) cites cases where forests contribute around half of people's income in certain regions of Africa and India.
80 TEEB, 2011b.
81 Klein *et al.*, 2007.
82 See, for example, two recent papers linking common pesticides to the decline of bee colonies, cited in Stokstad (2012).
83 TEEB, 2011b.
84 Turner *et al.*, 2012.

85 'Mother nature still a rich source of new drugs', *Reuters*, 19 March 2007. Online at http://www.reuters.com/article/2007/03/20/environment-drugs-nature-dc-idUSN1624228920070320

86 See Thompson Coon *et al.* (2011) for a review of evidence that exercise outdoors in natural environments is better for well-being than exercise indoors.

87 Mitchell and Popham, 2008.

88 EEA (2012b), citing a study by de Vries (2001).

89 Butler, 2011.

90 Trivedi *et al.*, 2008. The fires were worse than usual in 1997–1998 partly because of a drought affecting both Asia and South America as a result of the El Niño climatic cycle.

91 Gill *et al.*, 2007.

92 Costanza *et al.*, 1997.

93 Balmford and Bond, 2005.

94 Braat and ten Brink, 2008, Chapter 6, pp. 141–145.

95 Cost estimate from Eliasch (2008). This is compatible with a more recent estimate from UNEP (2011a) of costs of $29 billion per year to halve deforestation by 2030.

96 FAOSTAT online database for world food production index: 1961 = 36, 2010 = 110 (compared to 2005 = 100). See http://faostat.fao.org/site/612/DesktopDefault.aspx?PageID=612 (accessed 30 March 2012). Per capita food production increased by 38 per cent over the same period.

97 World Watch Institute, 2011.

98 World Agroforestry Centre, 2009.

99 World Watch Institute, 2011.

100 IAASTD, 2008.

101 Forum for the Future, 2010.

102 World Watch Institute, 2011: supplement: 'Innovations in action'. See also the website of the International Institute for Environment and Development at http://www.watershedmarkets.org/casestudies/Tanzania_Uluguru.htm#_ENREF_1 (accessed 8 September 2012).

103 Bayala *et al.*, 2011.

104 Gutteridge and Shelton, 1994, Chapter 7, section 5.

105 Davidson, 2012.

106 FAO, 2006. Sutton *et al.* (2011, p. 438) state that the IPCC estimate of 1.25 per cent conversion of N to N_2O in applied fertilisers is too low, with later studies reporting 2–2.5 per cent conversion or more. Crutzen *et al.* (2008), cited in UNEP (2009a), quote 3–5 per cent conversion.

107 Sutton *et al.*, 2011, p. 417.

108 Sutton *et al.*, 2011.

109 Davidson, 2012.

110 FAO, 2006, p. 103.

111 Sutton *et al.*, 2011, pp. 417–418.

112 FAO (2006, pp. 105–106) states that nitrogen uptake by crops ranges from 70 per cent of the applied nitrogen in fertilisers (best case) to 50 per cent (typical Europe) to 30 per cent (Asian rice). On average, 14 per cent is lost to air, so subtracting this from the previous figures implies that 15–55 per cent is lost to water.

113 Sutton *et al.*, 2011, pp. 388–389 and xxix.

114 Sutton *et al.*, 2011, p. 382.

115 Sutton *et al.*, 2011, p. 385.

116 Sustainable Development Commission, 2009.

117 Sutton *et al.*, 2011, p. xxviii.

118 Secretariat of the Convention on Biological Diversity, 2010.

119 FAO, 2006, pp. 26, 34 (figures for 2001). Land area is expressed as a percentage of ice-free land.

120 FAO, 2006.
121 Lal, 2004.
122 FAO, 2009a.
123 IFPRI, 2009.
124 FAO, 2009a, p. 52.
125 World Agroforestry Centre, 2009.
126 McCarthy *et al.*, 2011.
127 UNEP, 2008, p. 266.
128 UNEP, 2011a.
129 McCarthy *et al.*, 2011.
130 Karousakis, 2009.
131 Barr *et al.*, 2010.
132 Naughton-Treves and Day, 2012.
133 Rights and Resources Initiative, 2012.
134 See, for example, a plan by the Democratic Republic of the Congo to seek international funding for converting 1 million hectares of 'degraded forest' into plantations (http://news.mongabay.com/2011/0810-hance_congo-plantations.html). See also Barr *et al.* (2010) for details of how reforestation funds in Indonesia provided a perverse incentive for commercial plantations to replace 'degraded' natural forest.
135 'McKinsey accused of driving deforestation with poor REDD+ advice', *Business Green*, 7 April 2011. Online at http://www.businessgreen.com/bg/news/2041685/mckinsey-accused-driving-deforestation-poor-redd-advice. It should be noted that McKinsey & Company claimed that Greenpeace had misunderstood its report. See also http://news.mongabay.com/2012/0117-interview_czebiniak_greenpeace_redd.html#ixzz1oWKtFyal
136 Llanos and Feather (2011, p. 18) citing a study by the UK Hadley Centre.
137 Llanos and Feather, 2011, p. 22.
138 http://www.redd-monitor.org/2012/01/31/indigenous-communities-in-peru-condemn-the-further-adventures-of-an-australian-carbon-cowboy/
139 Dooley *et al.*, 2011.
140 See, for example, the REDD-Monitor website at http://www.redd-monitor.org/redd-an-introduction/ (accessed 16 April 2012).
141 Barr *et al.*, 2010.
142 http://www.emansion.gov.lr/press.php?news_id=1679
143 http://news.mongabay.com/2012/0228-hance_corruption_redd.html
144 Vidal, 2011; Zagema, 2011. The company involved, New Forests Company, claims that the residents were illegally occupying forest reserve land and that they left voluntarily. They accept no responsibility for the manner of the eviction, which was entirely organised by the government of Uganda, and state that they offered to pay compensation, but were prevented from doing so by the government.
145 Butler, 2011.
146 http://news.mongabay.com/2011/0822-hance_mabira.html
147 FAO, 2011a.
148 'Mexico's poor seek relief from tortilla shortage', *National Geographic Daily News*, 4 June 2008. Online at http://news.nationalgeographic.com/news/2008/06/080604-mexico-food_2.html
149 Rights and Resources Initiative, 2012.
150 Zagema, 2011, pp. 18–21.
151 Nuffield Council on Bioethics, 2011.
152 Campbell and Doswald, 2009, pp. 14–15: 15–22 per cent of forest species survive in oil palm plantations.
153 Campbell and Doswald, 2009, p. 12.
154 UNEP (2009a, p. 53), citing a paper by Beer *et al.* (2007).

155 UNEP (2009a, p. 55) citing Reinhardt *et al.* (2008).

156 TEEB (2011a, p. 93), citing a paper by Fargione *et al.* (2008).

157 UNEP, 2009a, p. 54.

158 Campbell and Doswald, 2009, p. 12.

159 Butler, 2011; Bertzky *et al.*. 2011.

160 UNEP (2009a, p. 55) citing Reinhardt *et al.* (2008).

161 World Watch Institute, 2011. Online at http://www.worldwatch.org/system/files/NtP-Innovations-in-Action.pdf

162 IEA Bioenergy, 2010, p. 15; WWF *et al.*, 2011, p. 163.

163 IEA, 2010b, p. 453.

164 Campbell and Doswald, 2009, p. 20.

165 Cotula *et al.*, 2011.

166 Woolf *et al.*, 2010.

167 Directive 2009/28/EC; FAO, 2011a.

168 Nuffield Council on Bioethics, 2011. See also European Commission Press Release IP/11/901, 19 July 2011: 'First EU sustainability schemes for biofuels get the go-ahead'. Online at http://europa.eu/rapid/pressReleasesAction.do?reference=IP/11/901&format=HTML&aged=0&language=EN&guiLanguage=en

169 Campbell and Doswald, 2009.

170 Marginal land is uncultivated land of low productivity, but it can be important for biodiversity. Degraded land has previously been cultivated, but has been damaged by soil erosion or nutrient depletion. Abandoned land may or may not be degraded, or it may be in the process of natural regeneration. See UNEP, 2009a, p. 75.

171 Campbell and Doswald, 2009.

172 Bowyer, 2010.

173 IEA Bioenergy, 2010.

174 McCarthy *et al.*, 2010.

175 FAO, 2011a.

176 OECD–FAO, 2011. This exceeds the projected population growth of 1 per cent per year, however, so food production per capita is still expected to grow.

177 FAO, 2009b.

178 Collette *et al.*, 2011.

179 FAO, 2006.

180 FAO, 2009b.

181 Sutton *et al.*, 2011, p. xxviii. For comparison, biological nitrogen fixation (e.g. by leguminous cover crops) can increase yields to 4–6 tonnes per hectare.

182 UNEP, 2011a.

183 OECD, 2012: this is one of the Aichi Biodiversity Targets agreed in 2010.

184 UNEP, 2011a. Governments have traditionally subsidised plantations by as much as 75 per cent, and global subsidies for plantations between 1994 and 1998 totalled $35 billion.

185 Rainforest Foundation UK *et al.*, 2012.

186 Trivedi *et al.*, 2008.

187 UNEP, 2011a.

188 As of March 2011, 17 projects were under way and a further 32 were being planned (UNEP Finance Initiative, 2011).

189 Rainforest Foundation UK *et al.*, 2012.

190 World Bank, 2011a.

191 Website of the Voluntary REDD+ Database: http://reddplusdatabase.org/ (accessed 23 April 2012). Other REDD-like initiatives include the Congo Basin Forest Fund supported by the UK government, and the Norwegian International Climate and Forest Initiative.

192 Friends of the Earth US, 2011.

193 Friends of the Earth US, 2011.

194 Rainforest Foundation UK *et al.*, 2012.
195 Ecosystem Marketplace, 2011.
196 For more on PINC, see Cranford *et al.* (2012).
197 Chomitz, 2007, p. 24.
198 Nelson and Chomitz (2011) found that protected areas in Latin America and Asia that allow access for sustainable use experienced 2 per cent less deforestation than strictly protected areas. Cranford and Mourato (2011) provide a case study from southern Peru of a two-stage scheme starting with education of the local community about the benefits of conservation. Persha *et al.* (2011) review 84 cases and find that although local access often leads to a loss of biodiversity, involving local communities in forest rule making significantly increases the chance of a win-win outcome where both local communities and wildlife benefit.
199 Naughton-Treves *et al.*, 2011.
200 Llanos and Feather, 2011.
201 World Resources Institute: see http://www.wri.org/stories/2011/05/restoring-forests-opportunity-africa
202 Naughton-Treves and Day, 2012; UNEP Finance Initiative, 2011.
203 Committee on World Food Security, 2012. See also the article in *The Guardian* at http://www.guardian.co.uk/global-development/2012/mar/14/negotiators-consensus-global-land-governance-guidelines (accessed 23 April 2012).
204 'REDD advances – slowly – at Durban': http://news.mongabay.com/2011/1214-redd_review_durban.html#ixzz1oWtINrLh
205 Moss and Nussbaum, 2011.
206 Ecosystem Marketplace, 2011.
207 'Kenya REDD project becomes first in Africa to win gold-level validation': http://news.mongabay.com/2009/1217-kenya.html#ixzz1ocVkPxVe; see also http://wildlifeworks.com; http://www.climate-standards.org/projects/files/taita_taveta_kenya/Rukinga_CCB_PDD_Ver_2_0.pdf
208 IEA, 2011c.
209 IEA, 2010b, p. 453.
210 Campbell and Doswald (2009), citing Ravindrath *et al.* (2009).
211 IEA, 2011c.
212 WWF *et al.*, 2011.
213 FAO, 2011c; Niggli *et al.*, 2009.
214 De Schutter, 2010.
215 Badgley *et al.*, 2007. Note that Niggli *et al.* (2007) cite higher yield reductions of typically 30–40 per cent when comparing the most recent data on organic and conventional production in temperate European countries. However, organic yields vary widely (e.g. from 44 per cent to 98 per cent of conventional yields for wheat), which implies considerable scope for improving yields through more research into the best methods of organic production.
216 Niggli *et al.*, 2009.
217 Pretty *et al.*, 2006.
218 Edwards *et al.*, 2010. See also Third World Network website: 'Is ecological agriculture productive?'. Online at http://www.twnside.org.sg/title2/susagri/susagri064.htm (accessed 25 April 2012). Numerous other examples of successful sustainable agriculture initiatives can be found in TEEB (2011a) and World Watch Institute (2011).
219 De Ponti *et al.*, 2012; Seufert *et al.*, 2012.
220 Fred Bahnson on the Worldwatch blogsite at http://blogs.worldwatch.org/nourishing theplanet/ending-the-hunger-season/ (accessed 24 April 2012).
221 De Schutter, 2010.
222 Collette *et al.*, 2011.
223 De Schutter, 2010. Half of cereal production goes to feed animals.

224 Friel *et al.*, 2009. Low levels of grazing can increase soil carbon levels because when grass is cropped, the roots shrink, releasing carbon into the soil. The grass and roots then regrow, absorbing more carbon from the air.
225 Randolph *et al.* (2007), cited in Friel *et al.* (2009).
226 Gowing and Palmer, 2008.
227 UNEP, 2011a.
228 Paul, 2012.
229 FAO, 2009a.
230 FAO, 2009a.

5 Secure and safe energy: adapting to Peak Oil

1 IEA website at http://www.iea.org/subjectqueries/keyresult.asp?KEYWORD_ID= 4103
2 BP (2011) for reserves and total production; IEA (2011a, p. 122) for the amount of crude oil. 'Proved recoverable reserves' are defined as the measurable oil which it is 90 per cent certain can be economically recovered with existing technology under current economic conditions. IEA (2011a) has 83.6 million barrels per day (mbd) total production – this is slightly larger than the BP estimate because it includes coal-to-oil (see under 'Unconventional oil').
3 See the website of the Association for the Study of Peak Oil and Gas at http://www.peakoil.net/headline-news/colin-campbell-and-100-months-of-peak-oil
4 Hubbert, 1956. Hubbert's best guess was for a peak in 1965, and his upper limit was for a peak in 1970.
5 Sorrell *et al.*, 2009.
6 Sorrell *et al.*, 2009.
7 Klemme and Ulmishek, 1991.
8 Sorkhabi, 2010.
9 Sorrell *et al.*, 2009; Robelius, 2007.
10 Robelius, 2007 and Figure 5.2 in this book.
11 IEA, 2011a, pp. 119–121: in 2010, 25 billion barrels were produced, 16 billion barrels of new reserves were discovered and existing reserves grew by 13 billion barrels.
12 Salameh, 2004.
13 IEA, 2010b, p. 150.
14 IEA, 2010b, p. 144.
15 WEC, 2010.
16 USGS, 2009; WEC, 2010.
17 WEC, 2010.
18 Converting oil shale to liquid fuel generates 1.2 to 1.75 times more greenhouse gas emissions than conventional oil production (Cleveland and O'Connor, 2010).
19 IEA, 2010b, p. 173.
20 See, for example, the websites of the Association for the Study of Peak Oil and Gas (http://www.peakoil.net), The Oil Drum (http://www.theoildrum.com) and the Oil Depletion Analysis Centre (http://www.odac-info.org/about-odac). Many of those warning of Peak Oil are ex-oil-company geologists: Hubbert was chief geologist for Shell; Colin Campbell (founder of ASPO) was vice president of Fina Exploration Norway Inc.; Chris Skrebowski was a long-term planner for BP; and Dr Richard Miller of the Oil Depletion Analysis Centre was a geochemist for BP.
21 One of the most vocal critics of the Peak Oil concept is Michael Lynch, an oil industry consultant (see http://www.energyseer.com/MikeLynch.html, accessed 30 August 2011). Other forecasts that take an optimistic view include those by CERA (Cambridge Energy Research Associates) – a US consultancy – and, up until recent years (2010 onwards), the International Energy Agency (IEA).

22 Robelius, 2007; ITPOES, 2010.
23 Sorrell *et al.*, 2009.
24 IEA, 2010b, p. 48: 'New Policies Scenario'.
25 Aleklett *et al.*, 2010.
26 BP, 2011. Proved recoverable reserves were 187 trillion cubic metres in 2010 and production was 3 tcm.
27 IEA, 2011d, p. 49.
28 IEA, 2011d, p. 54, Figure 2.5.
29 Energy Watch Group, 2009.
30 IEA, 2011d.
31 ITPOES (2010) reported that the utilisation of newly built US LNG terminals was down to just 8 per cent as a result of a massive increase in shale gas production.
32 European Parliament, 2011a.
33 BP, 2011.
34 BP, 2011. Average of prices in Europe, the United States and Japan.
35 Energy Watch Group (2007, pp. 11–12), based on total resource estimates from the German Federal Institute for Geosciences and Natural Resources (BGR), which fell from 13.4 trillion tonnes in 1980 to 6 trillion tonnes in 2005. The definition of 'total resources' here includes not only proved economically recoverable reserves, but also proved reserves that cannot currently be extracted economically, and hypothetical reserves that are thought to exist in regions with suitable geology, but are not yet proved to exist. Note that recent resource estimates from BGR are far higher, having more than tripled to 20.7 trillion tonnes in 2008, due mainly to the inclusion of large unproven resources in the United States, Russia and China. See Höök *et al.* (2010) who question the validity of this large increase. Estimates of proved global recoverable reserves over the last 10 years for which data are available (1998–2008/2009) have been roughly constant at around 900–1,000 million tonnes according to the BGR, and have fallen from around 1,000 million tonnes to 820 million tonnes according to the World Energy Council.
36 BP (2011) shows that coal production in terms of energy content peaked in the United States at 603 mtoe (million tonnes of oil equivalent) in 1998. Energy Watch Group (2007) predicts a global peak at around 2025 with production about 30 per cent above 2007 levels; and a peak in the United States between 2020 and 2030 (see figure on p. 7). Mohr and Evans (2009) predict a global peak between 2010 and 2048, with a best guess of 2034 (in tonnes of coal) and 2026 (in energy content of the coal). Höök *et al.* (2010) predict a global peak between 2020 and 2050.
37 WEC, 2010: reserves are 114 billion tonnes and annual production is 2.7 billion tonnes.
38 Lin and Liu, 2010.
39 WEC, 2010, Table 2.1.
40 IEA, 2011a, p. 92.
41 OECD (2007b) rated political stability in OPEC countries at 2.3 on a scale of 1 (good) to 3 (bad), compared with an average rating of 1.4 for OECD countries. The index was based on two of the six World Bank governance indicators: 'Political stability and absence of violence', and 'Regulatory quality'.
42 IEA, 2010b, p. 2.
43 IEA, 2011a, p. 93.
44 OECD, 2007b, pp. 77–82.
45 IEA, 2009a, pp. 120–122.
46 Seltmann, 2009.
47 The Yom Kippur War was the fourth Arab–Israeli conflict and was initiated in October 1973 when Egypt and Syria tried to drive Israel out of the land it had taken in the Golan Heights and Sinai in 1967. Other countries affected by the embargo were the Netherlands, Denmark, Portugal, South Africa, Rhodesia and Japan. The oil

embargo was finally relaxed in March 1974, following US negotiations that resulted in Israel partly withdrawing from the land it occupied in Sinai. See US Department of State website at http://history.state.gov/milestones/1969-1976/OPEC (accessed 20 September 2011).

48 The recession in the United States, UK and elsewhere cost OECD countries $350 billion in the money of the day, $1.1 trillion in 2003 prices, equivalent to 7 per cent of GDP (Hirsch *et al.*, 2005, p. 29). However, other factors also contributed to the recession, including the collapse of the Bretton Woods Agreement, the stock market crash that began in early 1973, the high cost of the Vietnam War, and the 1973 miners' strike in the UK.

49 Hirsch *et al.*, 2005, p. 29.

50 Reuters, 2008. The Strait of Hormuz carries around 40 per cent of the world's globally traded oil, 90 per cent of all oil exported from the Gulf countries (Saudi Arabia, Kuwait, Oman, Bahrain, Qatar and the United Arab Emirates) and a large amount of liquefied natural gas from Qatar, the world's biggest exporter.

51 IEA, 2009a, p. 118.

52 Hirsch *et al.*, 2005, Figure IV-1.

53 WEC, 2010.

54 Hall *et al.* (2009, p. 35) for US figures; Gagnon *et al.* (2009) for global figures. Gagnon *et al.* found a global EROI of 26:1 in 1992; they speculate that the reason for the increase to 35:1 in 1999 followed by the decrease to 18:1 in 2006 is linked to a decline in the number of exploration wells drilled in the late 1990s, and a steep increase in the 2000s.

55 Murphy and Hall, 2011, p. 68, Figure 16; Cleveland, 2005, p. 13. Note that calculating EROI is not easy, as little detailed information is available for the energy consumption of the production process (which should include both direct energy, e.g. fuel to power the drill; and indirect energy, e.g. energy needed to make drilling equipment and to make steel and concrete for well casings). Instead, researchers use the average energy intensity of that sector of the economy, expressed as energy consumed per unit of expenditure (GJ/$), and multiply this by the expenditure in a given year (correcting for inflation). However, this is a highly imperfect measure of actual energy use.

56 See 'An update on the energy return on Canadian natural gas', online at http://www. theoildrum.com/node/4376.

57 Cleveland, 2005, p. 12, Figure 6.

58 Tar sands estimate from Murphy and Hall (2010, p.109; EROI from various studies ranges from 2:1 to 4:1); oil shale estimate from Cleveland and O'Connor (2010; this includes internal energy); coal-to-liquids estimate from Cleveland (2005, p.12, Figure 6).

59 Aucott, 2011. This is an unpublished estimate. It assumes a 25-year production profile from each shale gas well, which may be an overestimate. Berman and Pittinger (2011) suggest that typical wells produce between 1 billion and 3 billion cubic feet rather than between 2.6 billion and 5 billion cubic feet, as assumed by Aucott. The EROI would be lower if this were the case – perhaps in the range of 35–50:1.

60 Robelius, 2007.

61 Robelius, 2007. Tar sand reserves at the time were estimated as 140 billion barrels (this has since increased to 170 billion barrels).

62 Nexen has installed a combined upgrading and gasification plant at Long Lake. See http://www.nexeninc.com/en/Operations/OilSands/LongLake/Technology.aspx (accessed 26 October 2011).

63 IEA, 2011a, p. 70.

64 IEA, 2011a, p. 103. Note that demand in 2010 (87 mbd) is higher than production (84 mbd according to the IEA; 82 mbd according to BP, 2011) because the difference is supplied by 'refinery processing gains': when the long chain molecules in heavy crude oil are 'cracked' into shorter molecules, the volume of the liquid increases.

65 ITPOES, 2010.
66 Sorrell *et al.*, 2009.
67 IEA, 2009a, p. 127.
68 ITPOES, 2010. See Aleklett *et al.* (2010) for a similar argument.
69 Salameh, 2010.
70 IEA, 2011b. Data for 2009.
71 See the IEA's baseline scenario (IEA, 2010a).
72 Ayres and Ayres, 2010, p. 5.
73 UNEP, 2011a, Chapter 7.
74 Cullen *et al.*, 2011.
75 The German standard for highly energy-efficient buildings. See Passivhaus Institut, Das Institut für Forschung und Entwicklung hocheffizienter Energieanwendung (http:// www.passiv.de).
76 Allwood, 2011.
77 Li, 2011.
78 'EU threatens mandatory energy efficiency targets': http://www.businessgreen.com/bg/news/2025551/eu-threatens-mandatory-energy-efficiency-targets
79 Adua (2010) found that behaviour was far more important than technical efficiency measures in reducing home energy consumption.
80 Jacobson and Delucchi (2009) estimate the accessible wind resource as 40–85 terawatts (TW) and the solar resource as 580 TW (excluding low wind areas, deep sea, high mountains and protected regions), compared to a current human demand of 12 TW. WEC (2010) points out that even if only 0.1 per cent of incoming solar energy could be converted at an efficiency of 10 per cent, we could supply four times the current world generating capacity of 3,000 GW. The total annual solar radiation falling on the earth is more than 7,500 times the world's total annual primary energy consumption of 450 exajoules (1 EJ = 10^{18} joules).
81 REN21, 2011, p. 17. See 'A note on units' in this book for a definition of final energy.
82 REN21, 2011, p. 76.
83 WWF *et al.*, 2011.
84 IPCC, 2011.
85 Jacobson and Delucchi, 2009, 2011.
86 Howarth *et al.*, 2011. As well as assuming that methane leakage and venting from natural gas production and distribution are higher than previous estimates, Howarth *et al.* use an updated figure for the global warming potential of methane: 33 times that of carbon dioxide, as opposed to the previous factor of 25 that was quoted in the 2007 IPCC guidelines and which is still used by most analysts. They also consider that in view of the urgent nature of the climate-change threat, it would be more appropriate to use 20-year global warming potentials instead of the commonly used 100-year potentials. As methane decays more quickly than carbon dioxide, its 20-year potential is considerably higher than its 100-year potential (105 compared to 33), so this gives a higher estimate for the life-cycle greenhouse gas emissions of natural gas, putting it on a par with emissions from coal and oil.
87 'France, Germany, Italy, Japan, UK and USA rated "high risk" for short-term energy security. Canada most secure – Maplecroft report', Maplecroft Press Release, 2 June 2011. Online at http://maplecroft.com/about/news/energy_security_2011.html (accessed 16 January 2012).
88 OECD, 2007b.
89 In the UK, the 'dash for gas' resulted partly from a move away from oil for power generation following the oil price shocks of the 1970s, but partly also from a desire to reduce dependence on coal in the power sector, following a serious supply disruption caused by the miners' strike of 1973. Switching from coal to gas was seen as a way of increasing the diversity of supply in the power sector, even though coal could be

viewed as a more secure fuel because its reserves-to-production ratio is higher than that of gas.

90 IEA, 2011a, p. 93.
91 IEA, 2011a, p. 93.
92 Energy Watch Group, 2009.
93 IEA, 2011b.
94 OECD, 2007b.
95 IEA, 2011a. The 450ppm Scenario corresponds to cumulative carbon emissions of 1.4 trillion tonnes between 2000 and 2050, giving a 50 per cent chance of keeping the average global temperature rise below 2 degrees centigrade.
96 See IEA, 2011a, Table 2.1.
97 OECD/IEA, 2010. Under the baseline scenario, carbon emissions from energy double by 2050.
98 Energy Saving Trust website at http://www.energysavingtrust.org.uk/Info/Our-calculations.
99 IEA, 2011a, p. 62.
100 OECD/FAO, 2011, p. 26.
101 Simms *et al.*, 2010.
102 ITPOES (2010), citing a report by Ofgem.
103 IEA, 2010b. Real 2008 dollars.
104 The 450ppm and BLUE Map Scenarios require a high carbon price that would effectively bring oil prices back up to a similar level to the baseline price. However, the carbon price would only be around half of the average amount currently added to the price of a barrel of oil in excise taxes by OECD governments.
105 Simms *et al.*, 2010, p. 71.
106 ITPOES, 2010.
107 OECD, 2007b.
108 OECD, 2008a.
109 IEA, 2009a, p. 124.
110 Li, 2011.
111 IMF, 2011, p. 90.
112 Reynolds, 1999.
113 Sorrell *et al.* (2009, p. 164), citing a paper by Cleveland (1991).
114 OECD, 2007b.
115 IMF, 2011.
116 IEA, 2011a, p. 95.
117 IPCC, 2011, Table 9.4. Prices in 1999–2001 dollars; poorest countries defined as those with *per capita* annual income of less than $300.
118 IEA (2011a, p. 227), with detail on gas and coal from the earlier 'Outlook' in IEA (2009a).
119 IEA, 2011a, p. 227.
120 Data for the United States from the website of the Mine Safety and Health Adminstration of the US Department of Labor, online at http://www.msha.gov/stats/centurystats/coalstats.asp. Tu Jianjin (2007) provides data for China, based on various editions of the *China Coal Industry Yearbook*, and explains that systems introduced to reduce accidents provide an incentive for under-reporting.
121 Tu Jianjin, 2007; US Department of Labor, 2009. The reduction in the Chinese death rate per tonne of coal is partly due to mechanisation improving the productivity rate.
122 US Department of Labor, 2010.
123 Robelius, 2007.
124 'Security raised in Malacca Strait after terror warning', Reuters, 2 March 2010. Online at http://www.reuters.com/article/2010/03/04/us-malacca-threat-idUSTRE623351 20100304 (accessed 16 January 2012).

125 San Sebastián and Hurtig, 2004.
126 ITPOES (2010) reports that production declines by 50–65 per cent in the first year after drilling, and the economic life of each well is no more than 10 years and often as little as 5 years. European Parliament (2011a) reports that wells can decline by 85 per cent in the first year.
127 European Parliament, 2011a.
128 Entrekin *et al.* (2011), for wells in parts of the Marcellus Shale in the United States.
129 European Parliament, 2011a; Entrekin *et al.*, 2011.
130 Osborn *et al.*, 2011.
131 See, for example, YouTube videos: 'Do not drink this water!', online at http://www.youtube.com/watch?v=4ApZkNsXfJE&NR=1 'Tap water on fire!', online at http://www.youtube.com/watch?v=qYJj-1jNOxE&NR=1 'Holy crap, the water's on fire!', online at http://www.youtube.com/watch?v=VEQMA0zwMM4 The 'Gasland' documentary trailer at http://www.youtube.com/watch?v=dZe1AeH0Qz8
132 European Parliament, 2011a.
133 Howarth *et al.*, 2011; European Parliament, 2011a.
134 Jiang *et al.* (2011) claim that shale gas emissions are only 11 per cent higher than conventional gas emissions on a well-to-burner basis and 3 per cent higher in terms of emissions per unit of electricity produced, but their analysis excludes methane leaks from fracking. IEA (2011d) claims that emissions from shale gas are between 3 per cent and 12 per cent higher than for conventional gas when the combustion stage is included. Neither of these estimates includes the impact of methane emissions escaping from the shale gas formation through overlying rocks.
135 IEA, 2012.
136 IEA, 2011a, p. 473. Figures are for 2009: 587 million from Sub-Saharan Africa; 290 million from India; 364 million from the rest of South Asia; 31 million from Latin America; and 21 million in the Middle East.
137 IEA, 2011a.
138 IPCC, 2011.
139 Slaski and Thurber, 2009.
140 'Eight19 looks to expand "pay as you go" solar across Africa', *Business Green*, 17 January 2012. Online at http://www.businessgreen.com/bg/news/2137460/eight19-looks-expand-pay-solar-africa
141 UNEP, 2008. The future of Grameen Shakti is currently (2012) in doubt as the government of Bangladesh forced the removal of its founder and director, Dr Mohammed Yunus in March 2011 and appointed a new chairman. There were rumours that Dr Yunus had angered the government with political criticism, and that the government might be trying to take over the bank. The move is being contested by Dr Yunus and the bank.
142 See http://www.ashdenawards.org/winners/toyola11 and http://www.ashden.org/files/Toyola%20winner.pdf
143 See http://www.ashdenawards.org/winners/CRELUZ10
144 IEA, 2011d, p. 271; UNDP, 2011a.
145 Meinshausen *et al.*, 2009; Allen *et al.*, 2009.
146 Farrell and Brandt, 2005. Brandt (2011) estimates emissions are 23 per cent higher.
147 IEA, 2012; Howarth *et al.*, 2011.
148 IEA, 2011d, 2012.
149 WEC (2010) lists 6.3 million tonnes of 'Identified Resources' of uranium which are recoverable at prices below $260 per kg U. Of these, 4 million tonnes are 'Reasonably Assured Resources' (roughly equivalent to 'proved resources' in oil terms) which could supply current reactor requirement for 62 years, and the remaining 2.3 million tonnes are 'Inferred Resources'. There are a further 10.4 million tonnes of undiscovered and speculative resources. Note that 'reactor requirements' exceed current ore production due to the contribution from secondary sources.

150 The countries that have exhausted their economically recoverable reserves are Germany, France, the Democratic Republic of the Congo, Gabon, Bulgaria, Tajikistan, Hungary, Romania, Spain, Portugal and Argentina (Energy Watch Group, 2006). However, high uranium prices have prompted renewed exploration activity in some of these countries recently (2012).

151 WEC (2010, p. 209), for reserves that can be produced at under $260 per kilogram.

152 Energy Watch Group (2006, pp. 30–31), citing Storm van Leeuwen *et al.* (2005). The figure of 0.02 per cent is for hard ores and 0.01 per cent is for soft ores. The figures have been disputed by a 2006 study by the University of Sydney (discussed in the same reference) which finds that for ore grades of 0.015 per cent (the average for Australia), the input energy of the nuclear fuel cycle is between 16 per cent and 40 per cent of the output electrical energy (assuming that both reactors and uranium-processing facilities are state-of-the-art).

153 WEC, 2010, p. 212.

154 World Nuclear Association website at http://www.world-nuclear.org/info/reactors.html

155 WEC, 2010, p. 212.

156 The website of the NGO WISE (http://wise-uranium.org/uisl.html) lists examples including Stráz pod Ralskem in the Czech Republic, where 235 million cubic metres of groundwater over an area of 28 square kilometres has been contaminated, threatening local water supplies (Andel and Pribán, 1996, pp. 113–135); the Cheshmata site in Bulgaria, where the measured groundwater pH is 2.2; and the Devladovo site in Ukraine where groundwater contamination from spills of sulphuric and nitric acid was spreading downstream at a speed of 53 metres per year.

157 OECD, 2007b, p. 41.

158 The treaty limits the possession of nuclear weapons to the five agreed countries (Russia, the United States, China, France and the UK) in return for a promise of assistance to states that want to develop a civil nuclear power programme, and also promises to progressively reduce the stockpile of nuclear weapons held by the five states (although progress on this front has been slow).

159 World Nuclear Association website at http://www.world-nuclear.org/info/inf75.html and http://www.world-nuclear.org/info/inf102.html (accessed 10 October 2011).

160 World Nuclear Association website at http://www.world-nuclear.org/info/cooling_power_plants_inf121.html (accessed 13 October 2011).

161 World Nuclear Association website at http://www.world-nuclear.org/info/cooling_power_plants_inf121.html (accessed 13 October 2011). Once-through cooling systems use around 90 m^3 per second for a 1,600 MW power plant, compared to 2 m^3 per second for a recirculating system. However, most of this water is returned unused – there are evaporation losses of only 0.4 litres per kWh generated, compared to losses of 1.8 litres per kWh for a recirculating system. Dry cooling systems are also possible, using air driven by fans, though these are unlikely to be adopted because they are expensive, do not work in hot weather and do not offer adequate emergency cooling in the event of a reactor shutdown.

162 'Olkiluoto EPR on schedule for 2012 completion', *World Nuclear News*, 30 November 2010. Online at http://www.world-nuclear-news.org/newsarticle.aspx?id=28896&terms=olkiluoto (accessed 13 October 2011).

163 See, for example, Sustainable Development Commission (2006).

164 World Nuclear Association website at http://www.world-nuclear.org/info/inf77.html (accessed 13 October 2011). The three reactors selected for prototypes are gas-cooled, sodium-cooled and lead-cooled fast reactors.

165 Wind speeds between 2.5 and 25 metres per second are best. Many modern turbines will not operate at speeds below 4 metres per second. The power delivered is proportional to the cube of the wind speed, so at high wind speeds there is a risk of overload.

354 *Notes*

166 IEA, 2008.
167 See IEA (2008) for a good overview.
168 IEA, 2007a
169 RAE, 2012.
170 Inage (2009, pp. 9, 16) suggests that wind power from a large region ramps down at about 3 per cent per minute, which can easily be matched with hydropower (ramps at 100 per cent per minute), pumped storage (30 per cent per minute), gas turbines (8 per cent per minute) or even coal power (4 per cent per minute).
171 Inage, 2009.
172 See Inage (2009) for an overview of storage technologies.
173 Holttinen *et al.*, 2009, p. 175.
174 IEA, 2011e. PowerCentsDC was a smart meter pilot programme in Washington, DC for 850 residential customers that ran from July 2008 to October 2009. Customer response to three different residential pricing options contributed to reducing peak demand by 4 per cent to 34 per cent in the summer and 2 per cent to 13 per cent in the winter.
175 See http://www.businessgreen.com/bg/news/2111599/evs-smarter-clean-energy-grid for a description of a trial project in Denmark to charge electric vehicles using surplus wind power.
176 IPCC, 2011, Chapter 8, p. 29.
177 Holttinen *et al.*, 2009, p. 180.
178 IEA, 2011e.
179 The study (IEA, 2011f) takes account of available balancing capacity (gas turbines, pumped storage), grid interconnections, demand management and storage capacity. It does not fully account for market factors (e.g. gate closure times), grid strength or the potential for variability to be reduced with a wide diversity of renewable sources.
180 Committee on Climate Change, 2011a, pp. 59–60; Gross *et al.*, 2006.
181 IPCC, 2011, Chapter 7, p. 568. Costs are in 2005 US dollars. Comprises 0.14–0.56 c$/kWh for short-term balancing (from Holttinen *et al.*, 2009, p. 171); 0.58–0.96 c$/kWh for long-term back-up (from Gross *et al.*, 2006) and 0–1.5 c$/kWh for adding transmission. Gross *et al.* (2006) found short- and long-term balancing costs of 0.5–0.8 p/kWh (compared to base costs of 3–5.5 p/kWh) for UK wind power.
182 IEA, 2011a, p. 191.
183 IPCC, 2011, Chapter 8, p. 27.
184 'Solar power generation world record set in Germany': http://www.guardian.co.uk/environment/2012/may/28/solar-power-world-record-germany
185 Excessive rainfall can also cause hydroelectric schemes to close down – this has been experienced in Guatemala in recent years.
186 IPCC, 2011, Chapter 9, p. 50.
187 Powers *et al.*, 2010.
188 FAO, 2011d, p. 89.
189 IEA, 2007b, p. 58.
190 Murphy and Allen, 2011.
191 Not all of these dams were for hydropower. See World Commission on Dams (2000). Online at http://www.unep.org/dams/WCD/
192 'After 30 years, secrets, lessons of China's worst dams burst accident surface', *People's Daily*, 1 October 2005. Online at http://english.people.com.cn/200510/01/eng2005 1001_211892.html (accessed September 2011).
193 IPCC, 2011.
194 IPCC (2011), citing a study by Edenhofer *et al.* (2010).
195 IEA, 2011a. The remaining 7 per cent is electricity and biofuels.
196 IEA, 2011f (balancing renewables), 2011e (smart grids); IPCC, 2011, Chapter 8.
197 According to UNEP (2011a, p. 272), the monopoly of energy companies on the supply

of electricity to the grid is a major barrier to the uptake of combined heat and power (CHP).

198 IPCC, 2011.

6 Less waste: a resource-efficient economy

1 The Hertwich and Peters (2009) 'Mobility' category has been split into direct use of transport fuel (shown here as 'personal transport fuel') and emissions from vehicle manufacture (added to the 'manufactured products' category). Their 'Shelter' category has been split into emissions related to fuel and electricity use in the home (shown as 'home energy'); and emissions from production of minerals, timber, iron and steel and other metals (added to the 'construction' category, even though some of these emissions, in fact, will relate to products other than building materials). Hertwich and Peters exclude land-use emissions, so this has been added using the EDGAR database for 2001 (category: CO_2, CH_4 and N_2O from biomass burning, which roughly equates to forest clearance for food production).

2 Hammond and Jones, 2011. Exact figures are 1.77 tCO_2/t steel for a 'typical world-wide' recycled content of 42 per cent, and 2.75 tCO_2/t steel for primary (virgin) steel production. For aluminium, 8.24 tCO_2/t Al for typical recycled content of 33 per cent; 11.5 tCO_2/t Al for primary Al.

3 IEA, 2011b. Figures are for 2009. 'Direct' use of coal and gas excludes that burnt in power stations to generate electricity.

4 See Figure 4.1 in this book.

5 Fischer-Kowalski, 2009.

6 Giljum *et al.*, 2009, p. 27.

7 Fischer-Kowalski, 2009.

8 UNEP, 2011c.

9 Allwood *et al.*, 2010. Steel accounts for 25 per cent of industrial carbon emissions; concrete 19 per cent, plastic 5 per cent, paper 4 per cent and aluminium 3 per cent.

10 Peters *et al.*, 2011.

11 Davis and Caldeira, 2010.

12 Brinkley and Less, 2010.

13 Peters *et al.* (2011), using data in sheets 5 and 6 of the accompanying spreadsheet on the *Proceedings of the National Academy of Sciences* (PNAS) website. Note that Brinkley and Less (2010) imply an even bigger difference, but their analysis is less accurate as it uses the carbon intensity of each national economy rather than looking at individual sub-sectors using input–output analysis.

14 IPCC (2007) estimates that a 50–85 per cent cut across all sectors is needed for a reasonable chance of limiting temperature increases to 2°C. Table 1 in Allwood *et al.* (2010) shows that if demand doubles by 2050, industrial emissions per unit output must be reduced – by 75 per cent for a target of a 50 per cent cut in total emissions or by 90 per cent for an 80 per cent target – in order to avoid unrealistic targets (i.e. almost a 100 per cent cut in emissions per unit output!) for other sectors.

15 Allwood *et al.*, 2010.

16 IPCC (2001) estimated potential carbon dioxide savings of 2.2 gigatonnes (Gt) from material efficiency measures compared to 2.6–3.3 Gt from energy efficiency.

17 WRAP, 2009. See point 12 in Executive Summary. The 28 per cent saving is for the 'Beyond Best Practice' scenario, and is equivalent to 8 per cent of total UK consumer emissions.

18 Kram *et al.*, 2001. Total emissions are 5 billion tonnes. The study is based on the predicted response to high carbon prices of €100–€200 per tonne.

19 Prognos, 2008. The savings of 244 $MtCO_2$e are additional to savings of 204 $MtCO_2$e already achieved in 2004 by existing levels of recycling and energy recovery. The

scenario assumes that recycling targets are increased to 60 per cent for municipal waste (currently 38 per cent) and 80 per cent for construction and demolition waste; that high-calorie waste is banned from landfill sites; and that market mechanisms are used to stimulate recycling and energy recovery. The potential for waste reduction is excluded.

20 Extraction of metal ores accounts for 7 per cent of global energy use (UNEP, 2011d).

21 Von Weizsäcker *et al.*, 2010, p. 46; steel from IPCC, 2007, Chapter 7, p. 459; glass from Enviros, 2003.

22 Friends of the Earth Europe, 2009. Municipal waste is that which is collected by local councils from households and businesses. A later study (ETC/SCP, 2011a, p. 52, Table 7.1) puts the savings at around 44 $MtCO_2e$.

23 ETC/SCP, 2011a, p. 25.

24 In the 18 EU countries that report waste electrical and electronic equipment recycling figures, 23 per cent is collected and 79 per cent of this is recycled (EEA, 2010b, p. 38), but little is collected elsewhere, bringing the global rate down to 15 per cent.

25 UNEP, 2011d. In a study on global recycling rates for 60 metals by the UNEP Resource Panel, it is estimated that only 18 have recycling rates of over 50 per cent, and 34 rare metals have recycling rates of less than 1 per cent.

26 OECD, 2010a.

27 See IEEP *et al.* (2010, pp. 125, 137), which note the importance of separate collection in achieving high-quality recyclates with a high market value, and preventing 'dumping' of low-quality waste outside the EU.

28 UK government guidance states that:

> Market research has shown that the value of any recyclate is significantly reduced if it is contaminated, even if the contaminant is another recyclate. It follows therefore that source separation of materials is the first step to maximising the value of recycling.
>
> *(DEFRA, 2005, p. 8)*

29 Dougherty Group LLC, 2006, p. 43.

30 Allwood *et al.*, 2010.

31 European Parliament, 2009, p. 94 and http://www.xerox.com/about-xerox/environment/recycling/enus.html

32 NISP, 2011.

33 Lombardi and Laybourn, 2007.

34 See European Commission (2011a) and the website of the Zero Waste Alliance (http://www.zerowaste.org/index.htm) for more information.

35 Envirowise, 2003.

36 'Reduced environmental impact from renewable tyre fillers', Chemistry Innovation website. Online at http://www.chemistryinnovation.co.uk/roadmap/sustainable/roadmap.asp?id=183

37 WSP Environmental Ltd, 2011.

38 European Commission, 2011a, p. 69, Table 5.2.

39 WRAP, 2009, p. 34.

40 Foster, 2011a.

41 WRAP, 2009, p. 34.

42 Meadows *et al.*, 1972, 2004.

43 Hertwich *et al.*, 2010, pp. 30–31.

44 Reynolds, 1999.

45 Cordell *et al.*, 2009.

46 OECD, 2008b, p. 42.

47 Seventy per cent of water is used for agriculture, with 10 per cent each for the domestic sector, industry and the energy sector (including power station cooling and fossil fuel extraction). However, the share used by industry is expected to grow to 20 per cent by

2030 (UNEP, 2011a, Chapter 7, p. 250). Water supply is energy-intensive – pumping, treatment and desalination require 5 per cent of the world's primary energy use.

48 Millennium Ecosystem Assessment, 2005, p. 1.

49 'Global human appropriation of net primary production': http://www.eoearth.org/article/Global_human_appropriation_of_net_primary_production_(HANPP), citing a study by Haberl *et al.* (2007).

50 Brown (2006, Chapter 1), citing an estimate by Paul MacCready.

51 Global Footprint Network at http://www.footprintnetwork.org/en/index.php/GFN/page/earth_overshoot_day/

52 Rare earth elements are the 15 lanthanides plus scandium and yttrium.

53 European Commission, 2011b.

54 European Commission (2008b, p. 20) describes the complaint about China's export restrictions (tariffs or quotas) on yellow phosphorous, bauxite, coke, fluorspar, magnesium, manganese, silicon metal, silicon carbide and zinc. See also Bradbury (2011) for details of restrictions on rare earth exports.

55 European Commission, 2008b, p. 10.

56 European Commission (2008b, p. 5): 'For example, just three producing companies now control about 75% of the seaborne trade in iron ore.'

57 OECD, 2008b.

58 Graedel, 2008.

59 Morley and Eatherley, 2008.

60 Armin Reller, Augsburg University, Germany, quoted in Cohen (2007).

61 Turner *et al.*, 2007.

62 European Commission, 2008b, 2010a, 2011b.

63 European Commission, 2008b

64 European Commission, 2008b.

65 AEA Technology, 2010. For aggregates, the risks are due to planning issues and transport costs.

66 Japan Oil, Gas and Metals National Corporation maintains stockpiles of chromium, nickel, manganese, cobalt, tungsten, vanadium and molybdenum equivalent to 42 days of standard consumption. See http://www.jogmec.go.jp/english/activities/stockpiling_metal/raremetals.html

67 European Commission, 2008b, p. 15.

68 See http://www.env.go.jp/recycle/3r/en/approach.html

69 WRAP, 2010.

70 UNEP, 2009b. Based on data from the Uppsala Conflict Data Program, for the period from 1946 to 2007.

71 Cohen, 2007.

72 UNEP, 2009b.

73 Millennium Ecosystem Assessment, 2005, p. 1.

74 Deininger and Byerlee, 2011.

75 International Aluminium Institute, 2007.

76 The Associated Press, 2010.

77 In 2007, with copper prices at a record-breaking $3.24 per lb, there were 40 deaths in Chilean copper mines; and in 2008, with prices still high, there were 43, compared to an average for the decade of 34. The safest year in the history of Chilean mining was 1999, when the average copper price fell to just 72 cents, its lowest level in over 10 years, a consequence of the Asian crisis. See Long (2010).

78 Giljum *et al.*, 2009.

79 'Peru declares state of emergency after two killed in mining protests', *The Guardian*, 29 May 2012 at http://www.guardian.co.uk/world/2012/may/29/peru-emergency-mining-protests

80 Figures are for 2008 and are taken from the International Labour Organization

Laborsta database, online at http://laborsta.ilo.org/default.html (accessed 16 November 2011), and from ILO, 2011a.

81 Survival International, 2011. See also Bates and Dale (2008, p. 121).

82 See Bates and Dale (2008) for a comprehensive survey of environmental impacts associated with resources imported to the EU.

83 Bates and Dale, 2008, pp. 5, 121. Brazil is the world's largest producer of iron ore and supplies 45 per cent of the EU-27 iron ore imports. Mining in the Amazon region has led indirectly to deforestation as new roads and railways have allowed access to loggers and farmers. The Guajá, Brazil's last hunter-gatherer nomadic people, have suffered evictions and habitat loss.

84 European Commission, 2011a.

85 EEA, 2010b. In the EU-27 in 2006, 16 tonnes of material *per capita* were used and 6 tonnes of waste *per capita* were generated (less than half of which was recycled).

86 Laybourn and Morrissey, 2009, p. 29.

87 WWF, 2006.

88 Murray, 2002, pp. 84–85.

89 See UNEP, 2011a, Chapter 8, p. 302.

90 EEA, 2010b, pp. 28–30. Of the 19 per cent of inspected waste shipments that were found to be in violation of EU law in 2008 and 2007, almost half were destined for Africa and Asia.

91 European Commission, 2011d, p. 16.

92 For example, the amount of hazardous waste generated in Europe increased by 15 per cent between 1997 and 2006 (IEEP *et al.*, 2010). It now accounts for 3 per cent of total waste generated.

93 See Millennium Ecosystem Assessment (2005, p. 117) for the percentage of untreated industrial waste discharges. See 'Trafigura found guilty of exporting toxic waste', BBC News 23 July 2010, online at http://www.bbc.co.uk/news/world-africa-10735255 for the Trafigura case.

94 EEA (2010b) estimates a 34 per cent reduction since 1990. European Commission (2010d) estimates a 39 per cent reduction in the EU-15 between 1990 and 2006 with a further 6 per cent reduction predicted for the period from 2006 to 2010. If the emission savings from recycling are also included, savings from waste management are estimated as 57 per cent between 1995 and 2008 (for municipal waste).

95 UNEP, 2011a.

96 Green Alliance, 2009.

97 See http://www.greenbiz.com/news/2011/02/08/european-parliament-backs-tough-e-waste-proposals

98 Cooper, 2004.

99 WRAP, 2011c.

100 WRAP, 2010.

101 UNEP, 2011a.

102 See http://www.tristramstuart.co.uk/FoodWasteFacts.html for this and other eye-opening food waste facts.

103 Benzie *et al.*, 2011.

104 WRAP, 2009.

105 Lee *et al.*, 2011. The slight discrepancy in the figures is due to rounding up.

106 European Commission, 2011d, p. 6, footnote 6.

107 UNEP, 2011a, p. 323.

108 WRAP, 2011d, p. 43.

109 Lee *et al.*, 2011.

110 Lee *et al.*, 2007.

111 EEA, 2010b.

112 Friends of the Earth Europe, 2009. The analysis does not include the costs of recycling

and does not include the market value of rarer metals such as those which could be recovered from waste electrical and electronic equipment.

113 http://www.nisp.org.uk (accessed 7 December 2011).
114 European Commission, 2011d, p. 6.
115 Laybourn and Morrisey (2009), plus NISP website (http://www.nisp.org.uk).
116 UNEP, 2010b.
117 For example, used catalytic converters from vehicles typically contain 2–5 grams of platinum group metals, giving a concentration of over 1,000ppm – more than ten times the concentration found in natural ores. See Buchert *et al.* (2009, p. 15).
118 AEA Technology, 2010.
119 EEA, 2010b.
120 IEEP *et al.*, 2010.
121 European Commission, 2008b, 2011b.
122 Herrndorf *et al.*, 2007.
123 European Commission, 2008b.
124 Foster, 2011b.
125 Murray, 2002, p. 106.
126 DEFRA and BIS, 2010.
127 Stahel, 2000.
128 See Safechem company website at http://www.dow.com/safechem/solutions_na/ complease.htm (accessed 15 November 2011).
129 Kingston, 2010.
130 Ericsson, 2011.
131 Wright, 2011.
132 Buchert *et al.*, 2009; IPCC, 2011, p. 29, Table 9.6. Many of these substances have no known substitutes; but in some cases, alternatives are being developed, such as nano-composites in place of rare earths in permanent magnets.
133 REN21, 2011, p. 42.
134 Zuser and Rechberger, 2011.
135 Giljum *et al.*, 2009.
136 EEA, 2010b; IEEP *et al.*, 2010; Hansen *et al.*, 2010; Reid and Miedzinski, 2008.
137 See Table 6.7 in this book.
138 EEA, 2010b, p. 22.
139 EEA, 2010b, p. 26.
140 Tukker *et al.*, 2006.
141 EEA, 2010c.
142 Best *et al.*, 2008.
143 Reisinger *et al.* (2009) note that the quality of trade statistics has been decreasing. For recommendations on methodology improvements, see OECD (2008b).
144 Von Weizsäcker *et al.*, 1998.
145 Von Weizsäcker *et al.*, 2010.
146 See EEA (2011d) for an overview and UNEP (2011c) for more detail on Germany.
147 European Commission, 2011d, p. 20, footnote 26.
148 Schmidt-Bleek, 2007; Ekins *et al.*, 2009.
149 World Resources Forum, 2009.
150 The Royal Society, 2012.
151 Fischer-Kowalski *et al.*, 2010.
152 Reisinger *et al.*, 2009; OECD, 2010b, p. 34.
153 OECD, 2008c.
154 Lee *et al.*, 2011.
155 IEEP *et al.*, 2010, Table 15.
156 US EPA, 2011.
157 IEEP *et al.*, 2010, p. 124.

158 European Parliament, 2009, p. 55.
159 Ekins *et al.*, 2009.
160 See, for example, OECD (2009a, p. 7) and Reid and Miedzinski (2008, p. 75).
161 European Parliament, 2011b.
162 Green Alliance, 2009. The proportion incinerated also increased.
163 See Murray (2002, pp. 110–128) for a powerful critique of the way in which UK waste strategy in the 1990s prioritised incinerators over recycling, with local authorities becoming tied into long-term contracts which divert recyclable waste to incinerators and resulting in one of the lowest recycling rates in Europe.
164 European Parliament, 2009. The WEEE Directive is 2002/96/EC. The Restriction on the Use of Hazardous Substances (RoHS) Directive (2002/95/EC) bans the use of certain heavy metals and flame retardants in electrical equipment.
165 IEEP *et al.*, 2010, pp. 153–154. There are also voluntary schemes in Canada, Australia and New Zealand. The EPR directives are 2006/66/EC (batteries), 2000/53/EC (end-of-life vehicles) and 2004/12/EC (packaging).
166 An estimated 1.2 billion mobile phones were sold in 2008; between 50 per cent and 80 per cent of these were replacement handsets. That equals between 51,000 and 82,000 tonnes of replacement chargers every year (GSMA, 2009).
167 European Parliament, 2009. See European Commission (2011a) for more details of how the Ecodesign Directive (2005/32/EC) could be extended.
168 A number of cities and countries have bans or restrictions on the use of plastic bags. See Cernansky (2010).
169 OECD, 2009a, p. 6; Reid and Miedzinski, 2008.
170 Reid and Miedzinski (2008) identify a lack of relevant education programmes as a key constraint on eco-innovation in Europe. House of Lords Science and Technology Committee (2008) notes that there is inadequate incorporation of sustainable design principles in educational curricula and professional training for designers and engineers in the UK.
171 However, beware of 'information overload' on product labels. Studies show that simple 'brand' type labels work best, such as the FAIRTRADE Mark or the Marine Stewardship Council (MSC) label for sustainable seafood.
172 World Bank, 2006, Appendix 3. Countries with negative wealth include Algeria, Bolivia, Burundi, Cameroon, Colombia, the Democratic Republic of the Congo, Ecuador, Ethiopia, Gabon, Ivory Coast, Malawi, Nigeria, Saudi Arabia, Sierra Leone, Sudan and Venezuela, amongst others. The report concludes that:

> It is striking that natural capital constitutes a quarter of total wealth in low-income countries, greater than the share of produced capital . . . Per capita, most low-income countries have experienced declines in both total and natural capital . . . Growth . . . will be illusory if it is based on mining soils and depleting fisheries and forest.
>
> *(World Bank, 2006, Appendix 3, p. vii)*

173 Also known as the 'commodity curse'. Exploitation of mineral commodities has a positive short-term effect on growth, but a negative long-term effect in African countries which have weak governance (Collier and Goderis, 2007).
174 Giljum *et al.*, 2009.
175 Giljum *et al.*, 2009, pp. 17–18.
176 UNEP (2010a) citing Roubini Global Economics (2009).
177 See Bates and Dale (2008) for a comprehensive list of potential policies to reduce environmental impacts from goods imported to the EU.

7 A stronger economy: long-term stability and prosperity

1 OECD, 2011a; UNEP, 2011a.
2 Strand and Toman, 2010.

3 ILO, 2011b.
4 Assuming that an energy slave uses 2,000 kilocalories per day (the average human intake of food). Average *per capita* primary energy use is 1.83 tonnes of oil equivalent per year (from the IEA website) which equates to 25 energy slaves (1 tonne of oil equivalent = 42 GJ net calorific value = 10,034 million calories).
5 OECD, 2010c.
6 UNEP, 2011a, Chapter 14; UNEP, 2008, p. 102.
7 Pollin *et al.*, 2009, p. 28.
8 In West Kalimantan (Indonesia), 200,000 hectares of biofuel plantation employ under 2,000 people, but 80,000 hectares provide employment and subsistence for 200,000 small farmers – almost 260 times the employment potential. In West Kalimantan, more than 5 million indigenous people, whose livelihoods are tied to intact forests, are at risk of displacement by oil palm expansion (UNEP, 2008).
9 Pollin *et al.*, 2009.
10 Study by ECOTEC for Friends of the Earth in 1997, cited in UNEP (2008, p. 171).
11 UNEP, 2008. New jobs were partly as a result of recycling fuel tax revenues to consumers, stimulating extra spending elsewhere in the economy.
12 UNEP, 2008, p. 167.
13 Walz, 2011.
14 Friends of the Earth, 2010.
15 A 1999 study of three US cities found that 79 jobs were required for every 100,000 tons of materials collected and sorted, and another 162 jobs for processing – a total of 241. This is ten times the job potential of waste disposal. A 1991 study found that recycling in the US state of Vermont generates between 550 and 2,000 jobs, compared with 150–1,100 for incineration and 50–360 for landfills. See UNEP (2008, p. 215).
16 UNEP, 2011a, Chapter 7, p. 28.
17 Morison *et al.*, 2005.
18 OECD, 2011a, p. 91. Data are for 2004 and are an unweighted average for 27 countries. Note that the figures for CO_2 intensity do not re-allocate emissions to the sectors consuming electricity.
19 Medhurst, 2009.
20 UNEP, 2008.
21 Fraunhofer Institute *et al.*, 2009.
22 IPCC (2007, Chapter 11), citing Jeeninga *et al.* (1999).
23 UNEP, 2008. In the United States, labour productivity tripled between 1950 and 2000, but energy productivity only increased slightly, and materials productivity stayed the same. In Germany, labour productivity increased by a factor of 3.5 between 1960 and 2000, whereas resource productivity doubled.
24 Spangenberg, 2009.
25 Innovas, 2009.
26 EEA, 2010c. Note that 30 per cent of 'eco-industry' employment was in the waste sector, which includes both high- and low-carbon activities.
27 Innovas, 2009.
28 Roland Berger Strategy Consultants, 2009.
29 REN21, 2011; OECD, 2011a.
30 UNEP, 2008.
31 Friends of the Earth, 2010.
32 Medhurst, 2009. Four of the eight studies reporting a negative impact were by the International Council for Capital Formation. Several of those reporting positive impacts were focusing only on one sector, e.g. renewable energy, and did not assess net changes in employment.
33 UNEP (2008), citing Dupressoir *et al.* (2007).
34 OECD, 2011a.

35 Jaeger *et al.*, 2011.
36 Pollin *et al.*, 2008.
37 Fraunhofer Institute *et al.*, 2009.
38 UNEP, 2008.
39 Innovas, 2009.
40 The Climate Group, 2011.
41 Friends of the Earth, 2010.
42 UNEP, 2011a.
43 Morison *et al.*, 2005.
44 Sustainable Development Commission, 2009.
45 Institut für Wirtschaft und Umwelt & AK Wien (2000), cited in Medhurst (2009), on the introduction of clean technology in Germany, Austria, Sweden, Spain and the Netherlands.
46 UNEP, 2008; Pernick *et al.*, 2010.
47 UNEP, 2008.
48 UNEP, 2008.
49 OECD (2011a) and IEA (2010a), using a discount rate of 10 per cent.
50 UNEP, 2011a, Synthesis Report, p. 24 and Chapter 7, p. 271. G2 scenario (extra 2 per cent green investment) compared to BAU2 scenario (extra 2 per cent normal investment). Note that the study assumes equal abatement costs in all sectors.
51 IPCC, 2007.
52 European Commission, 2010b.
53 European Commission, 2011c.
54 Lee *et al.*, 2011, Chapter 6.
55 IPCC, 2007, Working Group 3, Chapter 7, section10.
56 Sorrell (2007), citing a study by Worrell (2003).
57 Sorrell (2007), citing a study by Lovins and Lovins (1997).
58 Sustainable Development Commission, 2010.
59 European Commission, 2011c.
60 UNEP, 2011a, Chapter 7, p. 255.
61 Wilkinson *et al.* (2009) estimates that carbon reduction measures (such as insulation) in UK households could save eight premature cold-related deaths per million of population per year, as well as 107 premature deaths from particulate pollution.
62 WBCSD, 2010, p. 44.
63 OECD, 2011a.
64 UNEP, 2008.
65 OECD (2010d) finds that links between climate policy and innovation are reduced when policy is too lenient.
66 Roland Berger Strategy Consultants, 2009.
67 Innovas, 2009. Low-carbon goods and services form 48 per cent of sales; renewables 31 per cent; other environmental goods and services 21 per cent.
68 REN21, 2011.
69 UNEP, 2008.
70 'Boards urged to boost CSR skills in response to green shareholder resolutions', *Business Green*, 4 May 2011. Online at http://www.businessgreen.com/bg/news/2047132/boards-urged-boost-csr-skills-response-green-shareholder-resolutions
71 A.T. Kearney, 2009.
72 WBCSD, 2010.
73 'Leading companies call for sharper EU carbon cuts', WWF Press Release, 15 June 2011. Online at http://www.wwf.org.uk/what_we_do/press_centre/?5012/Leading-companies-call-for-sharper-EU-carbon-cuts
74 Bleischwitz *et al.* (2009) found that countries with a higher level of material productivity were more competitive. Oekom Research (undated, but using data from 2003) found

links between sustainability and credit ratings, and Bank Sarasin (Magyar, 2011) found links between sustainability and bond yields.

75 This argument is outlined in detail in Hargroves and Smith (2005).

76 Townshend *et al.* (2011), referring to the Framework Act on Low Carbon, Green Growth enacted in 2010.

77 Dupressoir *et al.*, 2007.

78 IPCC, 2007, Summary for policy makers, Figure SPM.11 and Table SPM.6. For concentrations of 1,000ppm CO_2e, the best estimate of temperature rise in 2090–2099 compared to 1980–1999 is 4°C, and the multi-century warming compared to pre-industrial levels would be 5–6°C.

79 OECD, 2009b.

80 Ernst & Young, 2011.

81 Porchia and Bigano (2008) show that including the external costs of pollution in energy prices would typically double the price of electricity generated from fossil fuels, making renewable energy sources such as wind, hydropower and biomass cheaper than conventional coal-, oil- and gas-fired generation. Epstein *et al.* (2011) show that including external costs would triple the price of electricity from coal-fired power plants. See also IPCC (2011, Chapter 9, p. 76, Chapter 10, section 10.6).

82 Medhurst (2009), citing McKinsey & Company (2009).

83 UNEP, 2011a.

84 Stern, 2006.

85 Stern (2006), cited in Jowit and Wintour (2008).

86 IPCC, 2011, Figure 10.11.

87 OECD, 2012.

88 IEA, 2010a. Investments of $46 trillion under the IEA's BLUE Map Scenario would be outweighed by fuel savings of $112 trillion between 2010 and 2050.

89 Medhurst, 2009.

90 McKinsey & Company, 2009.

91 European Commission, 2011c.

92 IEA, 2011a.

93 IEA, 2009a. Christiana Figueres, head of the UN Framework Convention on Climate Change, cites an estimate that the cost of restricting the temperature rise to 2°C rises by $1 trillion for each year of delay (interview with *The Guardian* newspaper – Harvey, 2011).

94 IEA, 2010b, p. 3.

95 OECD, 2012.

96 IEA, 2010b, p. 417.

97 Stern, 2006; OECD, 2012.

98 Ackerman, 2009.

99 IPCC (2007) cites a 2004 review of the effects of carbon and energy taxes on international competitiveness by Zhang and Baranzini, which concludes that 'competitive losses are not significant'. A 2003 study by Margolis and Walsh, cited in OECD (2009b), looked at 109 studies published between 1972 and 2002 and found that 54 reported a positive link between corporate responsibility (including environmental responsibility) and economic performance, 7 reported a negative relationship, 28 found a non-significant relationship and 20 had mixed findings.

100 See OECD (2010e) for the study on competitiveness. OECD (2010d, p. 31) quotes a 2007 study by Sherrington and Moran: 'reviews of a number of environmental and industrial regulations have shown that ex ante costs tend to exceed the ex post (or outturn) costs'. For quote on adverse competitiveness impacts, see OECD (2010e).

101 OECD (2010c), citing a study by Fankhauser *et al.* (2008).

102 OECD (2010e, p. 3), citing a study by Burniaux *et al.* (2010). Annex 1 countries are those that signed Annex 1 of the Kyoto Protocol: Europe, Russia, the United States, Canada, Australia and New Zealand.

103 European Commission, 2010c.

104 FPAG, 2008.

105 Defined as 21°C in the living room and 18°C in other rooms during daytime, and less at night.

106 DECC, 2010.

107 EPEE (undated), but using data for 2005.

108 DECC, 2011, p. 26.

109 Committee on Climate Change, 2011b.

110 For example, see the response of the G77 group of 131 developing countries to the UNEP report *Towards a Green Economy* (2011a):

> Green Economy should not imply conditionality to development assistance nor be used as a trade restriction. An outcome . . . should allow for expanded market access for products from developing countries while combating trade-distortive measures, such as subsidies in developed countries and 'green protectionism'.
>
> *(G77, 2011)*

111 OECD, 2009b, p. 88.

112 OECD, 2009b, p. 34.

113 See Christian Aid (2011) for a good overview of the potential for Sub-Saharan Africa to 'leapfrog' to a low-carbon future using technologies such as renewable energy and more efficient stoves, and how this can play a key role in poverty reduction.

114 UNEP, 2011a.

115 Hargroves and Smith (2005), citing *Sustainability and Cities* by Newman and Kenworthy (1999).

116 Von Weizsäcker *et al.*, 2010.

117 'Increased energy efficiency only fuels our appetite': http://www.guardian.co.uk/ sustainable-business/digital-efficiency-gains-fuel-increased-demand

118 Sorrell (2007, p. 64), citing Fouquet and Pearson (2006).

119 EEA, 2010c.

120 The case for 'backfire' is based partly on the theory that the availability of high-quality energy has been a major driver of economic growth.

121 Sorrell (2007) reviews over 500 studies and concludes that rebound is often over 50 per cent, but that the case for it always to exceed 100 per cent has not been proven, and that it varies depending on the technology and country. See also Maxwell *et al.* (2011, p. 11, Table 1.1) and Sorrell (2010, p. 3).

122 Druckman *et al.*, 2010.

123 Barker *et al.* (2009), assuming a direct rebound effect of 25 per cent for residential buildings, 10 per cent for transport, 5 per cent for service and low or zero for other sectors.

124 Simms *et al.*, 2010. Data on growth of world economy (previous paragraph) are from the World Bank Development Indicators databank, online at http://data.worldbank. org/data-catalog (accessed 17 September 2012). Global GDP in constant 2000 US$ (i.e. corrected for inflation to the year 2000) has doubled from $20 trillion in 1985 to $41 trillion in 2010. In current US$ (i.e. not corrected for inflation), the global economy increased by a factor of 5 over the same period, from $12 trillion to $63 trillion. IEA (2011a, p. 58) predicts that the world economy will more than double again in real terms, from $70 trillion (in constant 2010 US$) in 2009 to $173 trillion in 2035.

125 Solow was speaking at the Ely lecture to the American Economic Association in 1973, with a talk entitled 'The economics of resources or the resources of economics'. In response to the publication of *The Limits to Growth* (Meadows *et al.*, 1972), he argued that man-made capital could substitute for natural resources: 'If it is very easy to substitute other factors for natural resources, then there is, in principle, no "problem". The world can, in effect, get along without natural resources.'

126 Jackson, 2009, pp. 78–81. Also note that world carbon intensity has reversed its long-term decline in recent years, with a slight increase due to the rising use of coal for power generation by many countries.

127 Spangenberg, 2009.

128 UNEP, 2011c.

129 UNEP, 2008, p. 73.

130 See http://www.capandshare.org/

131 Joint Implementation (JI) and the Clean Development Mechanism (CDM) allow developed countries to achieve part of their emission reductions under the Kyoto Protocol by paying for emission reduction projects in other countries. These JI and CDM projects account for around 13 per cent of the value of permits traded in the EU carbon market, but verification costs can be 14–22 per cent of the revenue from the sale of carbon credits. There has been criticism that the schemes are geared towards obtaining cheap carbon credits rather than meeting the development needs of the host country (UNEP, 2008).

132 Pelletier, 2010.

133 Von Weizsäcker *et al.*, 2010, Chapter 9.

134 EEA, 2010c, p. 34.

135 Allcott, 2011.

136 Maxwell *et al.*, 2011.

137 IEA, 2009b.

138 UNEP, 2011a, Chapter 15. This is predicted to increase to 18 per cent of GDP by 2050 under business as usual.

139 OECD, 2009b, pp. 20, 63.

140 IEA, 2011a.

141 IEA, 2010b. Based on a survey of 37 OECD and non-OECD countries thought to represent over 95 per cent of global subsidized fossil-fuel consumption.

142 OECD (2011b) estimates $45–75 billion per year in production and consumption subsidies for 24 OECD countries.

143 The G20 agreement was made in September 2009. Subsidies are calculated using the 'price-gap' approach; i.e. the gap between world market prices and the price paid by consumers. This excludes subsidies that do not affect the market price, such as direct payments to low-income consumers, but in any case, these do not count as 'wasteful'. Oil-producing countries argue that subsidies in their countries should be calculated on the basis of production price, not world market price, which would greatly reduce the estimates of subsidies in oil-producing nations (for example, if oil trades at $100 per barrel but is produced at $20 per barrel and sold to consumers at $15 per barrel, the subsidy on a production price basis would be $5, rather than $85 on a market price basis) (IEA *et al.*, 2010).

144 OECD, 2010e, p. 43.

145 Pearce, 2003, pp. 9–30.

146 UNEP, 2011a.

147 IPCC, 2011, Chapter 9. In Indonesia, the 40 per cent lowest-income families receive only 15 per cent of subsidies.

148 IEA, 2011a, p. 7.

149 UNEP, 2011a.

150 UNEP, 2011a, Chapter 14, Box 7.

151 REN21, 2011.

152 OECD (2010e), citing a 2006 OECD study.

153 OECD, 2009b.

154 OECD (2010e) describes how this has been successfully implemented with a NO_x tax in Sweden. There were plans to adopt this scheme in the UK, as part of the Carbon Reduction Commitment, but this was eventually scrapped on the grounds of over-complexity.

155 OECD, 2009b, p. 90.
156 UNEP (2011a, Chapter 14) recommends border taxes for the cases where countries are acting unilaterally.
157 Von Weizsäcker *et al.*, 2010.
158 UNEP, 2011a, Chapter 15.
159 UNEP, 2008.
160 UNEP (2008), UNEP (2011a, Chapter 14) and Medhurst (2009) all report skill shortages in countries including the United States, the UK, Germany, Australia, Brazil and China. The US National Renewable Energy Laboratory has identified a shortage of skills and training as a leading barrier to renewable energy and energy efficiency growth; and in the UK, the Confederation of British Industry has expressed concern that sectors going green are struggling to find technical specialists, designers, engineers and electricians.
161 Cedefop, 2010.
162 Cedefop, 2010.
163 Townshend *et al.*, 2011.
164 Cedefop, 2010.
165 UNEP, 2008, p. 278.
166 DECC, 2010.
167 The Warm Front scheme is being phased out by 2013 to be replaced by the Green Deal.
168 DECC, 2011. Energy-saving policies include home energy-efficiency schemes and product efficiency standards.
169 NEA Cymru, 2011, p. 5.
170 Dresner and Ekins, 2005.
171 OECD (2009b), based on 2005 IEA data in tonnes CO_2e for all greenhouse gases.
172 OECD, 2009b, p. 204.
173 OECD, 2009b, p. 204.
174 Wilkins, 2002.
175 See http://www.climatefundsupdate.org/listing for a list of over 20 existing climate funds and data on donors, recipients and funding levels.
176 UNEP, 2008.
177 See Herrndorf *et al.* (2007) for a full overview of potential policies to promote greener production in developing countries that export goods to Europe.
178 Woodward and Simms, 2006.
179 Sachs, 2005.
180 The two West African entrepreneurs who founded Toyola, making energy-efficient stoves, found that the process of obtaining finance from international climate funds was very difficult. See 'Energy is central to development', Guardian Poverty Matters blog, posted by Madeline Bunting, 16 June 2011 at http://www.guardian.co.uk/global-development/poverty-matters/2011/jun/16/ashden-awards-renewable-energy
181 WBCSD, 2010.
182 Alcott, 2008.
183 Steinberger and Roberts, 2011.
184 Kuznets, 1934, p. 7; 1962, p. 67; Kennedy, 1968.
185 Costanza *et al.*, 2009.
186 Survey by GlobeScan Radar. See http://www.ethicalmarkets.com/2011/01/20/continued-public-support-for-going-%e2%80%98beyond-gdp%e2%80%99-global-poll/
187 See Stiglitz *et al.* (2009), European Commission (2009), Costanza *et al.* (2009) and the OECD 'Your Better Life Index' website at http://www.oecd.org/document/35/0,3746,en_2649_201185_47837411_1_1_1_1,00.html
188 World Bank, 2012a; European Commission, 2009.

189 Lawn, 2009.
190 Ecological tax reform (ETR) has been implemented at a modest level in a number of countries, including Germany where it created an estimated 250,000 jobs and cut carbon emissions by 2 per cent between 1999 and 2003.
191 See Sorrell (2010).
192 Pollitt *et al.* (2010) recommend that macro-economic models should widen their perspective to include environmental issues.
193 'China to slow GDP growth in bid to curb emissions': http://www.guardian.co.uk/environment/2011/feb/28/china-gdp-emissions
194 *Vision 2050* discussion workshop, cited by WBCSD (2010, p. 15).
195 Victor, 2008. Main results also available in Pollit *et al.* (2010).
196 O'Neill, 2008.
197 Bell, 2010.

8 Health and well-being: benefits of a low-carbon lifestyle

1 See Figure 6.1 in this book.
2 FAO, 2006.
3 McMichael *et al.*, 2007.
4 See Figure 2.1 in this book.
5 According to the Food and Agriculture Organization of the United Nations (FAO), 830 million people were undernourished in 2005–2007, the last period assessed, and this may have risen to 915 million in 2008 and passed 1,000 million in 2009. UN, 2010, p. 11.
6 Popkin, 2006.
7 World Health Organization, 2011b.
8 World Health Organization, 2009a.
9 Obesity is measured using the body mass index (BMI): weight (in kilograms) divided by the square of height (in metres). A BMI of 25 or more is generally classified as overweight, and 30 or more indicates obesity.
10 'Physical inactivity: a global public health problem': http://www.who.int/dietphysicalactivity/factsheet_inactivity/en/index.html
11 World Health Organization website: Global health observatory: Overweight. See http://www.who.int/gho/ncd/risk_factors/overweight_text/en/index.html (accessed 12 July 2011).
12 World Health Organization, 2008.
13 Sustainable Development Commission, 2009.
14 Sustainable Development Commission, 2007.
15 Popkin *et al.*, 2006.
16 ETC/SCP, 2011b, p. 59.
17 Pan *et al.* (2012) showed that an extra portion of red meat or processed meat per day increases the risk of death by 13 per cent or 20 per cent respectively.
18 JRC/IPTS, 2009.
19 World Health Organization, 2009a.
20 Scarborough *et al.*, 2010.
21 Pan *et al.*, 2012.
22 World Health Organization, 2009a.
23 For the UK, see Sustainable Development Commission (2009).
24 FAO, 2006.
25 Pimental and Pimental, 2003.
26 World Health Organization Europe and UNECE, 2009.
27 Thompson Coon *et al.*, 2011.
28 '10 facts on physical activity', World Health Organization website, November 2010.

Online at http://www.who.int/features/factfiles/physical_activity/en/index.html (accessed 12 July 2011).

29 World Health Organization, 2012. Online at http://www.who.int/mediacentre/factsheets/fs311/en/ (accessed 18 September 2012).

30 Craft and Perna, 2004.

31 World Health Organization, 2003b.

32 Woodcock *et al.*, 2009.

33 De Nazelle *et al.*, 2011.

34 Oja *et al.*, 2011.

35 Pucher *et al.*, 2010.

36 Woodcock *et al.*, 2009.

37 'Global strategy on diet, physical activity and health', World Health Organization website at http://www.who.int/dietphysicalactivity/childhood/en/

38 De Nazelle *et al.*, 2011.

39 De Hartog *et al.*, 2010. Rabl and de Nazelle (2011) found that the personal health benefits of shifting from car travel to cycling were €1,310/year, and the public health benefits from reduced pollution were €33/year, compared to losses of –€19/year from increased pollution exposure and –€53/year from increased accident risk.

40 De Nazelle *et al.* (2011), citing Gatersleben and Appleton (2007).

41 De Nazelle *et al.* (2011), citing Pucher and Buehler (2008).

42 Woodcock *et al.*, 2009.

43 Hendriksen *et al.*, 2010.

44 Vandermeulen *et al.*, 2011.

45 All data from the World Health Organization. See '10 facts about road safety', online at http://www.who.int/features/factfiles/roadsafety/02_en.html; World Health Organization, 2011c, 2009b.

46 De Nazelle *et al.*, 2011.

47 World Health Organization Europe and UNECE, 2009.

48 Pucher *et al.*, 2009.

49 World Health Organization Europe and UNECE, 2009.

50 Pucher *et al.*, 2009.

51 World Health Organization and JRC, 2011.

52 Of the EU agglomerations that have provided data to the Noise Observation and Information Service for Europe, 55 million of the population (51 per cent) are exposed to day–evening–night road noise levels over the recommended level of 55 decibels (dB(A)), compared to 3.6 million who are exposed to rail noise, 2.6 million to airport noise and 0.5 million to noise from industry. Thirty-six per cent are exposed to night road noise of over 50 dB(A). Data extracted from the database of the Noise Observation and Information Service for Europe (European Topic Centre on Land Use and Spatial Information, 2011).

53 World Health Organization Europe and JRC, 2011.

54 European Commission, 2011e, 2011f.

55 De Nazelle *et al.*, 2011.

56 However, note that although congestion may increase the risks of collisions, it may also reduce speeds so that accidents are less likely to result in serious injury.

57 Schrank *et al.*, 2010.

58 De Nazelle *et al.*, 2011.

59 For case studies and examples of local food initiatives in the UK, see https://www.makinglocalfoodwork.co.uk/about/fwm/index.cfm

60 For a long list of examples, see http://www.collaborativeconsumption.com/the-movement/snapshot-of-examples.php

61 Weintrobe, 2011.

62 Vestbro, 2010.

63 See Layard (2011) for a summary of the evidence.
64 Boyce *et al.*, 2010.
65 Easterlin and Angelescu, 2009.
66 Layard, 2009, p. 94.
67 Proto and Rustichini, 2011. Figure 1 (p. 25) shows life satisfaction increasing steadily to an income of about \$30,000 (UK) or \$40,000 (Germany) and then growing more slowly to an eventual peak at \$150,000–\$200,000 before declining, though there is considerable scatter at incomes over about \$200,000.
68 Kasser, 2002; James, 2007; Abdhallah and Thompson, 2008.
69 Layard, 2009, p. 94.
70 Kasser, 2002.
71 Layard, 2011; Wilkinson and Pickett, 2009.
72 Dunn *et al.*, 2008.
73 Howell and Hill, 2009.
74 For example, James, 2007.
75 EEA, 2010c.
76 Headey, 2008.
77 EEA, 2010c.
78 Ausubel and Grübler, 1995.
79 Keynes, 1963.
80 Nørgård, 2010.
81 Beder, 2000.
82 Beder, 2000.
83 Nørgård, 2010.
84 Nørgård, 2010.
85 Schneider, 2010.
86 UNEP, 2008, p. 82.
87 UNEP, 2008, p. 82.
88 UNEP, 2008, p. 82.
89 Baker, 2011.
90 Huppes and Huele, 2010.
91 EEA, 2010c.
92 DEFRA, 2008.
93 These findings arise from a series of UK studies including *I Will if You Will* (Sustainable Consumption Roundtable, 2006); Sustainable Development Commission, 2011; DEFRA, 2008; and Phillips and Rowley, 2011.
94 EEA, 2010c, Chapter 1.
95 Phillips and Rowley, 2011.
96 DEFRA, 2008.
97 Sustainable Consumption Roundtable, 2006.
98 Transition Network website at http://www.transitionnetwork.org/initiatives (accessed 10 August 2011).
99 Sustainable Consumption Roundtable, 2006.
100 Phillips and Rowley, 2011.
101 Puska, 2009.
102 'Environmentally effective food choices', Swedish National Food Administration/ Swedish Environmental Protection Agency. Online at http://www.slv.se/upload/ dokument/miljo/environmentally_effective_food_choices_proposal_eu_2009.pdf (accessed 10 August 2011).
103 Popkin, 2006.
104 Powell and Chaloupka, 2009.
105 EEA, 2010c.
106 Sustainable Consumption Roundtable, 2006.

107 Sustainable Consumption Roundtable, 2006.
108 Lister and Harnish, 2011.
109 Griffiths and Reeves, 2009, p. 19.
110 Griffiths and Reeves, 2009, p. 104.
111 Wilkinson and Pickett, 2009.
112 Compton and Kasser, 2009.
113 Compton and Kasser, 2009; Compton, 2010.

References

All online references were accessed 19 May 2012 unless stated otherwise.

Abdhallah, S. and Thompson, S. (2008) 'Psychological barriers to de-growth: values mediate the relationship between well-being and income', paper presented at the First International Conference on Economic Degrowth for Ecological Sustainability and Social Equity, Paris, 18–19 April. Online at http://events.it-sudparis.eu/degrowth conference/en/themes/

Ackerman, F. (2009) *Can We Afford the Future?* London: Zed Books.

Adua, L. (2010) 'To cool a sweltering earth: does energy efficiency improvement offset the climate impacts of lifestyle?', *Energy Policy*, 38(10): 5719–5732.

AEA Energy and Environment (2008) *Air Pollution in the UK 2007*. Harwell, UK: AEA Energy and Environment. Online at http://uk-air.defra.gov.uk/library/annualreport/view online?year=2007

AEA Technology (2010) *Review of the Future Resource Risks Faced by UK Business and an Assessment of Future Viability*. Report to DEFRA, London. Online at http://randd.defra.gov.uk

Ahmia, M. (ed.) (2011) *The Group of 77 at the United Nations: The Collected Documents of the Group of 77, Volume 4: Environment and Development*. New York: Oxford University Press.

Alcott, B. (2008) 'The sufficiency strategy: would rich-world frugality lower environmental impact?', *Ecological Economics*, 64(3): 770–786.

Aleklett, K. , Höök, M., Jakobsson, K., Lardelli, M., Snowden, S. and Söderbergh, B. (2010) 'The peak of the oil age – analyzing the world oil production reference scenario in World Energy Outlook 2008', *Energy Policy*, 38(3): 1398–1414.

Allcott, H. (2011) 'Social norms and energy conservation', *Journal of Public Economics*, 95(9–10): 1082–1095.

Allen, M., Frame, D., Huntingford, C., Jones, C., Lowe, J., Meinshausen, M. and Meinshausen, N. (2009) 'Warming caused by cumulative carbon emissions towards the trillionth tonne', *Nature*, 458(7242): 1163–1166.

Allwood, J. (2011) 'The physical basis for a low-carbon future economy', presentation at Imperial College London, 8 February. Online at https://workspace.imperial.ac.uk/ energyfutureslab/Public/Events/JMA%20Feb%202011.pdf

Allwood, J., Cullen, J. and Milford, R. (2010) 'Options for achieving a 50% cut in industrial carbon emissions by 2050', *Environmental Science and Technology*, 44(6): 1888–1894.

American Lung Association (2011) *State of the Air 2011*. Washington, DC: American Lung Association.

Andel, P. and Pribán, V. (1996) 'Environmental restoration of uranium mines and mills in the Czech Republic', in IAEA-TECDOC-865 *Planning for Environmental Restoration of Radioactively Contaminated Sites in Central and Eastern Europe, Vol.1: Identification and Characterization of Contaminated Sites.* Vienna: International Atomic Energy Agency (IAEA).

AQEG (2007) *Air Quality and Climate Change: A UK Perspective.* London: Air Quality Expert Group. Online at http://www.defra.gov.uk/environment/quality/air/air-quality/committees/aqeg/publish/

AQEG (2009) *Ozone in the United Kingdom.* London: Air Quality Expert Group. Online at http://www.defra.gov.uk/environment/quality/air/air-quality/committees/aqeg/publish/

AQEG (2011) *Road Transport Biofuels: Impact on UK Air Quality.* London: Air Quality Expert Group. Online at http://www.defra.gov.uk/publications/files/pb13464-road-transport-biofuels-110228.pdf

A.T. Kearney (2009) *Green Winners: The Performance of Sustainability-Focused Companies in the Financial Crisis.* Online at http://www.atkearney.com/index.php/News-media/companies-with-a-commitment-to-sustainability-tend-to-outperform-their-peers-during-the-financial-crisis.html

Aucott, M. (2011) 'Shale gas EROI: preliminary estimate suggests 70 or greater', *Energy Bulletin*, 23 June. Online at http://www.energybulletin.net/stories/2011-08-19/shale-gas-eroi-preliminary-estimate-suggests-70-or-greater

Ausubel, J. and Grübler, A. (1995) 'Working less and living longer: long term trends in working time and time budgets', *Technological Forecasting and Social Change*, 50(3): 113–131. Online at http://phe.rockefeller.edu/work_less/

Ayres, R. and Ayres, E. (2010) *Crossing the Energy Divide: Moving from Fossil Fuel Dependence to a Clean-Energy Future.* Upper Saddle River, NJ: Wharton School Publishing.

Badgley, C., Moghtader, J., Quintero, E., Zakem, E., Chappell, M. J., Avilés-Vázquez, K., Samulon, A. and Perfecto, I. (2007) 'Organic agriculture and the global food supply', *Renewable Agriculture and Food Systems*, 22(2): 86–108.

Baker, D. (2011) 'How to make short work of unemployment', *The Guardian*, 30 June. Online at http://www.guardian.co.uk/commentisfree/cifamerica/2011/jun/30/economic-policy-short-work

Bakkes, J. and Bosch, P. (eds) *et al.* (2008) *Background Report to the OECD Environmental Outlook to 2030. Overviews, Details, and Methodology of Model-Based Analysis.* Bilthoven: Netherlands Environmental Assessment Agency.

Balmford, A. and Bond, W. (2005) 'Trends in the state of nature and their implications for human well-being', *Ecology Letters*, 8(11): 1218–1234.

Barker, T., Dagoumas, A. and Rubin, J. (2009) 'The macroeconomic rebound effect and the world economy', *Energy Efficiency*, 2(4): 411–427.

Barr, C., Dermawan, A., Purnomo, H. and Komarudin, H. (2010) *Financial Governance and Indonesia's Reforestation Fund during the Soeharto and Post-Soeharto Periods, 1989–2009. A Political Economic Analysis of Lessons for REDD+.* Bogor, Indonesia: Center for International Forestry Research (CIFOR).

Bates, J. and Dale, N. (2008) *Environmental Impacts of Significant Resource Flows into the EU.* Report to the European Commission Environment Directorate. Harwell, UK: AEA Technology. Online at http://ec.europa.eu/environment/natres/pdf/env_impact.pdf

Bayala J., Kalinganire, A., Tchoundjeu, Z., Sinclair, F. and Garrity, D. (2011) *Conservation Agriculture with Trees in the West African Sahel – A Review.* ICRAF Occasional Paper No. 14. Nairobi: World Agroforestry Centre.

Beder, S. (2000) *Selling the Work Ethic: From Puritan Pulpit to Corporate PR.* Melbourne, Vic.: Scribe.

Bell, K. (2010) 'Degrowth: what can we learn from Cuba?', paper presented at the Second International Conference on Economic Degrowth for Ecological Sustainability and

Social Equity, Barcelona, 26–29 March. Online at http://www.barcelona.degrowth. org/Posters-download.113.0.html

Bell, M., Davis, D. and Fletcher, T. (2004) 'A retrospective assessment of mortality from the London smog episode of 1952: the role of influenza and pollution', *Environmental Health Perspectives*, 112(1): 6–8. Online at http://ehp.niehs.nih.gov/members/2003/6539/6539. html

Benyus, J. (1997) *Bio-mimicry: Innovation Inspired by Nature.* New York: William Morrow and Co.

Benzie, M., Harvey, A., Burningham, K., Hodgson, N. and Siddiqi, A. (2011) *Vulnerability to Heat Waves and Drought: Case Studies of Adaptation to Climate Change in South-West England.* York, UK: Joseph Rowntree Foundation.

Berman, A. and Pittinger, L. (2011) 'U.S. shale gas: less abundance, higher cost', *The Oil Drum*, 5 August. Online at http://www.theoildrum.com/node/8212

Bertzky, M., Kapos, V. and Scharlemann, J. (2011) *Indirect Land Use Change from Biofuel Production: Implications for Biodiversity.* Peterborough, UK: Joint Nature Conservation Committee.

Best, A., Giljum, S., Simmons, C., Blobel, D., Lewis, K., Hammer, M., Cavalieri, S., Lutter, S. and Maguire, C. (2008) *Potential of the Ecological Footprint for Monitoring Environmental Impacts from Natural Resource Use: Analysis of the Potential of the Ecological Footprint and Related Assessment Tools for Use in the EU's Thematic Strategy on the Sustainable Use of Natural Resources.* Report to the European Commission, DG Environment. Online at http://ec.europa.eu/ environment/natres/pdf/footprint.pdf

Bleischwitz, R., Bahn-Walkowiak, B., Onischka, M., Röder, O. and Steger, S. (2009) *The Relation between Resource Productivity and Competitiveness.* Report to the European Commission, Project ENV.G.1/ETU/2007/0041. Online at http://ec.europa.eu/environment/ enveco/economics_policy/pdf/part2_report_comp.pdf

Bollen, J., Hers, S. and van der Zwaan, B. (2010) 'An integrated assessment of climate change, air pollution, and energy security policy', *Energy Policy*, 38(8): 4021–4030.

Bollen, J., Brink, C., Eerens, H. and Manders, T. (2009a) *Co-Benefits of Climate Policy.* PBL Report No. 500116005. Bilthoven: Netherlands Environmental Assessment Agency. Online at http://www.pbl.nl/en/publications/2009/Co-benefits-of-climate-policy.html

Bollen, J., Guay, B., Jamet, S. and Corfee-Morlot, J. (2009b) 'Co-benefits of climate change mitigation policies: literature review and new results', OECD Economics Department Working Paper No. 693. Paris: OECD.

Bowyer, C. (2010) *Anticipated Indirect Land Use Change Associated with Expanded Use of Biofuels in the EU – An Analysis of Member State Performance.* London: Institute for European Environmental Policy (IEEP).

Boyce, C., Brown, G. and Moore, S. (2010) 'Money and happiness: rank of income, not income, affects life satisfaction', *Psychological Science*, 21(4): 471–475. Online at www.vrg. cf.ac.uk/Files/2010_boyce_brown_moore.pdf

BP (2011) *BP Statistical Review of World Energy June 2011.* Online at http://www.bp.com/ statisticalreview

Braat, L. and ten Brink, P. (eds) (2008) *The Cost of Policy Inaction: The Case of Not Meeting the 2010 Biodiversity Target.* A study for the European Commission by Alterra, Wageningen UR in co-operation with the Institute for European Environmental Policy and other partners. Online at http://ecologic.eu/2363

Bradbury, D. (2011) 'Searching for a solution to the rare-earth problem', *Business Green*, 28 January. Online at http://www.businessgreen.com/bg/analysis/1940308/searching-solution-rare-earth

Brandt, A. (2011) *Upstream Greenhouse Gas (GHG) Emissions from Canadian Oil Sands as a Feedstock for European Refineries*. Stanford, CA: Stanford University, Department of Energy Resources Engineering.

Brandt, J., Silver, J., Christensen, J., Andersen, M., Bønløkke, J., Sigsgaard, T., Geels, C., Gross, A., Hansen, A., Hansen, K., Hedegaard, G., Kaas, E. and Frohn, L. (2011) *Assessment of Health–Cost Externalities of Air Pollution at the National Level Using the EVA Model System*. CEEH Scientific Report No. 3. Roskilde: Centre for Energy, Environment and Health. Online at http://www.ceeh.dk/CEEH_Reports/Report_3/ CEEH_Scientific_Report3.pdf

Brinkhoff, T. (2009) *The Principal Agglomerations of the World*. Online at http://www.citypopulation.de/world/Agglomerations.html

Brinkley, A. and Less, S. (2010) *Carbon Omissions: Consumption-Based Accounting for International Carbon Emissions*. London: Policy Exchange. Online at http://www.policyexchange.org.uk/publications/category/item/carbon-omissions-consumption-based-accounting-for-international-carbon-emissions?category_id=24

Brinkman, N., Wang, M., Weber, T. and Darlington, T. (2005) *Well-to-Wheels Analysis of Advanced Fuel/Vehicle Systems – A North American Study of Energy Use, Greenhouse Gas Emissions, and Criteria Pollutant Emissions*. Argonne National Laboratory, General Motors Corporation and Air Improvement Resource, Inc. Online at http://www.transportation.anl.gov/pdfs/TA/339.pdf

Brown, L. (2006) *Plan B 2.0: Rescuing a Planet under Stress and a Civilization in Trouble*. Washington, DC: Earth Policy Institute. Online at http://www.earth-policy.org/index.php?/books/pb2/pb2ch1_ss2

Buchert, M., Schüler, D. and Bleher, D. (2009) *Critical Metals for Future Sustainable Technologies and Their Recycling Potential*. Nairobi: United Nations Environment Programme.

Butler, R. (2011) 'Greening the world with palm oil?'. Online at http://news.mongabay.com/2011/0126-palm_ oil.html

Campbell, A. and Doswald, N. (2009) *The Impacts of Biofuel Production on Biodiversity: A Review of the Current Literature*. Cambridge, UK: UNEP-WCMC.

Campbell, A., Kapos, V., Scharlemann, J., Bubb, P., Chenery, A., Coad, L., Dickson, B., Doswald, N., Khan, M., Kershaw, F. and Rashid, M. (2009) *Review of the Literature on the Links between Biodiversity and Climate Change: Impacts, Adaptation and Mitigation*. Montreal: Secretariat of the Convention on Biological Diversity.

Canadell, J., Le Quéréc, C., Raupach, M., Field, C., Buitenhuis, E., Ciais, P., Conway, T., Gillett, N., Houghton, R. and Marland, G. (2007) 'Contributions to accelerating atmospheric CO_2 growth from economic activity, carbon intensity, and efficiency of natural sinks', *Proceedings of the National Academy of Sciences of the USA*, 104(47): 18866–18870.

Cardis, E., Krewski, D., Boniol, M. *et al.* (2006) 'Briefing document: the cancer burden from Chernobyl in Europe', World Health Organization, International Agency for Research on Cancer. Online at http://www.iarc.fr/en/media-centre/pr/2006/IARCBriefing Chernobyl.pdf

Carrington, D. (2012) 'Sun, sewage and algae: a recipe for success?', *The Guardian*, 6 March. Online at http://www.guardian.co.uk/environment/damian-carrington-blog/2012/mar/06/sewage-algae-biofuels-energy

Cedefop (2010) *Skills for Green Jobs. European Synthesis Report*. Luxembourg: Publications Office of the European Union.

Cernansky, R. (2010) 'How many cities have a ban on plastic bags?'. Online at http://planet green.discovery.com/work-connect/how-many-cities-have-a-ban-on-plastic-bags.html

Chomitz, K. (2007) *At Loggerheads? Agricultural Expansion, Poverty Reduction, and Environment in the Tropical Forests*. Washington, DC: World Bank.

Christian Aid (2011) *Low-Carbon Africa: Leapfrogging to a Green Future*. London: Christian Aid.

Clancy, L., Goodman, P., Sinclair, H. and Dockery, D. (2002) 'Effect of air-pollution control on death rates in Dublin, Ireland: an intervention study', *The Lancet*, 360(9341): 1210–1214.

Clarke, R. (1990) 'The 1957 Windscale accident revisited' in R. C. Ricks and S. A. Fry (eds) *The Medical Basis for Radiation Accident Preparedness*. New York: Elsevier, pp. 281–289.

Cleveland, C. (2005) 'Net energy from oil and gas extraction in the United States, 1954–1997', *Energy*, 30(5): 769–782. Online at http://www.eoearth.org/article/Net_Energy:_Concepts,_Issues,_and_Case_Studies_(collection)

Cleveland, C. and O'Connor, P. (2010) 'An assessment of the energy return on investment (EROI) of oil shale', report to Western Resource Advocates, Boulder, Colorado. Online at http://www.westernresourceadvocates.org/land/pdf/oseroireport.pdf

Cohen, D. (2007) 'Earth's natural wealth: an audit', *New Scientist*, 23 May: 34–41.

Collette, L., Hodgkin, T., Kassam, A., Kenmore, P., Lipper, L., Nolte, C., Stamoulis, K. and Steduto, P. (2011) *Save and Grow: A Policymaker's Guide to the Sustainable Intensification of Smallholder Crop Production*. Rome: Food and Agriculture Organization of the United Nations.

Collier, P. and Goderis, B. (2007) *Commodity Prices, Growth, and the Natural Resource Curse: Reconciling a Conundrum*. Working Paper 2007-15, Centre for the Study of African Economies, University of Oxford. Online at http://www.csae.ox.ac.uk/workingpapers/pdfs/2007-15text.pdf (accessed 7 September 2012).

Committee on Climate Change (2011a) *The Renewable Energy Review*. London: Committee on Climate Change.

Committee on Climate Change (2011b) *Household Energy Bills – Impacts of Meeting Carbon Budgets*. London: Committee on Climate Change.

Committee on the Significance of International Transport of Air Pollutants, Board on Atmospheric Sciences and Climate, National Research Council of the National Academies (2009) *Global Sources of Local Pollution: An Assessment of Long-Range Transport of Key Air Pollutants to and from the United States*. Washington, DC: The National Academies Press. Online at http://books.nap.edu/openbook.php?record_id=12743&page=R1

Committee on World Food Security (2012) *Voluntary Guidelines on the Responsible Governance of Tenure of Land, Fisheries and Forests in the Context of National Food Security*. Rome: Food and Agriculture Organization of the United Nations.

Compton, T. (2010) *Common Cause: The Case for Working with Our Cultural Values*. Godalming, UK: Worldwide Fund for Nature.

Compton, T. and Kasser, T. (2009) *Meeting Environmental Challenges: The Role of Human Identity*. Totnes, UK: Green Books.

Cooper, O. *et al.* (2010) 'Increasing springtime ozone mixing ratios in the free troposphere over western North America', *Nature*, 463: 344–348.

Cooper, T. (2004) 'Inadequate life? Evidence of consumer attitudes to product obsolescence', *Journal of Consumer Policy*, 27(4): 421–449.

Cordell, D., Jangert, D-O. and White, S. (2009) 'The story of phosphorus: global food security and food for thought', *Global Environmental Change*, 19: 292–305.

Corvalan, C., Hales, S., McMichael, A., Butler, C., Campbell-Lendrum, D., Confalonieri, U., Leitner, K., Lewis, N., Patz, J., Polson, K., Scheraga, J., Woodward, A. and Younes, M. (2005) *Ecosystems and Human Well-Being: Health Synthesis, Millennium Ecosystem Assessment*. Geneva: World Health Organization.

Costanza, R., Hart, M., Posner, S. and Talberth, J. (2009) *Beyond GDP: The Need for New Measures of Progress*. The Pardee Papers No. 4. Boston, MA: Boston University. Online at www.bu.edu/pardee/publications/pardee-paper-004-beyond-gdp/

Costanza, R., d'Arge, R., de Groot, R., Farber, S., Grasso, M., Hannon, B., Limburg, K., Naeem, S., O'Neill, R. V., Paruelo, J., Raskin, R. G., Sutton, P. and van den Belt, M. (1997) 'The value of the world's ecosystem services and natural capital', *Nature*, 387: 253–260.

Cotula, L., Finnegan, L. and McQueen, D. (2011) *Biomass Energy: Another Driver of Land Acquisitions?* London: International Institute for Environment and Development.

Cox, T., Glover, J., van Tassel, D., Cox, C. and De Haan, L. (2006) 'Prospects for developing perennial grain crops', *BioScience*, 56(8): 649–659.

Craft, L. and Perna, F. (2004) 'The benefits of exercise for the clinically depressed', *The Primary Care Companion to the Journal of Clinical Psychiatry*, 6(3): 104–111.

Cranford, M. and Mourato, S. (2011) 'Community conservation and a two-stage approach to payments for ecosystem services', *Ecological Economics*, 71: 89–98.

Cranford, M., Leggett, M., Oakes, N. and Parker, C. (2012) *The Little Biodiversity Finance Book 2012*. Oxford: Global Canopy Programme.

Crouse D., Peters, P., van Donkelaar, A., Goldberg, M., Villeneuve, P., Brion, O. *et al.* (2012) 'Risk of nonaccidental and cardiovascular mortality in relation to long-term exposure to low concentrations of fine particulate matter: a Canadian national-level cohort study', *Environmental Health Perspectives*, 120(5): 708–714.

Cullen, J., Allwood, J. and Borgstein, E. (2011) 'Reducing energy demand: what are the practical limits?', *Environmental Science and Technology*, 45(4): 1711–1718.

Daily Yomiuri Online (2011) 'Accidents add to expense of N-power: damage may cost 1.1 yen per kilowatt-hour'. Online at http://www.yomiuri.co.jp/dy/national/T111025005767.htm

Davidson, E. (2012) 'Representative concentration pathways and mitigation scenarios for nitrous oxide', *Environmental Research Letters*, 7: 024005.

Davis, S. and Caldeira, K. (2010) 'Consumption-based accounting of CO_2 emissions', *Proceedings of the National Academy of Sciences*, 107(12): 5687–5692.

DECC (2010) *Annual Report on Fuel Poverty Statistics 2010*. London: Department of Energy and Climate Change.

DECC (2011) *Estimated Impacts of Energy and Climate Change Policies on Energy Prices and Bills: November 2011*. London: Department of Energy and Climate Change. Online at http://www.decc.gov.uk/en/content/cms/meeting_energy/aes/impacts/impacts.aspx

DEFRA (2005) *Guidance for Waste Collection Authorities on the Household Waste Recycling Act 2003*. London: Department for Environment, Food and Rural Affairs.

DEFRA (2008) *A Framework for Pro-Environmental Behaviours*. London: Department for Environment, Food and Rural Affairs.

DEFRA (2010) *Air Pollution: Action in a Changing Climate*. London: Department for Environment, Food and Rural Affairs.

DEFRA and BIS (2010) *Less Is More: Business Opportunities in Waste Management*. London: Department for Environment, Food and Rural Affairs.

De Hartog, J., Boogaard, H., Nijland, H. and Hoek, G. (2010) 'Do the health benefits of cycling outweigh the risks?', *Environmental Health Perspectives*, 118(8): 1109–1116.

Deininger, K. and Byerlee, D. (2011) *Rising Global Interest in Farmland: Can It Yield Sustainable and Equitable Benefits?* Washington, DC: World Bank.

De Nazelle, A., Nieuwenhuijsen, M., Antó, J. *et al.* (2011) 'Improving health through policies that promote active travel: a review of evidence to support integrated health impact assessment', *Environment International*, 37: 766–777.

De Ponti, T., Rijk, B. and van Ittersum, M. (2012) 'The crop yield gap between organic and conventional agriculture', *Agricultural Systems*, 108: 1–9.

De Schutter, O. (2010) *Report Submitted by the Special Rapporteur on the Right to Food, Olivier De Schutter*. United Nations General Assembly, 20 December. Online at http://www. srfood.org/images/stories/pdf/officialreports/20110308_a-hrc-1649_agroecology_en.pdf

Dooley, K., Griffiths, T., Martone, F. and Ozinga, S. (2011) *Smoke and Mirrors: A Critical Assessment of the Forest Carbon Partnership Facility*. FERN and Forest Peoples' Programme. Online at http://www.redd-monitor.org/wordpress/wp-content/uploads/2011/03/smokeandmirrorsinternet.pdf

Dougherty Group LLC (2006) *Materials Recovery Facilities. Comparison of Efficiency and Quality*. Banbury, UK: Waste and Resources Action Programme.

Dresner, S. and Ekins, P. (2005) *Climate Change and Fuel Poverty*. London: Policy Studies Institute, University of Westminster.

Druckman, A., Chitnis, M., Sorrell, S. and Jackson, T. (2010) 'An investigation into the rebound and backfire effects from abatement actions by UK households', RESOLVE Working Paper 05-10. Guildford, UK: University of Surrey.

Dunn, E., Aknin, L. and Norton, M. (2008) 'Spending money on others promotes happiness', *Science*, 319(5870): 1687–1688.

Dupressoir, S. *et al.* (2007) *Climate Change and Employment: Impact on Employment in the European Union-25 of Climate Change and CO_2 Emission Reduction Measures by 2030*. Brussels: European Trade Union Confederation (ETUC), Instituto Sindical de Trabajo, Ambiente y Salud (ISTAS), Social Development Agency (SDA), Syndex and Wuppertal Institute.

Easterlin, R. and Angelescu, L. (2009) *Happiness and Growth the World Over: Time Series Evidence on the Happiness–Income Paradox*. IZA Discussion Paper No. 4060. Bonn, Germany: Institute for the Study of Labor.

EC/JRC-PBL (2011) *Emission Database for Global Atmospheric Research (EDGAR), Release Version 4.2*. European Commission, Joint Research Centre (JRC)/Netherlands Environmental Assessment Agency (PBL). Online at http://edgar.jrc.ec.europa.eu/overview.php?v=42

Ecosystem Marketplace (2011) *State of the Forest Carbon Market 2011: From Canopy to Currency* (Authors: D. Diaz, K. Hamilton and E. Johnson). Washington, DC: Forest Trends Association. Online at http://www.forest-trends.org/publication_details.php?publicationID=2963

Edwards, S., Berhan, T., Egziabher, G. and Araya, H. (2010) *Successes and Challenges in Ecological Agriculture: Experiences from Tigray, Ethiopia*. Rome: Food and Agriculture Organization of the United Nations.

EEA (2003) *Air Pollution by Ozone in Europe in Summer 2003 – Overview of Exceedances of EC Ozone Threshold Values during the Summer Season April–August 2003 and Comparisons with Previous Years*. Topic Report 3/2003. Copenhagen: European Environment Agency, European Topic Centre on Air and Climate Change. Online at http://www.eea.europa.eu/publications/topic_report_2003_3

EEA (2005) *The European Environment: State and Outlook 2005*. Copenhagen: European Environment Agency.

EEA (2006) *Air Quality and Ancillary Benefits of Climate Change Policies*. EEA Technical Report 4/2006. Copenhagen: European Environment Agency.

EEA (2010a) *The European Environment: State and Outlook 2010. Thematic Assessment of Air Pollution*. Copenhagen: European Environment Agency.

EEA (2010b) *The European Environment: State and Outlook 2010. Material Resources and Waste*. Copenhagen: European Environment Agency.

EEA (2010c) *The European Environment: State and Outlook 2010. Consumption and Environment.* Copenhagen: European Environment Agency.

EEA (2011a) *NEC Directive Status Report 2010.* EEA Technical Report 3/2011. Copenhagen: European Environment Agency.

EEA (2011b) *Air Quality in Europe: 2011 Report.* EEA Technical Report 12/2011. Copenhagen: European Environment Agency.

EEA (2011c) *Air Pollution Impacts from Carbon Capture and Storage (CCS).* EEA Technical Report 14/2011. Copenhagen: European Environment Agency.

EEA (2011d) *Resource Efficiency in Europe: Policies and Approaches in 31 EEA Member and Cooperating Countries.* Copenhagen: European Environment Agency.

EEA (2012a) *Atmospheric Greenhouse Gas Concentrations (CSI 013) – Assessment Published Jan 2012.* Copenhagen: European Environment Agency. Online at http://www.eea.europa.eu/data-and-maps/indicators/atmospheric-greenhouse-gas-concentrations-2/assessment

EEA (2012b) *Forests, Health and Climate Change.* Copenhagen: European Environment Agency. Online at http://www.eea.europa.eu/articles/forests-health-and-climate-change

Ekins, P., Meyer, B. and Schmidt-Bleek, F. (2009) *Reducing Resource Consumption: A Proposal for Global Resource and Environmental Policy.* GWS Discussion Paper 2009/5. Osnabrück, Germany: Gesellschaft für Wirtschaftliche Strukturforschung.

Eliasch, J. (2008) *Climate Change: Financing Global Forests.* The Eliasch Review. London: The Stationery Office.

Ellison, D., Futter, M. and Bishop, K. (2012) 'On the forest cover–water yield debate: from demand- to supply-side thinking', *Global Change Biology*, 18(3): 806–820.

Energy Watch Group (2006) *Uranium Resources and Nuclear Energy.* Aachen, Germany: Energy Watch Group. Online at http://www.energywatchgroup.org/fileadmin/global/pdf/EWG_Report_Uranium_3-12-2006ms.pdf

Energy Watch Group (2007) *Coal: Resources and Future Production.* Aachen, Germany: Energy Watch Group. Online at http://www.energywatchgroup.org/fileadmin/global/pdf/EWG_Report_Coal_10-07-2007ms.pdf

Energy Watch Group (2009) 'Natural gas reserves: a false hope', *Sun and Wind Energy*, 12: 16–19. Online at http://www.energywatchgroup.org/fileadmin/global/pdf/2009_SWE_12_Natural_Gas_Seltmann.pdf

Entrekin, S., Evans-White, M., Johnson, B. and Hagenbuch, E. (2011) 'Rapid expansion of natural gas development poses a threat to surface waters', *Frontiers in Ecology*, 9(9): 503–511.

Environmental Protection UK and LACORS (2009) *Biomass and Air Quality Guidance for Local Authorities: England and Wales.* Brighton: Environmental Protection UK. Online at http://www.environmental-protection.org.uk/biomass/

Enviros (2003) *Glass Recycling – Life Cycle Carbon Dioxide Emissions. Report to British Glass Manufacturers' Confederation.* Sheffield: Enviros. Online at http://www.britglass.org.uk/lca-report-investigates-environmental-benefit-recycling-glass%C2%A0

Envirowise (2003) 'Over £1 million of efficiency savings achieved by an electroplating company', Case Study 389. Didcot, UK: Envirowise. Online at http://www.wrap.org.uk/sites/files/wrap/CS389.pdf

EPEE Project (undated) *Definition and Evaluation of Fuel Poverty in Belgium, Spain, France, Italy and the United Kingdom.* Work Package 2, Deliverable 7. Online at http://www.fuel-poverty.eu/files/WP2_D7_en.pdf

Epstein, P., Buonocore, J., Eckerle, K., Hendryx, M., Stout, B., Heinberg, R., Clapp, R., May, B., Reinhart, N., Ahern, M., Doshi, S. and Glustrom, L. (2011) 'Full cost accounting for the life cycle of coal', *Annals of the New York Academy of Sciences*, 1219: 73–98.

Ericsson (2011) 'Ericsson and Novatium bring simple and convenient computing for the masses through the cloud', Ericsson Press Release, 11 February. Online at http://www.ericsson.com/news/1488296

Ernst & Young (2011) *Ernst & Young UK Solar PV Industry Outlook: The UK 50kW to 5MW Solar PV Market*. Online at http://www.oursolarfuture.org.uk/coalition-government-derail-50kw-solar-as-ey-report-launched-today-confirms-its-potential/

ETC/SCP (2011a) *Projections of Municipal Waste Management and Greenhouse Gases*. ETC/SCP Working Paper 4/2011. Online at http://scp.eionet.europa.eu/wp/2011wp4

ETC/SCP (2011b) *Progress in Sustainable Consumption and Production in Europe. Indicator-Based Report*. Copenhagen: European Topic Centre on Sustainable Consumption and Production.

EU Climate Change Expert Group (2008) *The 2°C Target: Reference Document. Background on Impacts, Emission Pathways, Mitigation Options and Costs*. Online at: http://ec.europa.eu/clima/policies/international/negotiations/future/docs/brochure_2c_en.pdf

European Commission (2008a) *Annex to the Impact Assessment on the Package of Implementation Measures for the EU's Objectives on Climate Change and Renewable Energy for 2020*. SEC(2008)85 Vol. II. Online at http://ec.europa.eu/energy/renewables/doc/sec_2008_85-2_ia_annex.pdf

European Commission (2008b) *The Raw Materials Initiative – Meeting Our Critical Needs for Growth and Jobs in Europe*. Communication from the Commission and Staff Working Document accompanying the communication. Brussels: European Commission. Online at http://ec.europa.eu/enterprise/non_energy_extractive_industries/raw_materials.htm

European Commission (2009) *GDP and Beyond: Measuring Progress in a Changing World*. COM (2009) 433 final. Brussels: European Commission. Online at http://www.beyond-gdp.eu

European Commission (2010a) *Critical Raw Materials for the EU*. Report of the Ad-Hoc Working Group on Defining Critical Raw Materials. Brussels: European Commission. Online at http://ec.europa.eu/enterprise/e_i/news/article_10514_en.htm

European Commission (2010b) *EUROPE 2020: A Strategy for Smart, Sustainable and Inclusive Growth*. Brussels: European Commission.

European Commission (2010c) *Analysis of Options to Move beyond 20% Greenhouse Gas Emission Reductions and Assessing the Risk of Carbon Leakage*. COM (2010) 265 final. Brussels: European Commission.

European Commission (2010d) *Staff Working Document Accompanying the Communication on the Thematic Strategy on the Prevention and Recycling of Waste*. Brussels: European Commission. Online at: http://ec.europa.eu/environment/waste/strategy.htm

European Commission (2011a) *Analysis of the Key Contributions to Resource Efficiency*. Brussels: European Commission.

European Commission (2011b) *Tackling the Challenges in Commodity Markets and on Raw Materials*. COM(2011) 25. Brussels: European Commission.

European Commission (2011c) *Roadmap for Moving to a Competitive Low-Carbon Economy by 2050*. Brussels: European Commission. Online at http://ec.europa.eu/clima/policies/roadmap/index_en.htm

European Commission (2011d) *Roadmap to a Resource-Efficient Europe*. COM(2011) 571. Brussels: European Commission.

European Commission (2011e) *Report from the Commission to the European Parliament and the Council on the Implementation of the Environmental Noise Directive*. Brussels: European Commission.

European Commission (2011f) *Commission Staff Working Paper: Impact Assessment. Accompanying Document to the White Paper. Roadmap to a Single European Transport Area – Towards a Competitive and Resource-Efficient Transport System*. SEC(2011) 359 final. Brussels: European Commission.

European Parliament (2009) *Study on Eco-Innovation: Putting the EU on the Path to a Resource and*

Energy Efficient Economy. Brussels: European Parliament Committee on Industry, Research and Energy.

European Parliament (2011a) *Impacts of Shale Gas and Shale Oil Extraction on the Environment and on Human Health*. Brussels: European Parliament.

European Parliament (2011b) 'MEPs demand better e-waste management', Press Release, 3 February. Brussels: European Parliament.

European Topic Centre on Land Use and Spatial Information (2011) *Population Exposure to Noise from Different Sources in Europe*. Spreadsheet containing data reported by member states up to 30 June 2010. Online at http://noise.eionet.europa.eu/index.html (accessed 19 July 2011).

ExternE (2005) *ExternE (Externalities of Energy)*. Brussels: European Commission. Online at http://www.externe.info/

Fairlie, S. (2010) *Meat: A Benign Extravagance*. East Meon, Hampshire, UK: Permanent Publications.

FAO (2001) *Global Forest Resources Assessment 2000*. Rome: Food and Agriculture Organization of the United Nations.

FAO (2006) *Livestock's Long Shadow: Environmental Issues and Options*. Rome: Food and Agriculture Organization of the United Nations.

FAO (2009a) *Food Security and Agricultural Mitigation in Developing Countries: Options for Capturing Synergies*. Rome: Food and Agriculture Organization of the United Nations.

FAO (2009b) *Global Agriculture towards 2050*. Rome: Food and Agriculture Organization of the United Nations.

FAO (2011a) *Agricultural Outlook 2011–2020*. Rome: Food and Agriculture Organization of the United Nations.

FAO (2011b) *Global Forest Land-Use Change from 1990 to 2005: Initial Results from a Global Remote Sensing Survey*. Rome: Food and Agriculture Organization of the United Nations.

FAO (2011c) *Organic Agriculture and Climate Change Mitigation*. A Report of the Round Table on Organic Agriculture and Climate Change. Rome: Food and Agriculture Organization of the United Nations.

FAO (2011d) *State of the World's Forests 2011*. Rome: Food and Agriculture Organization of the United Nations.

FAOSTAT (2012) Online statistics of the Food and Agriculture Organization of the United Nations. Online at http://faostat.fao.org/

Farrell, A. and Brandt, A. (2005) 'Scraping the bottom of the barrel: CO_2 emissions consequences of a transition to low-quality and synthetic petroleum resources', Carnegie Mellon Climate Decision Making Center, Energy and Resources Group, University of California, Berkeley. Online at http://cdmc.epp.cmu.edu/docs/pub/Farrell_Brandt.pdf

Fischedick, M., Esken, A., Pastowski, A., Schüwer, D., Supersberger, N., Nitsch, J., Viebahn, P., Bandi, A., Zuberbühler, U. and Edenhofer, O. (2008) *RECCS: Comparison of Renewable Energy Technologies (RE) with Carbon Capture and Storage (CCS) Regarding Structural, Economic, and Ecological Aspects*. Berlin: Federal Ministry for the Environment, Nature Conservation and Nuclear Safety (BMU). Online at www.wupperinst.org/en/ccs

Fischer-Kowalski, M. (2009) 'Future scenarios of global materials flows', presentation to the World Resources Forum, Davos, 14–16 September. Online at http://www.world resourcesforum.org/archive/wrf-2009/workshop-presentations

Fischer-Kowalski, M., Krausmann, F., Steinberger, J. and Ayres, R. (2010) 'Towards a low carbon society: setting targets for a reduction of global resource use', Social Ecology Working Paper 115. Vienna: Institute of Social Ecology. Online at www.uni-klu.ac.at/socec/downloads/WP115_web.pdf

Forum for the Future (2010) *The Future Climate for Development*. London: Forum for the Future.

Foster, P. (2011a) 'Mechanical engineering students demonstrate a user-recyclable laptop', *Green IT Review*, 3 February. Online at http://www.thegreenitreview.com/2011/02/mechanical-engineering-students.html

Foster, P. (2011b) 'Automated consumer electronics buy-back kiosk launched in the US', *Green IT Review*, 24 February. Online at http://www.thegreenitreview.com/2011/02/automated-consumer-electronics-buy-back.html

FPAG (2008) *Fuel Poverty Advisory Group: Seventh Annual Report*. London: FPAG. Online at http://www.decc.gov.uk/en/content/cms/about/partners/public_bodies/fpag/fpag.aspx

Fraunhofer Institute (2011) *Wind Energy Report Germany 2010*. Kassel, Germany: Fraunhofer Institute for Wind Energy and Energy System Technology. Online at http://windmonitor.iwes.fraunhofer.de/wind/download/Windenergie_Report_2010_en.pdf

Fraunhofer Institute, Ecofys, Energy Economics Group, Rütter and Partner, SEURECO and Lithuanian Energy Institute (2009) *The Impact of Renewable Energy Policy on Economic Growth and Employment in the European Union*. Brussels: European Commission.

Friel, S., Dangour, A., Garnett, T., Lock, K., Chalabi, Z., Roberts, I., Butler, A., Butler, C., Waage, J., McMichael, A. and Haines, A. (2009) 'Public health benefits of strategies to reduce greenhouse-gas emissions: food and agriculture', *The Lancet*, 374(9706): 2016–2025.

Friends of the Earth (2010) *More Jobs, Less Waste*. London: Friends of the Earth.

Friends of the Earth Europe (2009) *Gone to Waste: The Valuable Resources that European Countries Bury and Burn*. Brussels: Friends of the Earth Europe.

Friends of the Earth US (2011) 'Issue brief: REDD in California's cap and trade: undermining environmental and financial market integrity'. Washington, DC: Friends of the Earth. Online at http://libcloud.s3.amazonaws.com/93/d9/7/637/Issue_Brief_California_Air_Resources_Board_REDD.pdf

Fthenakis, V., Kim, H. C. and Alsema, E. (2008) 'Emissions from photovoltaic life cycles', *Environmental Science and Technology*, 42(6): 2168–2174.

G77 (2011) 'Message from the Group of 77 and China to the informal interactive thematic debate of the General Assembly on "Green economy: a pathway to sustainable development", delivered by Mr Marcos Stancanelli of the Permanent Mission of Argentina to the United Nations (New York, 2 June 2011)', United Nations. Online at http://g77.org/statement/getstatement.php?id=110602

Gagnon, N., Hall, C. and Brinker, L. (2009) 'A preliminary investigation of energy return on energy investment for global oil and gas production', *Energies*, 2: 490–503.

Ghirga, G. (2010) 'Cancer in children residing near nuclear power plants: an open question', *Italian Journal of Pediatrics*, 36: 60.

Gibson, L., Lee, T. M., Koh, L. P., Brook, B., Gardner, T., Barlow, J., Peres, C., Bradshaw, C., Laurance, W., Lovejoy, T. and Sodhi, N. (2011) 'Primary forests are irreplaceable for sustaining tropical biodiversity', *Nature*, 478: 378–381.

Giljum, S., Hinterberger, F., Bruckner, M., Burger, E., Frühmann, J., Lutter, S., Pirgmaier, E., Polzin, C., Waxwender, H., Kernegger, L. and Warhurst, M. (2009) *Overconsumption? Our Use of the World's Natural Resources*. Vienna: Sustainable Europe Research Institute.

Gill, S., Handley, J., Ennos, A. and Pauleit, S. (2007) 'Adapting cities for climate change: the role of green infrastructure', *Built Environment*, 33(1): 115–133. Online at http://www.fs.fed.us/ccrc/topics/urban-forests/docs/Gill_Adapting_Cities.pdf

Global Footprint Network (2010) *2010 Data Tables*. Online at http://www.footprintnetwork.org/en/index.php/GFN/page/footprint_data_and_results

Gowing, J. and Palmer, M. (2008) 'Sustainable agricultural development in sub-Saharan

Africa: the case for a paradigm shift in land husbandry', *Soil Use and Management*, 24(1): 92–99.

Graedel, T. (2008) 'Are non-renewable resources critical?', presentation at European Commission Green Week, Brussels, 3–6 June. Online at http://www.unep.org/resource panel/Portals/24102/PDFs/Presentation_Graedel_Greenweek2.pdf

Green Alliance (2009) *Landfill Bans and Restrictions in the EU and US.* Research report to DEFRA. London: Department for Environment, Food and Rural Affairs. Online at randd.defra.gov.uk

Griffiths, S. and Reeves, R. (eds) (2009) *Well-Being: How to Lead the Good Life and What Government Should Do to Help*. London: Social Market Foundation.

Gross, R., Heptonstall, P., Anderson, D., Green, T., Leach, M. and Skea, J. (2006) *The Costs and Impacts of Intermittency: An Assessment of the Evidence on the Costs and Impacts of Intermittent Generation on the British Electricity Network*. London: UK Energy Research Centre. Online at http://www.ukerc.ac.uk/Downloads/PDF/06/0604Intermittency/0604Intermittency Report.pdf

GSMA (2009) 'Mobile industry unites to drive universal charging solution for mobile phones', *GSMA News*, 17 February.

Gutteridge, R. and Shelton, M. (1994) *Forage Tree Legumes in Tropical Agriculture*. Wallingford, Oxon, UK: CAB International. Online at http://www.fao.org/ag/AGP/AGPC/doc/Publicat/Gutt-shel/x5556e00.htm#Contents

Haberl, H., Erb, K-H., Krausmann, F., Gaube, V., Bondeau, A., Plutzar, C., Gingrich, S., Lucht, W. and Fischer-Kowalski, M. (2007) 'Quantifying and mapping the human appropriation of net primary production in earth's terrestrial ecosystems', *Proceedings of the National Academy of Sciences of the USA*, 104(31): 12942–12947.

Hall, C., Balogh, S. and Murphy, D. (2009) 'What is the minimum EROI that a sustainable society must have?', *Energies*, 2(1): 25–47.

Hamerschlag, K. (2011) *Meat Eater's Guide to Climate Change and Health*. Environmental Working Group. Online at http://www.ewg.org/meateatersguide

Hammond, G. and Jones, C. (2011) Inventory of Carbon and Energy (ICE) Database. University of Bath, UK. Available on request from http://www.bath.ac.uk/mech-eng/research/sert/

Hansen, J., Sato, M., Kharecha, P., Beerling, D., Berner, R., Masson-Delmotte, V., Pagani, M., Raymo, M., Royer, D. and Zachos, J. (2008) 'Target atmospheric CO_2: where should humanity aim?', *Open Atmospheric Science Journal*, 2: 217–231.

Hansen, M. S., McKinnon, D. L. and Watson, D. (2010) *Sustainable Consumption and Production Policies – A Toolbox for Practical Use*. Wuppertal, Germany: SWITCH-Asia Network Facility, UNEP/Wuppertal Institute Collaborating Centre on Sustainable Consumption and Production (CSCP).

Hargroves, K. and Smith, M. (2005) *The Natural Advantage of Nations*. London: Earthscan.

Harvey, F. (2011) 'Global warming crisis may mean world has to suck greenhouse gases from air', *The Guardian*, 5 June. Online at http://www.guardian.co.uk/environment/2011/jun/05/global-warming-suck-greenhouse-gases

Headey, B. (2008) 'Life goals matter to happiness: a revision of set-point theory', *Social Indicators Research*, 86(2): 213.

HEI (1999) *Diesel Emissions and Lung Cancer: Epidemiology and Quantitative Risk Assessment*. Cambridge, MA: Health Effects Institute.

Hendriksen, I., Simons, M., Galindo Garre, F. and Hildebrandt, V. (2010) 'The association between commuter cycling and sickness absence', *Preventative Medicine*, 51(2): 132–135.

Herrndorf, M., Kuhndt, M. and Fisseha, T. (2007) *Raising Resource Productivity in Global Value*

Chains – Spotlights on International Perspectives and Best Practice. Wuppertal, Germany: UNEP/Wuppertal Institute Collaborating Centre on Sustainable Consumption and Production. Online at http://www.scp-centre.org/fileadmin/content/files/publications/CSCP-2008-RessourceProductivity-GlobalValueChains.pdf

Hertwich, E. and Peters, G. (2009) 'Carbon footprint of nations: a global, trade-linked analysis', *Environmental Science and Technology*, 43(16): 6414–6420.

Hertwich, E., van der Voet, E., Suh, S., Tukker, A., Huijbregts, M., Kazmierczyk, P., Lenzen, M., McNeely, J. and Moriguchi, Y. (2010) *Assessing the Environmental Impacts of Consumption and Production: Priority Products and Materials*. A Report to the International Panel for Sustainable Resource Management. Nairobi: United Nations Environment Programme.

Hirsch, R., Bezdek, R. and Wendling, R. (2005) *Peaking of World Oil Production: Impacts, Mitigation and Risk Management*. Report to US Department of Energy. Online at http://www.netl.doe.gov/publications/others/pdf/Oil_Peaking_NETL.pdf

Holland, M. (2009) *The Co-Benefits to Health of a Strong EU Climate Change Policy*. Report for Climate Action Network Europe, Health and Environment Alliance and WWF. Online at http://assets.panda.org/downloads/co_benefits_to_health_report_ september_2008. pdf

Holland, M., Hunt, A., Wagner, A., Hurley, F., Miller, B., Mathur, R. and Ramaprasad, A. (2011a) *Quantification and Monetisation of the Co-Benefits from Control of Regional Air Pollutants*. Brussels: European Commission ClimateCost project.

Holland, M., Amann, M., Heyes, C., Rafaj, P., Schöpp, W., Hunt, A. and Watkiss, P. (2011b) 'The reduction in air quality impacts and associated economic benefits of mitigation policy. Summary of results from the EC RTD ClimateCost Project', in P. Watkiss (ed.) *The ClimateCost Project. Final Report. Volume 1: Europe*. Stockholm: Stockholm Environment Institute. Online at http://www.climatecost.cc/images/Policy_Brief_ master_REV_WEB_medium_.pdf

Holttinen, H. *et al*. (2009) *Design and Operation of Power Systems with Large Amounts of Wind Power*. IEA Wind Task 25, Final Report, Phase One 2006–08. Vuorimiehentie, Finland: VTT. Online at http://www.vtt.fi/inf/pdf/tiedotteet/2009/T2493.pdf

Höök, M., Zittel, W., Schindler, J. and Aleklett, K. (2010) 'Global coal production outlooks based on a logistic model', *Fuel*, 89(11): 3546–3558. Online at http://www.diva-portal.org/smash/get/diva2:329110/FULLTEXT01

House of Lords Science and Technology Committee (2008) *Waste Reduction Volume 1: Report*. London: The Stationery Office. Online at http://www.publications.parliament.uk/pa/ld200708/ldselect/ldsctech/163/163.pdf

Howarth, R., Santoro, R. and Ingraffea, A. (2011) 'Methane and the greenhouse-gas footprint of natural gas from shale formations', *Climatic Change*, 106(4): 679–690.

Howell, G. and Hill, R. (2009) 'The mediators of experiential purchases: determining the impact of psychological needs satisfaction and social comparison', *The Journal of Positive Psychology*, 4(6): 511–522.

Hubbert, M. K. (1956) 'Nuclear energy and the fossil fuels', paper presented at the Spring Meeting of the Southern District, American Petroleum Institute, San Antonio, Texas, 7– 9 March. Online at http://www.hubbertpeak.com/hubbert/

Huppes, G. and Huele, R. (2010) 'Degrowth with an ageing population: increasing leisure for improving the environment. The key role of pensions and their funding', presentation at the 2nd International Conference on Economic Degrowth for Ecological Sustainability and Social Equity, Barcelona, 26–29 March. Online at http://www.barcelona.degrowth.org/Oral-presentations-slides.116.0.html

IAASTD (2008) *Agriculture at a Crossroads: Global Report*. International Assessment of Agricultural Knowledge, Science and Technology for Development, edited by B. MacIntyre, H. R. Herren, J. Wakhungu and R.T. Watson. Washington, DC: Island Press.

IEA (2007a) *IEA Energy Technology Essentials – Nuclear Power*. Paris: International Energy Agency. Online at http://www.iea.org/techno/essentials4.pdf

IEA (2007b) *Contribution of Renewables to Energy Security*. Paris: International Energy Agency.

IEA (2008) *Empowering Variable Renewables: Options for Flexible Electricity Systems*. Paris: International Energy Agency.

IEA (2009a) *World Energy Outlook 2009*. Paris: International Energy Agency.

IEA (2009b) 'The impact of the financial and economic crisis on global energy investment', background paper for G8 energy ministers' meeting May 2010. Paris: International Energy Agency.

IEA (2010a) *Energy Technology Perspectives 2010*. Paris: International Energy Agency.

IEA (2010b) *World Energy Outlook 2010*. Paris: International Energy Agency.

IEA (2011a) *World Energy Outlook 2011*. Paris: International Energy Agency.

IEA (2011b) *Key World Energy Statistics 2010*. Paris: International Energy Agency.

IEA (2011c) *Technology Roadmap: Biofuels for Transport*. Paris: International Energy Agency.

IEA (2011d) *World Energy Outlook 2011 Special Report: Are We Entering a Golden Age of Gas?* Paris: International Energy Agency.

IEA (2011e) *Technology Roadmap: Smart Grids*. Paris: International Energy Agency.

IEA (2011f) *Harnessing Variable Renewables: A Guide to the Balancing Challenge*. Paris: International Energy Agency.

IEA (2012) *Golden Rules for a Golden Age of Gas*. Paris: International Energy Agency.

IEA Bioenergy (2010) *Bioenergy, Land Use Change and Climate Change Mitigation*. Paris: International Energy Agency.

IEA Bioenergy (2011) *Algae as a Feedstock for Biofuels*. Paris: International Energy Agency.

IEA Data Services (2012) Estimated RD&D Budgets by Region. Online at http://iea.org/stats/rd.asp

IEA, OECD, OPEC and World Bank (2010) *Analysis of the Scope of Energy Subsidies and the Suggestions for the G-20 Initiative*. Paris: International Energy Agency.

IEEP, Ecologic, Arcadis, Umweltbundesamt, Bio Intelligence Services and VITO (2010) *Final Report – Supporting the Thematic Strategy on Waste Prevention and Recycling*. Report to the European Commission. Online at http://ec.europa.eu/environment/waste/strategy.htm

IFPRI (2009) *Millions Fed: Proven Successes in Agricultural Development*. Edited by D. Spielman and R. Pandya-Lorch. Washington, DC: International Food Policy Research Institute.

ILO (2011a) *ILO Introductory Report: Global Trends and Challenges on Occupational Safety and Health*. XIX World Congress on Safety and Health at Work, Istanbul, 11–15 September. Geneva: International Labour Organization.

ILO (2011b) *Global Employment Trends 2011*. Geneva: International Labour Organization. Online at http://www.ilo.org/global/publications/books/WCMS_150440/lang—en/index.htm

IMF (2011) *World Economic Outlook April 2011. Tensions from the Two-Speed Recovery: Unemployment, Commodities, and Capital Flows*. Washington, DC: International Monetary Fund.

IMF (2012) IMF Commodity Price Indices. Online at http://www.imf.org/external/np/res/commod/index.aspx

Inage, S. (2009) *Prospects for Large-Scale Energy Storage in Decarbonised Power Grids*. IEA Working Paper. Paris: International Energy Agency. Online at http://www.iea.org/papers/2009/energy_storage.pdf

Innovas (2009) *Low-Carbon and Environmental Goods and Services: An Industry Analysis*. Report to

UK Department for Business, Enterprise and Regulatory Reform. Winsford, UK: Innovas Solutions Ltd. Online at http://www.defra.gov.uk/statistics/environment/ green-economy/scptb16-lcegs/

International Aluminium Institute (2007) *Life Cycle Assessment of Aluminium: Inventory Data for the Worldwide Primary Aluminium Industry. Year 2005 Update.* London: International Aluminium Institute. Online at http://www.world-aluminium.org

IPCC (2001) *Third Assessment Report: Climate Change 2001. Working Group III Report: Mitigation of Climate Change.* Geneva: Intergovernmental Panel on Climate Change.

IPCC (2005) *Special Report on Carbon Dioxide Capture and Storage.* Geneva: Intergovernmental Panel on Climate Change.

IPCC (2007*) Fourth Assessment Report: Climate Change 2007. Working Group III Report: Mitigation of Climate Change.* Geneva: Intergovernmental Panel on Climate Change.

IPCC (2011) *IPCC Special Report on Renewable Energy Sources and Climate Change Mitigation* (O. Edenhofer, R. Pichs?Madruga, Y. Sokona, K. Seyboth, P. Matschoss, S. Kadner, T. Zwickel, P. Eickemeier, G. Hansen, S. Schlömer and C. von Stechow [eds]). Cambridge, UK and New York: Cambridge University Press. Online at http://srren.ipcc-wg3.de/ report

ITPOES (2010) *The Oil Crunch: A Wake-Up Call for the UK Economy.* Second report of the UK Industry Task Force on Peak Oil and Energy Security. London: Ove Arup and Partners Ltd. Online at http://peakoiltaskforce.net/download-the-report/2010-peak-oil-report/

IUCN (2012) Red List Version 2011.2. Online at http://www.iucnredlist.org

Jackson, T. (2009) *Prosperity without Growth: Economics for a Finite Planet.* London: Earthscan.

Jacobson, M. (2002) 'Control of fossil-fuel particulate black carbon and organic matter, possibly the most effective method of slowing global warming', *Journal of Geophysical Research,* 107(D19): 4410.

Jacobson, M. and Delucchi, M. (2009) 'A path to sustainable energy by 2030', *Scientific American,* November: 58–65.

Jacobson, M. and Delucchi, M. (2011) 'Providing all global energy with wind, water, and solar power, part I: technologies, energy resources, quantities and areas of infrastructure, and materials', *Energy Policy,* 39(3): 1154–1169.

Jaeger, C., Paroussos, L., Mangalagiu, D., Kupers, R., Mandel, A. and David Tàbara, J. (2011) *A New Growth Path for Europe: Generating Prosperity and Jobs in the Low-carbon Economy: Synthesis Report.* A study commissioned by the German Federal Ministry for the Environment, Nature Conservation and Nuclear Safety. Potsdam, Germany: European Climate Forum. Online at http://www.european-climate-forum.net/fileadmin/ecf-documents/Press/A_New_Growth_Path_for_Europe__Synthesis_Report.pdf

James, O. (2007) *Affluenza.* London: Vermilion.

Jiang, M., Griffin, W. M., Hendrickson, C., Jaramillo, P., Van Briesen, J. and Venkatesh, A. (2011) 'Life cycle greenhouse gas emissions of Marcellus shale gas', *Environmental Research Letters,* 6(3): 034014.

Jowit, J. and Wintour, P. (2008) 'Cost of tackling global climate change has doubled, warns Stern', *The Guardian,* 26 June. Online at http://www.guardian.co.uk/environment/ 2008/jun/26/climatechange.scienceofclimatechange

JRC/IPTS (2009) *Environmental Impacts of Diet Changes in the EU.* JRC Scientific and Technical Reports, EUR23783EN. Seville: Joint Research Centre, Institute for Prospective Technological Studies. Online at http://ftp.jrc.es/EURdoc/JRC50544.pdf

Karlsson, G. P., Akselsson, C., Hellsten, S. and Karlsson, P. E. (2011) 'Reduced European emissions of S and N – effects on air concentrations, deposition and soil water chemistry in Swedish forests', *Environmental Pollution,* 159(12): 3571–3582.

Karousakis, K. (2009) *Promoting Biodiversity Co-Benefits in REDD*. OECD Environment Working Papers No. 11. Paris: OECD Publishing.

Kasser, T. (2002) *The High Price of Materialism*. Cambridge, MA: The MIT Press.

Kennedy, R. F. (1968) Speech at the University of Kansas, 18 March.

Keynes, J. M. (1963) 'Economic possibilities for our grandchildren', in *Essays in Persuasion*. New York: W.W. Norton & Company Ltd, pp. 358–373.

Kingston, E. (2010) 'Don't buy it – hire it', *The Ecologist, Newsletter 11*, May: 4–5.

Klein, A.-M., Vaissiere, B., Cane, J., Steffan-Dewenter, I., Cunningham, S., Kremen, C. and Tscharntke, T. (2007) 'Importance of pollinators in changing landscapes for world crops', *Proceedings of the Royal Society B: Biological Sciences*, 274(1608): 303–313.

Klemme, H. and Ulmishek, G. (1991) 'Effective petroleum source rocks of the world: stratigraphic distribution and controlling depositional factors', *AAPG Bulletin*, 75(12): 1809–1851. Summary online at http://www.searchanddiscovery.com/documents/Animator/klemme2.htm

Kram, T., Gielen, D., Bos, A., de Feber, M., Gerlagh, T., Groenendaal, B., Moll, H., Bouwman, M., Daniels, D., Worrell, E., Hekkert, M., Joosten, L., Groenewegen, P. and Goverse, T. (2001) *The MATTER Project Final Report: Integrated Energy and Materials Systems Engineering for GHG Emission Mitigation*. Paris: International Energy Agency. Online at http://www.iea-etsap.org/web/markal/matter/main.html

Krausmann, F., Gingrich, S., Eisenmenger, N., Erb, K-H., Haberl, H. and Fischer-Kowalski, M. (2009) 'Growth in global materials use, GDP and population during the 20th century', *Ecological Economics*, 68(10): 2696–2705. Dataset is online at http://www.uni-klu.ac.at/socec/inhalt/3108.htm

Kravitz, B., Robock, A., Oman, L., Stenchikov, G. and Marquardt, A. B. (2009) 'Sulfuric acid deposition from stratospheric geoengineering with sulfate aerosols', *Journal of Geophysical Research*, 114: D14109.

Kuznets, S. (1934) *National Income, 1929–1932*. Senate document No. 124, 73rd Congress, 2nd session.

Kuznets, S. (1962) 'How to judge quality', *The New Republic*, 20 October: 67.

Lal, R. (2004) 'Soil carbon sequestration impacts on global climate change and food security', *Science*, 304(5677): 1623–1627.

Lal, R. (2008) 'Carbon sequestration', *Philosophical Transactions of the Royal Society*, B363: 815–830. Online at http://rstb.royalsocietypublishing.org/content/363/1492/815.abstract

Lawn, P. (2009) *Environment and Employment: A Reconciliation*. London and New York: Routledge.

Layard, R. (2009) 'Afterword: the greatest happiness principle – its time has come', in S. Griffiths and R. Reeves (eds) *Well-Being: How to Lead the Good Life and What Government Should Do to Help*. London: Social Market Foundation, pp. 92–106.

Layard, R. (2011) *Happiness: Lessons from a New Science*, 2nd edition. London: Penguin Books.

Laybourn, P. and Morrisey, M. (2009) *National Industrial Symbiosis Programme: The Pathway to a Low Carbon Sustainable Economy*. Birmingham, UK: National Industrial Symbiosis Programme (NISP). Online at www.nisp.org.uk (under 'Publications').

Lee, P., Walsh, B. and Smith, P. (2007) *Quantification of the Business Benefits of Resource Efficiency: A Report to the Department for Environment, Food and Rural Affairs*. Aylesbury, UK: Oakdene Hollins. Online at randd.defra.gov.uk

Lee, P., MacGregor, A. and Willis, P. (2011) *Further Benefits of Business Resource Efficiency*. A *Report to the Department for Environment, Food and Rural Affairs*. Aylesbury, UK: Oakdene Hollins. Online at randd.defra.gov.uk

Lewis, S. *et al.* (2009) 'Increasing carbon storage in intact African tropical forests', *Nature*, 457: 1003–1006.

Li, M. (2011) *Peak Energy and the Limits to Global Economic Growth: Annual Report 2011*. Independent research report, University of Utah. Online at http://www.econ.utah.edu/ ~mli/Annual%20Reports/Annual%20Report%202011.pdf

Liddicoat, C., Schapel, A., Davenport, D. and Dwyer, E. (2010) *Soil Carbon and Climate Change*. PIRSA Discussion Paper. Adelaide, SA: Rural Solutions SA.

Lin, B. and Liu, J. H. (2010) 'Estimating coal production peak and trends of coal imports in China', *Energy Policy*, 38(1): 512–519.

Lister, K. and Harnish, T. (2011) *The State of Telework in the U.S. How Individuals, Business, and Government Benefit*. Telework Research Network. Online at http://www.workshifting.com/ downloads/downloads/Telework-Trends-US.pdf

Llanos, R. E. and Feather, C. (2011) *The Reality of REDD+ in Peru: Between Theory and Practice. Indigenous Amazonian Peoples' Analyses and Alternatives*. Lima, Puerto Maldonado and Satipo (all in Peru) and Moreton in Marsh, UK: AIDESEP, FENAMAD, CARE and FPP. Online at http://www.forestpeoples.org/sites/fpp/files/publication/2011/11/reality-redd-peru-between-theory-and-practice-website-english-low-res.pdf

Lombardi, R. and Laybourn, P. (2007) *Industrial Symbiosis in Action. Report on the Third International Industrial Symbiosis Research Symposium, Birmingham, England, August 5–6, 2006*. New Haven, CT: Yale School of Forestry and Environmental Studies.

Long, G. (2010) 'How safe are Chile's copper mines?', *BBC News*, 5 October. Online at http://www.bbc.co.uk/news/world-latin-america-11467279

Luyssaert, S. *et al.* (2008) 'Old-growth forests as global carbon sinks', *Nature*, 455: 213–215.

McCarthy, N., Lipper, L. and Branca, G. (2011) *Climate-Smart Agriculture: Smallholder Adoption and Implications for Climate Change Adaptation and Mitigation*. Rome: Food and Agriculture Organization of the United Nations.

McKinsey & Company (2009) *Pathways to a Low-Carbon Economy, Version 2 of the Global Greenhouse Gas Abatement Cost Curve*. Online at https://solutions.mckinsey.com/ ClimateDesk/default.aspx

McMichael, A., Powles, J., Butler, C. and Uauy, R. (2007) 'Food, livestock production, energy, climate change, and health', *The Lancet*, 370(9594): 1253–1263. Online at http://www.thelancet.com/journals/lancet/article/PIIS0140-6736(07)61256-2/fulltext

Magyar, B. (2011) *Sustainable Fulfilment of Sovereign Obligations – Sustainability and Performance of Sovereign Bonds*. Dubai: Bank Sarasin.

Massmann, C., Kosinska, I., Pessina, G., Brunner, P., Buschmann, H., Brandt, B., Neumayer, S. and Daxbeck, H. (2009) 'Report chapter with description of three what-if scenarios of waste treatment policies and their interplay with the macro-economic scenarios', Deliverable D5-3 of the FP6 project *FORWAST Overall Mapping of Physical Flows and Stocks of Resources to Forecast Waste Quantities in Europe and Identify Life-Cycle Environmental Stakes of Waste Prevention and Recycling*. Brussels: European Commission. Online at http://forwast.brgm.fr/results_deliver.asp

Maxwell, D., Owen, P., McAndrew, L., Muehmel, K. and Neubauer, A. (2011) *Addressing the Rebound Effect*. Report for the European Commission DG Environment. Online at http://ec.europa.eu/environment/eussd/pdf/rebound_effect_report.pdf

Mayor of London (2002) *50 Years On: The Struggle for Air Quality in London since the Great Smog of 1952*. London: Greater London Authority. Online at http://legacy.london.gov.uk/ mayor/environment/air_quality/docs/50_years_on.pdf

Meadows, Donella, Meadows, Dennis and Randers, J. (2004) *Limits to Growth: The 30 Year Update*. White River Junction, VT: Chelsea Green Publishing.

Meadows, Donella, Meadows, Dennis, Randers, J. and Behrens, W. III (1972) *The Limits to Growth*. New York: Universe Books.

Mearns, E. (2010) 'The Chinese coal monster – running out of puff?', *The Oil Drum*, 20 November. Online at http://europe.theoildrum.com/node/7123

Medhurst, J. (2009) *The Impacts of Climate Change on European Employment and Skills in the Short to Medium-Term: A Review of the Literature. Final Report (Volume 2)*. Report by GHK to the European Commission. London: GHK.

Meinshausen, M., Meinshausen, N., Hare, W., Raper, S., Frieler, K., Knutti, R., Frame, D. and Allen, M. (2009) 'Greenhouse gas emission targets for limiting global warming to 2°C', *Nature*, 458: 1158–1162.

Miles, L., Kapos, V. and Dunning, E. (2010) *Ecosystem Services from New and Restored Forests: Tool Development*. Multiple Benefits Series 5. Prepared on behalf of the UN–REDD Programme. Cambridge, UK: UNEP World Conservation Monitoring Centre.

Millennium Ecosystem Assessment (2005) *Ecosystems and Human Well-Being: Synthesis*. Washington, DC: Island Press.

Minnemeyer, S., Laestadius, L., Sizer, N., Saint-Laurent, C. and Potapov, P. (2011) *A World of Opportunity for Forest and Landscape Restoration*. World Resources Institute, IUCN and South Dakota State University. Online at http://www.wri.org/map/global-map-forest-landscape-restoration-opportunities

Mitchell, R. and Popham, F. (2008) 'Effect of exposure to natural environment on health inequalities: an observational population study', *The Lancet*, 372(9650): 1655–1660.

Mohr, S. and Evans, G. (2009) 'Forecasting coal production until 2100', *Fuel*, 88(11): 2059–2067.

Morison, J., Hine, R. and Pretty, J. (2005) 'Survey and analysis of labour on organic farms in the U.K. and the Republic of Ireland', *International Journal of Agricultural Sustainability*, 3(1): 24–43.

Morley, N. and Eatherley, D. (2008) *Material Security: Ensuring Resource Availability for the UK Economy*. Oakdene Hollins for the Resource Efficiency Knowledge Transfer Network. Chester, UK: C-Tech Innovation.

Moss, N. and Nussbaum, R. (2011) *A Review of Three REDD+ Safeguard Initiatives*. Geneva: Forest Carbon Partnership Facility and UN-REDD.

Murphy, C. F. and Allen, D. T. (2011) 'Energy-water nexus for mass cultivation of algae', *Environmental Science & Technology*, 45(13): 5861–5868.

Murphy, D. J. and Hall, C. A. S. (2010) 'Year in review – EROI or energy return on (energy) invested', *Annals of the New York Academy of Sciences* (Special Issue: Ecological Economics Reviews), 1185: 102–118.

Murphy, D. and Hall, C. (2011) 'Energy return on investment, peak oil and the end of economic growth', *Annals of the New York Academy of Sciences*, 1219: 52–72.

Murray, R. (2002) *Zero Waste*. London: Greenpeace Environmental Trust. Online at http://www.greenpeace.org.uk/media/reports/the-environmental-trust-zero-waste

NAEI (National Atmospheric Emissions Inventory) (2008) *UK Emissions of Air Pollutants 1970 to 2006*. Didcot, UK: AEA Technology. Online at www.naei.org.uk

National Academy of Engineering and National Research Council (2008) *Energy Futures and Urban Air Pollution, Challenges for China and the United States*. Washington, DC: The National Academies Press.

Naughton-Treves, L. and Day, C. (eds) (2012) *Lessons about Land Tenure, Forest Governance and REDD+. Case Studies from Africa, Asia and Latin America*. Madison, WI: UW-Madison Land Tenure Center. Online at www.rmportal.net/landtenureforestsworkshop

Naughton-Treves, L., Alex-Garcia, J. and Chapman, C. (2011) 'Lessons about parks and poverty from a decade of forest loss and economic growth around Kibale National Park, Uganda', *Proceedings of the National Academy of Sciences*, 108(34): 13919–13924.

NEA Cymru (2011) *Reducing the Impact of Fuel Poverty on the Health Services*. Swansea: NEA Cymru.

Nelson, A. and Chomitz, K. M. (2011) 'Effectiveness of strict vs. multiple use protected areas in reducing tropical forest fires: a global analysis using matching methods', *PLoS ONE*, 6(8): e22722.

Nemet, G., Holloway, T. and Meier, P. (2010) 'Implications of incorporating air-quality co-benefits into climate change policymaking', *Environmental Research Letters*, 5(1): 014007. Online at http://www.iop.org/EJ/article/1748-9326/5/1/014007/erl10_1_014007.html

Netherlands Environmental Assessment Agency (2010) *Rethinking Global Biodiversity Strategies: Exploring Structural Changes in Production and Consumption to Reduce Biodiversity Loss*. The Hague/Bilthoven: Netherlands Environmental Assessment Agency.

New Economics Foundation (2009) *The Happy Planet Index 2.0: Why Good Lives Don't Have to Cost the Earth*. London: New Economics Foundation. Online at http://www.happyplanetindex.org/learn/download-report.html

Newman, J., Zillioux, E., Newman, C., Denny, C., Colverson, P., Hill, K., Warren-Hicks, W. and Marynowski, S. (2009) *Comparison of Reported Effects and Risks to Vertebrate Wildlife from Six Electricity Generation Types in the New York/New England Region*. Albany, NY: New York State Energy Research and Development Authority.

Niggli, U., Schmid, H. and Fliessbach, A. (2007) *Organic Farming and Climate Change*. Geneva: International Trade Centre (UNCTAD/WTO) and Research Institute of Organic Agriculture (FiBL).

Niggli, U., Fliessbach, A., Hepperly, P. and Scialabba, N. (2009) *Low Greenhouse Gas Agriculture: Mitigation and Adaptation Potential of Sustainable Farming Systems*. Rome: Food and Agriculture Organization of the United Nations.

NISP (2011) 'A new industry in the limelite' and 'New life for old polystyrene' case studies. Birmingham, UK: National Industrial Symbiosis Programme. Online at http://www.nisp.org.uk/case_study_index.aspx

Nørgård, J. (2010) 'Sustainable degrowth through more amateur economy', presentation at the 2nd International Conference on Economic Degrowth for Ecological Sustainability and Social Equity, Barcelona, 26–29 March. Online at http://www.barcelona.degrowth.org/Oral-presentations-slides.116.0.html

Nowak, D., Crane, D. and Stevens, J. (2006) 'Air pollution removal by urban trees and shrubs in the United States', *Urban Forestry & Urban Greening*, 4: 115–123.

Nuffield Council on Bioethics (2011) *Biofuels: Ethical Issues*. London: Nuffield Council on Bioethics.

O'Connor, D., Zhai, F., Aunan, K., Bernsten, T. and Vennemo, H. (2003) *Agricultural and Human Health Impacts of Climate Policy in China: A General-Equilibrium Analysis with Special Reference to Guangdong*. OECD Technical Paper 206. Paris: OECD Development Centre.

OECD (2007a) *OECD Environmental Performance Review of China*. Paris: OECD.

OECD (2007b) *Energy Security and Climate Policy: Assessing Interactions*. Paris: OECD. Online at http://www.iea.org/textbase/nppdf/free/2007/energy_security_climate_policy.pdf

OECD (2008a) 'Oil dependence: is transport running out of affordable fuel?', Discussion paper 2008-5. Paris: OECD.

OECD (2008b) *Measuring Material Flows and Resource Productivity. Synthesis Report*. Paris: OECD.

OECD (2008c) *Recommendation of the Council on Resource Productivity*. Paris: OECD.

OECD (2009a) 'Sustainable manufacturing and eco-innovation: towards a green economy', Policy Brief, June. Paris: OECD.

OECD (2009b) *The Economics of Climate Change Mitigation: Policies and Options for Global Action beyond 2012*. Paris: OECD.

OECD (2010a) *Summary of SMM Case Studies*. Summary Paper No. 1 for the OECD Global Forum on Environment, Focusing on Sustainable Materials Management, 25–27 October, Mechelen, Belgium. Paris: OECD Environment Directorate.

OECD (2010b) *Setting and Using Targets for Sustainable Materials Management: Opportunities and Challenges*. Policy Report 2 for the OECD Global Forum on Environment, Focusing on Sustainable Materials Management, 25–27 October, Mechelen, Belgium. Paris: OECD Environment Directorate.

OECD (2010c) *Green Jobs and Skills: The Local Labour Market Implications of Addressing Climate Change*. Paris: OECD.

OECD (2010d) *Linkages between Environmental Policy and Competitiveness*. OECD Environment Working Papers No. 13. Paris: OECD Publishing.

OECD (2010e) *Addressing International Competitiveness in a World of Non-Uniform Carbon Pricing: Lessons from a Decade of OECD Analysis*. Paris: OECD.

OECD (2011a) *Towards Green Growth*. Paris: OECD.

OECD (2011b) *Inventory of Estimated Budgetary Support and Tax Expenditures for Fossil Fuels*. Paris: OECD. Online at http://www.oecd.org/dataoecd/40/35/48805150.pdf

OECD (2012) *OECD Environmental Outlook to 2050: The Costs of Inaction*. Paris: OECD.

OECD/EEA (2011) OECD/EEA database on instruments used for environmental policy and natural resources management. Online at http://www2.oecd.org/ecoinst/queries/index.htm (accessed 29 June 2011).

OECD–FAO (2011) *OECD–FAO Agricultural Outlook 2011–2020*. Paris: OECD.

OECD/IEA (2010) *Sustainable Production of Second-Generation Biofuels*. Paris: International Energy Agency.

Oekom Research (undated*)* 'The link between sustainability performance and credit standing of governmental bonds based on the example of Oekom Research's country rating', online at http://www.oekom-research.com/homepage/english/performance_countries.pdf (accessed 23 December 2011).

Oja, P., Titze, S., Bauman, A., de Geus, B., Krenn, P., Reger-Nash, B. and Kohlberger, T. (2011) 'Health benefits of cycling: a systematic review', *Scandinavian Journal of Medicine and Science in Sports*, 21(4): 496–509.

O'Neill, D. (2008) 'De-growth or recession?', paper presented at the First International Conference on Economic Degrowth for Ecological Sustainability and Social Equity, Paris, 18–19 April. Online at http://events.it-sudparis.eu/degrowthconference/en/themes/index.php

Osborn, S., Vengosh, A., Warner, N. and Jackson, R. (2011) 'Methane contamination of drinking water accompanying gas-well drilling and hydraulic fracturing', *Proceedings of the National Academy of Sciences*, 108(20): 8172–8176.

Page, N. J. and Creasey, S. C. (1975) 'Ore grade, metal production and energy', *Journal of Research of the U.S. Geological Survey*, 3(10): 9–13.

Pan, A., Sun, Q., Bernstein, A., Schulze, M., Manson, J., Stampfer, M., Willett, W. and Hu, F. (2012) 'Red meat consumption and mortality: results from 2 prospective cohort studies', *Archives of Internal Medicine*, 172(7): 555–563.

Pan, Y. *et al.* (2011) 'A large and persistent carbon sink in the world's forests, 1990–2007', *Science*, 333(6045): 988–993.

Parfitt, J. (2006) 'Summary of LCA Review including carbon issues', presentation at Making the Most of LCA Thinking, 23 November, Savoy Place, London. Banbury, UK: Waste and Resources Action Programme.

Partnership for Clean Indoor Air (2011) *Test Results of Cookstove Performance*. Washington, DC: PCIA. Online at http://www.pciaonline.org/files/Test-Results-Cookstove-Performance.pdf

Paul, H. (2012) 'Why we should continue to oppose the inclusion of agriculture in the climate negotiations', *EcoNexus*, February. Online at http://unfccc.int/resource/docs/2012/smsn/ngo/164.pdf

Pearce, D. (2003) 'Environmentally harmful subsidies: barriers to sustainable development', in *Environmentally Harmful Subsidies: Policy Issues and Challenges*. Paris: OECD.

Pehnt, M. (2006) 'Dynamic life cycle assessment (LCA) of renewable energy technologies', *Renewable Energy*, 31: 55–71.

Pelletier, N. (2010) 'Of laws and limits: an ecological economic perspective on redressing the failure of contemporary global environmental governance', *Global Environmental Change*, 20(2): 220–228.

Pernick, R., Wilder, C. and Winnie, T. (2010) *Clean Tech Job Trends 2010*. Clean Edge Inc. Online at http://www.cleanedge.com

Persha, L., Agrawal, A. and Chhatre, A. (2011) 'Social and ecological synergy: local rulemaking, forest livelihoods, and biodiversity conservation', *Science*, 331: 1606–1608. Online at http://illinois.academia.edu/AshwiniChhatre/Papers/930244/Social_and_Ecological_Synergy_Local_Rulemaking_Forest_Livelihoods_and_Biodiversity_Conservation

Peters, G., Minx, J., Weber, C. and Edenhofer, O. (2011) 'Growth in emission transfers via international trade from 1990 to 2008', *Proceedings of the National Academy of Science of the United States of America*, 108(21): 8903–8908.

Phillips, R. and Rowley, S. (2011) *Bringing it Home: Using Behavioural Insights to Make Green Living Policy Work*. London: Green Alliance.

Pilling, M. (2005) 'Air quality and climate change', presentation to the UK Meteorological Office. Online at www.airquality.co.uk/reports/cat12/0505101604_Mike_Pilling.pdf

Pimentel, D. and Pimentel, M. (2003) 'Sustainability of meat-based and plant-based diets and the environment', *American Journal of Clinical Nutrition*, 78(3): 660S–663S.

Platt, B. and Seldman, N. (2000) *Wasting and Recycling in the United States 2000*. Athens, GA: Grass Roots Recycling Network. Online at http://www.grrn.org/page/zero-waste-resources

Pleijel, H. (ed.) (2009) *Air Pollution and Climate Change: Two Sides of the Same Coin?* Stockholm: Swedish Environmental Protection Agency.

Pollin, R., Heintz, J. and Garrett-Peltier, H. (2009) *The Economic Benefits of Investing in Clean Energy: How the Economic Stimulus Program and New Legislation Can Boost U.S. Economic Growth and Employment*. Washington, DC: Center for American Progress.

Pollin, R., Garrett-Peltier, H., Heintz, J., and Scharber, H. (2008) *Green Recovery: A Program to Create Good Jobs and Start Building a Low-Carbon Economy*. Washington, DC: Center for American Progress.

Pollitt, H., Barker, A., Barton, J. *et al.* (2010) *A Scoping Study on the Macroeconomic View of Sustainability*. Final Report for the European Commission, DG Environment. Cambridge Econometrics and the Sustainable Europe Research Institute. Online at http://ec.europa.eu/environment/enveco/studies_modelling/index.htm

Popkin, B. (2006) 'Global nutrition dynamics: the world is shifting rapidly toward a diet linked with non-communicable diseases', *American Journal of Clinical Nutrition*, 84(2): 289–298.

Popkin, B., Kim, S., Rusev, E., Du, S. and Zizza, C. (2006) 'Measuring the full economic costs of diet, physical activity and obesity-related chronic diseases', *Obesity Reviews*, 7(3): 271–293.

Porchia, R. and Bigano, A. (2008) *Development of a Set of Full Cost Estimates of the Use of Different Energy Sources and Its Comparative Assessment in EU Countries*. Report by FEEM (Fondazione Eni Enrico Mattei) to the European Commission as part of the CASES project (Costs Assessment for Sustainable Energy Markets). Brussels: European Commission. Online at http://www.feem-project.net/cases/documents/deliverables/D_06_1%20part2%2008_09.pdf

Powell, L. and Chaloupka, F. (2009) 'Food prices and obesity: evidence and policy implications for taxes and subsidies', *The Milbank Quarterly*, 87(1): 229–257.

Powers, S., Dominguez-Faus, R. and Alvarez. P. J. J. (2010) 'The water footprint of biofuel production in the USA', *Biofuels*, 1(2): 255–260.

Pretty, J., Noble, A., Bossio, D., Dixon, J., Hine, R., Penning de Vries, F. and Morison, J. (2006) 'Resource-conserving agriculture increases yields in developing countries', *Environmental Science and Technology*, 40(4): 1114–1119.

Prior, T., Giurco, D., Mudd, G., Mason, L. and Behrisch, J. (2010) 'Resource depletion, Peak Minerals and the implications for sustainable resource management', Institute for Sustainable Futures, University of Technology, Sydney, Australia. Paper presented at the International Society for Ecological Economics (ISEE) 11th Biennial Conference, Oldenburg/Bremen, Germany, 22–25 August. Online at http://www.isf.uts.edu.au/publications/prioretal2010resourcedepletion.pdf

Prognos (2008) *Resource Savings and CO_2 Reduction Potential in Waste Management in Europe and the Possible Contribution to CO_2 Reduction Potential in 2020*. Berlin: Prognos AG.

Proto, E. and Rustichini, A. (2011) *The Hump-Shaped Relation between Income and Life Satisfaction*. Warwick University Working Paper. Online at http://www2.warwick.ac.uk/fac/soc/economics/staff/academic/proto/workingpapers/happiness_may_10.pdf

Pucher, J., Dill, J. and Handy, S. (2009) 'Infrastructure, programs, and policies to increase bicycling: an international review', *Preventive Medicine*, 50: S106–S125. Online at http://policy.rutgers.edu/faculty/pucher/Pucher_Dill_Handy10.pdf

Pucher, J., Buehler, R., Bassett, D. and Dannenberg, A. (2010) 'Walking and cycling to health: a comparative analysis of city, state, and international data', *American Journal of Public Health*, 100(10): 1986–1992.

Puska, P. (2009) 'Fat and heart disease: yes we can make a change – the case of North Karelia (Finland)', *Annals of Nutrition and Metabolism*, 54 (Supplement 1): 33–38.

Rabl, A. and de Nazelle, A. (2011) 'Benefits of shift from car to active transport', *Transport Policy*, 19(1): 121–131.

RAE (2012) *Heat: Degrees of Comfort? Options for Heating Homes in a Low-Carbon Economy*. London: Royal Academy of Engineering. Online at www.raeng.org.uk/heat

Rainforest Foundation UK *et al.* (2012) *Civil Society Submission to the AWG-LCA: Views on Modalities and Procedures for Financing Results-Based Actions and Considering Activities Related to Decision 1/C.P.16, Paragraphs 68, 69, 70 and 72*. Online at http://www.fern.org/sites/fern.org/files/Civil%20Society%20LCA%20REDD%2B%20submission.pdf

Ramanathan, V. and Carmichael, G. (2008) 'Global and regional climate changes due to black carbon', *Nature Geoscience*, 1(4): 221–227.

Reid, A. and Miedzinski, M. (2008) *Sectoral Innovation Watch in Europe: Eco-Innovation*. Brussels: Europe-INNOVA.

Reisinger, H., Eisenmenger, N., Ferguson, J., Kanthak, J., Finocchiaro, G., Donachie, G., Maguire, C., Arto, I. and Rotzetter, C. (2009) *Material Flow Analysis for Resource Policy Decision Support*. Position Paper of the Interest Group on the Sustainable Use of Natural Resources on the Needs for Further Development of MFA-Based Indicators. Copenhagen: European Environment Agency.

REN21 (2011) *Renewables 2011 Global Status Report*. Paris: REN21. Online at www.ren21.net

Reuters (2008) 'FACTBOX: the Strait of Hormuz, Iran and the risk to oil', 7 January. Online at http://www.reuters.com/article/2008/01/07/us-iran-hormuz-oil-idUSL 0715889 720080107

Reynolds, D. (1999) 'The mineral economy: how prices and costs can falsely signal decreasing scarcity', *Ecological Economics*, 31(1): 155–166. Online at http://www.oilcrisis.com/reynolds/MineralEconomy.htm

Rights and Resources Initiative (2012) *Turning Point: What Future for Forest Peoples and Resources in the Emerging World Order?* Washington, DC: Rights and Resources Initiative.

Robelius, F. (2007) *Giant Oil Fields – The Highway to Oil: Giant Oil Fields and their Importance for Peak Oil*. Published Doctoral Thesis, Department of Nuclear and Particle Physics, Uppsala University. Online at http://urn.kb.se/resolve?urn=urn:nbn:se:uu:diva-7625

Robock, A. (2008) '20 reasons why geo-engineering may be a bad idea', *Bulletin of Atomic Scientists*, 64(2): 14–18.

Rockström, J. *et al.* (2009) 'Planetary boundaries: exploring the safe operating space for humanity', *Ecology and Society*, 14(2): 32.

Roland Berger Strategy Consultants (2009) *GreenTech Atlas 2.0*. Online at http://www.rolandberger.com/company/press/releases/519-press_archive2009_sc_content/GreenTech_Atlas_2_0.html

Sachs, J. (2005) *The End of Poverty*. London: Penguin Books.

Salameh, M. (2004) 'How realistic are OPEC's proved oil reserves?', *Petroleum Review*, 58(691): 26–29. Online at http://www.aspo-australia.org.au/References/OPEC%20reserves%20Petroleum%20Review%20Aug-04%20SALAMEH.pdf

Salameh, M. (2010) 'An impending oil crunch by 2015?', USAEE-IAEE Working Paper No. 10-054. World Bank Oil Market Consultancy Service. Online at http://ssrn.com/abstract=1715853

San Sebastián, M. and Hurtig, A-K. (2004) 'Oil exploitation in the Amazon basin of Ecuador: a public health emergency', *Pan Am Journal of Public Health*, 15(3): 205–211. Online at http://publications.paho.org/english/TEMA_San_bastian.pdf

Scarborough, P., Clarke, D., Wickramasinghe, K. and Rayner, M. (2010) *Modelling the Health Impacts of the Diets Described in 'Eating the Planet' Published by Friends of the Earth and Compassion in World Farming*. Oxford: British Heart Foundation Health Promotion Research Group, Department of Public Health, University of Oxford.

Schmidt-Bleek, F. (2007) 'Future beyond Climate Change', Position Paper 08/01. Carnoules, France: Factor 10 Institute.

Schneider, F. (2010) 'Degrowth of production and consumption capacities for social justice, wellbeing and ecological sustainability', paper presented at the 2nd International Conference on Economic Degrowth for Ecological Sustainability and Social Equity, Barcelona, 26–29 March. Online at http://www.barcelona.degrowth.org/fileadmin/content/documents/Proceedings/Schneider.pdf

Schrank, D., Lomax, T. and Turner, S. (2010) *Urban Mobility Report 2010*. Texas Transportation Institute. Online at http://tti.tamu.edu/publications/catalog/record/ ?id=36580

Scialabba, N. E-H. and Müller-Lindenlauf, M. (2010) 'Organic agriculture and climate change', *Renewable Agriculture and Food Systems*, 25(2): 158–169.

Scottish Executive (2006) *Review of Greenhouse Gas Life Cycle Emissions, Air Pollution Impacts and Economics of Biomass Production and Consumption in Scotland*. Edinburgh: Scottish Executive. Online at http://www.scotland.gov.uk/Resource/Doc/149415/0039781.pdf

Secretariat of the Convention on Biological Diversity (2010) *Global Biodiversity Outlook 3*. Montreal: Convention on Biological Diversity.

Selin, N., Wu, S., Nam, K., Reilly, J., Paltsev, S., Prinn, R. and Webster, M. (2009) 'Global health and economic impacts of future ozone pollution', *Environmental Research Letters*, 4(4): 044014.

Seltmann, T. (2009) 'Coal is also becoming scarce', *Sun and Wind Energy*, 10: 36–38. Review by the Energy Watch Group. Online at http://www.energywatchgroup.org/fileadmin/global/pdf/2009_SWE_10_Coal_Seltmann.pdf

SEPA (2007) *Report on State of the Environment in China 2006*. Beijing: State Environmental Protection Administration.

SERI (Sustainable Europe Research Institute) (2011) *Global Resource Use*. Slide taken from http://www.materialflows.net/images/stories/global_resource_use.zip

Sermage-Faure, C.,, Laurier, D., Goujon-Bellec, S., Chartier, M., Guyot-Goubin, A., Rudant, J., Hémon, D. and Clavel, J. (2012) 'Childhood leukaemia around French nuclear power plants – the Geocap study, 2002–2007', *International Journal of Cancer*, 131(5): E769–E780.

Seufert, V., Ramankutty, N. and Foley, J. (2012) 'Comparing the yields of organic and conventional agriculture', *Nature*, 485: 229–232. Online at http://www.nature.com/nature/journal/vaop/ncurrent/full/nature11069.html

Shackley, S. and Sohi, S. (2011) *An Assessment of the Benefits and Issues Associated with the Application of Biochar to Soil*. Edinburgh: UK Biochar Research Centre.

Shindell, D. *et al.* (2012) 'Simultaneously mitigating near-term climate change and improving human health and food security', *Science*, 335(6065): 183–189.

Shindell, D., Faluvegi, G., Walsh, M., Anenberg, S. C., van Dingenen, R., Muller, N. Z., Austin, J., Koch, D. and Milly, G. (2011) 'Climate, health, agricultural and economic impacts of tighter vehicle-emission standards', *Nature Climate Change*, 1: 59–66.

Simmons, M. (2007) 'Is the world's supply of oil and gas peaking?', presentation at International Petroleum Week, London, 13 February. Online at http://www.docstoc.com/docs/15563812/1-IS-THE-WORLD-SUPPLY-OF-OIL-_-GAS-PEAKING-by-Matthew-R-Simmons

Simms, A., Johnson, V. and Chowla, P. (2010) *Growth Isn't Possible*. London: New Economics Foundation.

Sitch, S., Cox, P., Collins, W. and Huntingford, C. (2007) 'Indirect radiative forcing of climate change through ozone effects on the land-carbon sink', *Nature*, 448: 791–794.

Slaski, X. and Thurber, M. (2009) 'Cookstoves and obstacles to technology adoption by the poor', Working Paper No. 89, Program on Energy and Sustainable Development, Stanford University. Online at http://pesd.stanford.edu/publications/cookstoves_and_obstacles_to_technology_adoption_by_the_poor/

Sorkhabi, R. (2010) 'Why so much oil?', *GeoExPro*, 7(1). Online at http://www.geoexpro.com/article/Why_So_Much_Oil_in_the_Middle_East/58d94fc1.aspx

Sorrell, S. (2007) *The Rebound Effect: An Assessment of the Evidence for Economy-Wide Energy Savings from Improved Energy Efficiency*. London: UK Energy Research Centre. Online at http://www.ukerc.ac.uk/Downloads/PDF/07/0710ReboundEffect/0710ReboundEffectReport.pdf

Sorrell, S. (2010) 'Energy, growth and sustainability: five propositions', SPRU Working Paper No. 185. Brighton: University of Sussex. Online at http://www.mdpi.com/2071-1050/2/6/1784/

Sorrell, S., Speirs, J., Bentley, R., Brandt, A. and Miller, R. (2009) *Global Oil Depletion: An Assessment of the Evidence for a Near-Term Peak in Global Oil Production*. London: UK Energy Research Centre. Online at http://www.ukerc.ac.uk/support/Global%20Oil%20Depletion

Sovacool, B. (2010) 'A critical evaluation of nuclear power and renewable electricity in Asia', *Journal of Contemporary Asia*, 40(3): 393–400.

Spangenberg, J. (2009) 'The growth discourse, growth policy and sustainable development: two thought experiments', *Journal of Cleaner Production*, 18(6): 561–566.

Stahel, W. (2000) *From Manufacturing to Services – A Strategic Report*. Conches, Switzerland: Product Life Institute.

Steinberger, J. and Roberts, J. T. (2011) 'From constraint to sufficiency: the decoupling of energy and carbon from human needs, 1975–2005', *Ecological Economics*, 70(2): 425–433.

Stern, N. (2006) *Stern Review: The Economics of Climate Change*. London: HM Treasury. Online at http://webarchive.nationalarchives.gov.uk/+/http://www.hm-treasury.gov.uk/sternreview_index.htm

Stiglitz, J., Sen, A. and Fitoussi, J-P. (2009). *Report by the Commission on the Measurement of Economic Performance and Social Progress*. Commission on the Measurement of Economic Performance and Social Progress. Online at www.stiglitz-sen-fitoussi.fr/en/index.htm

Stohl, A., Seibert, P., Wotawa, G., Arnold, D., Burkhart, J., Eckhardt, S., Tapia, C., Vargas, A. and Yasunari, T. (2011) 'Xenon-133 and caesium-137 releases into the atmosphere from the Fukushima Dai-ichi nuclear power plant: determination of the source term, atmospheric dispersion, and deposition', *Atmospheric Chemistry and Physics*, 11: C28319–C28394.

Stokstad, E. (2012) 'Field research on bees raises concern about low-dose pesticides', *Science*, 335(6076): 1555. Online at http://www.sciencemag.org/content/335/6076/1555

Storm van Leeuwen, J. W. and Smith, P. (2005) *Nuclear Power – the Energy Balance*. Online at http://www.stormsmith.nl

Strand, J. and Toman, M. (2010) '*Green Stimulus*', *Economic Recovery, and Long-Term Sustainable Development*. Policy Research Working Paper No. 5163. Washington, DC: World Bank.

Sundseth, K., Pacyna, J., Pacyna, E., Munthe, J., Belhaj, M. and Astrom, S. (2010) 'Economic benefits from decreased mercury emissions: projections for 2020', *Journal of Cleaner Production*, 18: 386–394.

Survival International (2011) *The Uncontacted Indians of Peru and Brazil*. Online at http://www.survivalinternational.org/tribes/isolatedperu and http://www.survival international.org/tribes/uncontacted-brazil

Sustainable Consumption Roundtable (2006) *I Will If You Will: Towards Sustainable Consumption*. London: Sustainable Development Commission and National Consumer Council.

Sustainable Development Commission (2006) *The Role of Nuclear Power in a Low-Carbon Economy*. London: Sustainable Development Commission. Online at http://www.sd-commission.org.uk/publications/downloads/SDC-NuclearPosition-2006.pdf

Sustainable Development Commission (2007) *Healthy Futures 5: Sustainable Transport and Active Travel*. London: Sustainable Development Commission.

Sustainable Development Commission (2009) *Setting the Table: Advice to Government on Priority Elements of Sustainable Diets*. London: Sustainable Development Commission. Online at http://www.sd-commission.org.uk/publications.php?id=1033

Sustainable Development Commission (2010) *Becoming the 'Greenest Government Ever': Achieving Sustainability in Operations and Procurement*. London: Sustainable Development Commission.

Sustainable Development Commission (2011) *Making Sustainable Lives Easier*. London: Sustainable Development Commission.

Sutton, M., Howard, C., Erisman, J. W., Billen, G., Bleeker, A., Grennfelt, P., van Grinsven, H. and Grizzetti, B. (eds) (2011) *The European Nitrogen Assessment: Sources, Effects and Policy Perspectives*. Cambridge, UK: Cambridge University Press.

Tallis, M., Taylor, G., Sinnett, D. and Freer-Smith, P. (2011) 'Estimating the removal of atmospheric particulate pollution by the urban tree canopy of London, under current and future environments', *Landscape and Urban Planning*, 103(2): 129–138.

Tan, J. (2008) 'Commentary: people's vulnerability to heat wave', *International Journal of Epidemiology*, 37(2): 318–320.

TEEB (2009) *TEEB Climate Issues Update, September 2009*. Online at http://www.unep.ch/etb/ebulletin/pdf/TEEB-ClimateIssuesUpdate-Sep2009.pdf (accessed 28 August 2012).

TEEB (2010) *The Economics of Ecosystems and Biodiversity: Ecological and Economic Foundations*. London: Earthscan.

TEEB (2011a) *The Economics of Ecosystems and Biodiversity for Local and Regional Policy Makers*. London: Earthscan.

TEEB (2011b) *The Economics of Ecosystems and Biodiversity for National and International Policy Makers*. London: Earthscan.

The Associated Press (2010) 'Accident brings scrutiny to Chile's mining system', 26 August. Online at http://www.guardian.co.uk/world/feedarticle/9238348

The Chernobyl Forum (2005) *Chernobyl's Legacy: Health, Environmental and Socio-Economic Impacts*. Vienna: IAEA.

The Climate Group (2011) *India's Clean Revolution*. Online at http://www.theclimategroup.org/_assets/files/Indias-Clean-Revolution-Report-March-2011.pdf

The Royal Society (2012) *People and the Planet*. London: The Royal Society.

Thompson Coon, J., Boddy, K., Stein, K., Whear, R., Barton, J. and Depledge, M. (2011) 'Does participating in physical activity in outdoor natural environments have a greater effect on physical and mental wellbeing than physical activity indoors? A systematic review', *Environmental Science and Technology*, 45(5): 1761–1772.

Townshend, T., Fankhauser, S., Matthews, A., Feger, C., Liu, J. and Narciso, T. (2011) *GLOBE Climate Legislation Study*. London: GLOBE (Global Legislators Organisation for a Balanced Environment) and Grantham Research Institute at the London School of Economics. Online at http://www.globe-europe.eu/images/stories/toto/globe climlegitstudy.pdf

Trivedi, M., Papageorgiou, S. and Moran, D. (2008) *What Are Rainforests Worth?* Global Canopy Programme.

Trumper, K., Bertzky, M., Dickson, B., van der Heijden, G., Jenkins, M. and Manning, P. (2009) *The Natural Fix? The Role of Ecosystems in Climate Mitigation. A UNEP Rapid Response Assessment*. Cambridge, UK: United Nations Environment Programme, World Conservation Monitoring Centre.

Tu Jianjin (2007) 'Coal mining safety: China's Achilles' heel', *China Security*, 3(2): 36–53. Online at http://chinasecurity.us/index.php?option=com_content&view=article&id=180&Itemid=8

Tukker, A., Huppes, G., Guinée, J., Heijungs, R., de Koning, A., van Oers, L., Suh, S., Geerken, T., Van Holderbeke, M., Jansen, B. and Nielsen, P. (2006) *Environmental Impact of Products (EIPRO): Analysis of the Life Cycle Environmental Impacts Related to the Final Consumption of the EU-25*. EUR 22284 EN. Brussels: European Commission Joint Research Centre.

Turner, K., Morse-Jones, S. and Fisher, B. (2007) *Perspectives on the 'Environmental Limits' Concept*. A research report by CSERGE for the Department for Environment, Food and Rural Affairs. London: DEFRA.

Turner, W. R., Brandon, K., Brooks, T. M., Gascon, C., Gibbs, H. K., Lawrence, K. S., Mittermeier, R. A. and Selig, E. R. (2012) 'Global biodiversity conservation and the alleviation of poverty', *BioScience* 62(1): 85–92.

UN (2010) *The Millennium Development Goals Report 2010*. New York: United Nations.

UNDP (2011a) *One Digester Plus Three Renovations: Biogas Plants for Rural China*. New York: UNDP.

UNDP (2011b) *Human Development Report 2011*. New York: UNDP. Data online at http://hdrstats.undp.org/en/indicators/103106.html (accessed 23 January 2012).

UNEP (2008) *Green Jobs: Towards Decent Work in a Sustainable, Low-Carbon World*. Washington, DC: Worldwatch Institute.

UNEP (2009a) *Towards Sustainable Production and Use of Resources: Assessing Biofuels*. International Panel for Sustainable Resource Management. Nairobi: United Nations Environment Programme.

UNEP (2009b) *From Conflict to Peacebuilding: The Role of Natural Resources and the Environment*. Nairobi: United Nations Environment Programme.

UNEP (2010a) *Assessing the Environmental Impacts of Consumption and Production: Priority Products and Materials*. Nairobi: United Nations Environment Programme.

UNEP (2010b) *Metal Stocks in Society: Scientific Synthesis*. Nairobi: United Nations Environment Programme.

UNEP (2011a) *Towards a Green Economy: Pathways to Sustainable Development and Poverty Eradication*. Nairobi: United Nations Environment Programme.

UNEP (2011b) *Near-Term Climate Protection and Clean Air Benefits: Actions for Controlling Short-Lived Climate Forcers*. Nairobi: United Nations Environment Programme.

UNEP (2011c) *Decoupling Natural Resource Use and Environmental Impacts from Economic Growth*. Nairobi: United Nations Environment Programme.

UNEP (2011d) *Recycling Rates of Metals*. Nairobi: United Nations Environment Programme.

UNEP Finance Initiative (2011) *REDDy Set Grow*. Geneva: United Nations Environment Programme Finance Initiative.

UNEP-WCMC (2008) *Carbon and Biodiversity: A Demonstration Atlas*. Cambridge, UK: UNEP-WCMC.

UNEP/WMO (2011) *Integrated Assessment of Black Carbon and Tropospheric Ozone*. Nairobi: United Nations Environment Programme.

US Department of Labor: Mine Safety and Health Administration (2009) *Injury Experience in Coal Mining, 2008*. Online at http://www.msha.gov/Stats/Part50/Yearly%20IR's/2008/Coal%20Publication-2008.pdf

US Department of Labor: Mine Safety and Health Administration (2010) *Preliminary Regulatory Economic Analysis for Lowering Miners' Exposure to Respirable Coal Mine Dust Including Continuous Personal Dust Monitors. Proposed Rule (RIN 1219-AB64)*. Online at http://www.msha.gov/REGS/REA/CoalMineDust2010.pdf

US Energy Information Administration (2012) *International Energy Statistics: Petroleum Reserves*. Online at http://www.eia.gov/cfapps/ipdbproject/iedindex3.cfm?tid=5&pid=57&aid=6&cid=CG9,&syid=1980&eyid=2011&unit=MST

US EPA (2011) Pay-As-You-Throw website. Online at http://www.epa.gov/solidwaste/conserve/tools/payt/tools/ssintro.htm

USGS (2009) *An Estimate of Recoverable Heavy Oil Resources of the Orinoco Oil Belt, Venezuela*. Washington, DC: US Geological Survey. Online at http://pubs.usgs.gov/fs/2009/3028/pdf/FS09-3028.pdf

Vandermeulen, V., Verspecht, A., Vermeire, B., Van Huylenbroeck, G., Gellynck, X. *et al.* (2011) 'The use of economic valuation to create public support for green infrastructure investments in urban areas', *Landscape and Urban Planning*, 103: 198–206.

Van der Voet, E., van Oers, L., Moll, S., Schütz, H., Bringezu, S., de Bruyn, S., Sevenster, M. and Warringa, G. (2005). *A Policy Review on Decoupling. Development of Indicators to Assess*

Decoupling of Economic Development and Environmental Pressure in the EU-25 and AC-3 Countries. Brussels: European Commission, DG Environment.

Venter, O., Meijaard, E., Possingham, H., Dennis, R., Sheil, D., Wich, S., Hovani, L. and Wilson, K. (2009) 'Carbon payments as a safeguard for threatened tropical mammals', *Conservation Letters*, 2(3): 123.

Vestbro, D. U. (2010) 'Saving by sharing – collective housing for sustainable lifestyles', presentation at the 2nd International Conference on Economic Degrowth for Ecological Sustainability and Social Equity, Barcelona, 26–29 March. Online at http://www.barcelona.degrowth.org/Oral-presentations-slides.116.0.html

Victor, P. (2008) *Managing without Growth: Slower by Design, not Disaster.* Cheltenham: Edward Elgar.

Vidal, J. (2011) 'Ugandan farmer: "My land gave me everything. Now I'm one of the poorest"', *The Guardian*, 22 September. Online at http://www.guardian.co.uk/environment/2011/sep/22/uganda-farmer-land-gave-me-everything

Vlachokostas, Ch., Nastis, S. A., Achillas, Ch. *et al.* (2010) 'Economic damages of ozone air pollution to crops using combined air quality and GIS modelling', *Atmospheric Environment*, 44(28): 3352–3361.

Von Hippel, F. (2011) 'The radiological and psychological consequences of the Fukushima Daiichi accident', *Bulletin of the Atomic Scientists*, 67(5): 27–36.

Von Weizsäcker, E., Lovins, A. and Hunter Lovins, L. (1998) *Factor Four: Doubling Wealth, Halving Resource Use.* London: Earthscan.

Von Weizsäcker, E., Hargroves, C., Smith, M., Desha, C. and Stasinopoulos, P. (2010) *Factor Five: Transforming the Global Economy through 80% Improvements in Resource Productivity.* London: Earthscan.

Wakeford, R. (2007) 'The Windscale reactor accident – 50 years on', *Journal of Radiological Protection*, 27(3): 211–215.

Walz, R. (2011) 'Employment and structural impacts of material efficiency strategies: results from five case studies', *Journal of Cleaner Production*, 19(8): 805–815.

WBCSD (2010) *Vision 2050: The New Agenda for Business.* Geneva: World Business Council for Sustainable Development.

WEC (2004) *Comparison of Energy Systems Using Life Cycle Assessment.* London: World Energy Council.

WEC (2010) *2010 Survey of Energy Resources.* London: World Energy Council. Online at http://www.worldenergy.org/documents/ser_2010_report_1.pdf

WEC (2012) *World Energy Perspective: Nuclear Energy One Year after Fukushima.* London: World Energy Council.

Weintrobe, L. (2011) 'Q&A round-up: Rachel Botsman on collaborative consumption', *The Guardian*, 19 August. Online at http://www.guardian.co.uk/sustainable-business/best-bits-rachel-botsman-collaborative-consumption

Wenig M., Spichtinger, N., Stohl, A., Held, G., Beirle, S., Wagner, T., Jahne, B. and Platt, U. (2002) 'Intercontinental transport of nitrogen oxide pollution plumes', *Atmospheric Chemistry and Physics Discussions*, 2(6): 2151–2165.

West, J., Fiore, A., Horowitz, L. and Mauzerall, D. (2006) 'Global health benefits of mitigating ozone pollution with methane emission controls', *Proceedings of the National Academy of Sciences*, 103(11): 3988–3993.

Wilkins, G. (2002) *Technology Transfer for Renewable Energy: Overcoming Barriers in Developing Countries.* London: Earthscan.

Wilkinson, P., Smith, K., Davies, M. *et al.* (2009) 'Public health benefits of strategies to reduce greenhouse-gas emissions: household energy', *The Lancet*, 374(9705): 1917–1929.

Wilkinson, R. and Pickett, K. (2009) *The Spirit Level: Why More Equal Societies Almost Always Do Better*. London: Penguin Books.

Williams, M. (2007) 'UK air quality in 2050 – synergies with climate change policies', *Environmental Science & Policy*, 10(2): 169–175.

Williams, M. (2009) 'Environmental sustainability – policy aspects of climate change and air quality', presentation at ISPRA (Institute for Environmental Protection and Research) Conference on Environment Including Global Change, Palermo, 4–9 October.

Woodcock, J., Edwards, P., Tonne, C., Armstrong, B. G., Ashiru, O., Banister, D., Beevers, S., Chalabi, Z., Chowdhury, Z., Cohen, A., Franco, O. H., Haines, A., Hickman, R., Lindsay, G., Mittal, I., Mohan, D., Tiwari, G., Woodward, A. and Roberts, I.. (2009) 'Public health benefits of strategies to reduce greenhouse-gas emissions: overview and implications for policy makers', *The Lancet*, 374(9707): 2104–2114.

Woodward, D. and Simms, A. (2006) *Growth Isn't Working: The Unbalanced Distribution of Benefits and Costs from Economic Growth*. London: New Economics Foundation.

Woolf, D., Amonette, J., Street-Perrott, F. A., Lehmann, J. and Joseph, S. (2010) 'Sustainable biochar to mitigate global climate change', *Nature Communications*, 1, Article No. 56.

World Agroforestry Centre (2009) *Creating an Evergreen Agriculture in Africa for Food Security and Environmental Resilience*. Nairobi: World Agroforestry Centre.

World Bank (2004) *Sustaining Forests: A Development Strategy*. Washington, DC: World Bank.

World Bank (2006) *Where Is the Wealth of Nations? Measuring Capital for the 21st Century*. Washington, DC: World Bank.

World Bank (2011a) *State and Trends of the Carbon Market 2011*. Washington, DC: World Bank.

World Bank (2011b) Commodity Price Indices. Online at http://blogs.worldbank.org/prospects/category/tags/commodity-prices

World Bank (2012a) *A Smarter GDP: Factoring Natural Capital into Economic Decision-Making*. Report of the WAVES project (Wealth Accounting and the Valuation of Ecosystem Services). Online at http://siteresources.worldbank.org/EXTSDNET/Resources/A_Smarter_GDP.pdf

World Bank (2012b) Data: Indicators: GDP per capita. Online at http://data.worldbank.org/indicator/NY.GDP.PCAP.CD

World Bank and SEPA (2007) *Cost of Pollution in China: Economic Estimates of Physical Damages*. World Bank and State Environmental Protection Agency, People's Republic of China. Online at http://go.worldbank.org/FFCJVBTP40

World Commission on Dams (2000) *Dams and Development: A New Framework for Decision-Making*. London: Earthscan.

World Health Organization (2003a) *Health Aspects of Air Pollution with Particulate Matter, Ozone and Nitrogen Dioxide*. Copenhagen: World Health Organization Regional Office for Europe.

World Health Organization (2003b) *Investing in Mental Health*. Geneva: World Health Organization.

World Health Organization (2008) *2008–2013 Action Plan for the Global Strategy for the Prevention and Control of Non-Communicable Diseases: Prevent and Control Cardiovascular Diseases, Cancers, Chronic Respiratory Diseases and Diabetes*. Geneva: World Health Organization.

World Health Organization (2009a) *Global Health Risks: Mortality and Burden of Disease Attributable to Selected Major Risks*. Geneva: World Health Organization. Online at http://www.who.int/evidence/bod

World Health Organization (2009b) *Global Status Report on Road Safety*. Geneva: World Health Organization.

World Health Organization (2011a) *Fact Sheet No. 313. Air Quality and Health (updated September 2011)*. Geneva: World Health Organization.

World Health Organization (2011b) *World Health Statistics 2011*. Geneva: World Health Organization.

World Health Organization (2011c) *Global Plan for the Decade of Action on Road Safety 2011–2020*. Geneva: World Health Organization.

World Health Organization (2012) *Fact Sheet No. 311. Obesity and Overweight*. Geneva: World Health Organization.

World Health Organization Europe and JRC (2011) *Burden of Disease from Environmental Noise: Quantification of Healthy Life Years Lost in Europe*. Geneva: World Health Organization.

World Health Organization Europe and UNECE (2009) 'Key messages for Transport Health and Environment Pan European Programme (THE PEP) toolbox: what works?'. Geneva: World Health Organization. Online at http://www.healthytransport.com/athena/site/file_database/key_messages.pdf

World Nuclear Association website. Online at http://www.world-nuclear.org/info/inf75.html (accessed 10 October 2011).

World Resources Forum (2009) *Declaration of the World Resources Forum Sept 16 2009*. Online at http://www.worldresourcesforum.org/archive/documents-links/wrf-declaration

World Watch Institute (2011) *State of the World 2011: Innovations that Nourish the Planet*. Washington, DC: World Watch Institute.

WRAP (2006) *Environmental Benefits of Recycling – An International Review of Life Cycle Comparisons for Key Materials in the UK Recycling Sector*. Banbury, UK: Waste and Resources Action Programme.

WRAP (2009) *Meeting the UK Climate Change Challenge: The Contribution of Resource Efficiency*. WRAP Project EVA128. Banbury, UK: Stockholm Environment Institute, University of Durham Business School and Waste and Resources Action Programme.

WRAP (2010) *Securing the Future – The Role of Resource Efficiency*. Banbury, UK: Waste and Resources Action Programme.

WRAP (2011a) *Water and Carbon Footprint of Household Food Waste in the UK*. Banbury, UK: Waste and Resources Action Programme.

WRAP (2011b) *Benefits of Re-Use: Case Studies*. Banbury, UK: Waste and Resources Action Programme.

WRAP (2011c) *The Courtauld Commitment Phase 2: First Year Progress Report*. Banbury, UK: Waste and Resources Action Programme.

WRAP (2011d) *Evaluation Methodology Statement 2008–2011*. Banbury, UK: Waste and Resources Action Programme.

WRI (2010) *Earth Trends: Energy and Resources – Energy Consumption: Total Energy Consumption per Capita*. World Resources Institute. Online at http://earthtrends.wri.org/text/energy-resources/variable-351.html

Wright, M. (2011) 'Success means telling people to buy less', *The Guardian*, 7 November. Online at http://www.guardian.co.uk/sustainable-business/blog/patagonia-closed-loop-green-technology?INTCMP=SRCH

WSP Environmental Ltd (2011) *GlassRite: Wine – Monitoring of Savings 2010/11*. Banbury, UK: Waste and Resources Action Programme.

WWF (2006) *Counting Consumption: CO_2 Emissions, Material Flows and Ecological Footprint of the UK by Region and Devolved Country*. Godalming, UK: WWF-UK.

WWF, Ecofys and OMA (2011) *The Energy Report: 100% Renewable Energy by 2050*. Gland, Switzerland: WWF International.

Yablokov, A., Nesterenko, V. and Nesterenko, A. (2009) 'Chernobyl: consequences of the catastrophe for people and the environment', *Annals of the New York Academy of Sciences*, 1181.

Zagema, B. (2011) *Land and Power: The Growing Scandal Surrounding the New Wave of Investments in Land*. Oxford: Oxfam International.

Zuser, A. and Rechberger, H. (2011) 'Considerations of resource availability in technology development strategies: the case of photovoltaics', *Resources, Conservation and Recycling*, 56(1): 56–65.

Index

CDM *see* clean development mechanism
certification schemes 221, 237, 273
CH_4 *see* methane
Chilean copper mine accident 195
China
  acidification 15
  air pollution 18, 45
  biogas 152
  carbon exports 175
  coal mining accidents 146
  costs of obesity 289
  fossil fuel demand 123–4, 137, 144–5
  green jobs 131–2
  health co-benefits 37–41
  informal recycling 236
  ozone damage 16–17
  rare earth metals 188
  reducing growth targets 282
  resource efficiency 191
  see also Sloping Land Conversion Program
choice editing 311–12
CHP *see* combined heat and power
clean coal technologies *see* coal
clean development mechanism (CDM) 41, 59, 98–100
climate change targets 1
climate-smart agriculture 79, 86
closed-loop manufacture 183
cloud forests 72
CO_2e *see* carbon dioxide equivalent
coal
  clean coal technologies 22–3
  mining accidents 146, 148
  pollution from 11, 146
  see also FBC; IGCC
co-housing 301
Colombia 72, 237, 299
combined heat and power (CHP) 23–4, 171
commodity prices 186–7
community benefits 300–1, 319–21
  from active travel 299, 300–1
  from forest carbon schemes 102–4
  from local food 287, 300–1
competitiveness 239–42, 250–1, 266–8
compost 61, 108–9, 200–1
conflict, due to fuel shortages 137; *see also* resource wars
congestion 299–300
conservation tillage 60–1, 80, 84–7, 95, 110–11
conspicuous consumption 303
construction materials, recycling 210

consumerism 305–6; *see also* buy-less work-less lifestyles
contour planting 61, 81–2
cooking stoves 42, 46–7, 150–1; *see also* traditional biofuels
cost savings 139–40, 201–6, 237–9, 254–7, 295
cover crops 61, 80, 84–7, 107
cradle-to-cradle 183
crop yields 67, 85, 94–6, 106–12
Cuba 282–3
cycling 291–300

dash to gas 13, 24, 137
debt cancellation 221
Deepwater Horizon accident 146–8
deforestation 68–9, 84, 90, 97, 101
  carbon loss due to 58
  costs of 77–8
  soil erosion and flood risk due to 71–3
demand management 164
desertification 73, 80
developing countries 252–4, 271–5
  air pollution 18, 37–42, 44
  energy access 150–2
  energy import costs 144–5
  health co-benefits 37–42
  organic agriculture 107–8
  resource use and trade 209–11, 220–1
  waste disposal problems 199–200
  see also GDP of the poor
diesel, switching to from gasoline 45
diet, low-carbon 286–7, 289–91, 308, 311–12; *see also* meat and dairy produce, eating less
dioxins 12–13, 43, 46, 199
driving less *see* active travel

Easterlin Paradox 303
ecoATM 205
eco-design 183–4, 218
ecological footprint 187–8, 213, 246
ecological rucksack 213
Ecological Tax Reform 233, 271, 281
ecosystem services 77–8, 187, 244
eco-tourism 74–5
Ecuador 72, 75, 148
electric vehicles 30
electronic waste 179, 204–6, 210, 218, 236
Eliasch Review 60
emissions trading 41, 98–101, 260–3, 266–8, 271–2
end-of-pipe pollution control 43–5
energy access 150